Case Studies in Systems Biology

Pavel Kraikivski

Editor

Case Studies in Systems Biology

 Springer

Editor
Pavel Kraikivski
Academy of Integrated Science
Division of Systems Biology
Virginia Polytechnic Institute and State
University
Blacksburg, VA, USA

ISBN 978-3-030-67744-2   ISBN 978-3-030-67742-8   (eBook)
https://doi.org/10.1007/978-3-030-67742-8

This Springer imprint is published by the registered company Springer Nature Switzerland AG
The registered company address is: Gewerbestrasse 11, 6330 Cham, Switzerland

Preface

The purpose of this book is to provide a representative set of case studies that can be used for lectures, seminars, computational labs, and project presentations in Systems Biology and Computational Biology classes.

The aim is not to present all systems biology-related topics, but to provide a broad variety of current and well-established research case studies in these programs. The case studies demonstrate how mathematical modeling, computational informatics, and data analysis are used to explain and predict the dynamic behavior of biological systems.

The content of this book reflects the curriculums for Systems Biology and Computational Cell Biology courses that are taught in junior and senior undergraduate, and graduate, classes at Virginia Tech. Additionally, the case studies can be used in any interdisciplinary science or engineering curriculum that includes studies of principles governing the behavior of complex biological systems. However, this book does not aim to define Systems Biology, or to introduce the basic features of mathematical models, reaction networks, and biological systems. We assume these have already been discussed and covered in prerequisite courses, or that additional literature sources are available to cover these areas. Thus, our collection of studies will be used to supplement other basic course books. A case study from this book should be sufficient for lecture preparation, computational lab assignments, or as an individual assignment selected by a student to make a class presentation of the case study.

Case studies are organized as independent chapters and all are structured using a similar template, starting with a description of the biological system that highlights the most interesting aspects of the system's dynamic behavior. Then, the relevant details with regard to the cell physiology and molecular mechanisms of the system are explained, followed by an explanation of how the details are put together into a mathematical model representing the system, and how the model can be analyzed and simulated. After this, the results section presents the essential results of the model simulations and compares these results to the known facts of cell physiology. The discussion section summarizes results and explains what predictions can be made using the model, and how the model can be extended. After the discussion section, the chapter provides exercises that can be used for in-class activities or homework assignments. For case studies that use computational software to perform the mathematical modeling, the corresponding code, and instructions on how to run the code are provided in the Method section. Each case study ends with a list of relevant literature references for further study of the corresponding biological system.

The modular organization of this book supports addition of new case studies as independent chapters, while leaving the existing materials unchanged. It is also easy to update individual case studies, and/or substitute existing chapters with more novel and relevant systems biology case studies. Any single chapter will be sufficient for a project or presentation and can be used as the study source, independent of other book chapters.

Although the dynamic behavior of biological systems is often described by complex mathematical models using numerous mathematical equations, in this book we present "concise" models that capture the most essential physiological characteristics of the described systems or its components. For example, the model of mitotic oscillations is much simpler than a comprehensive model of the cell cycle, yet the former can well illustrate the principles governing cell cycle regulation. This type of model can be described and studied in one or two lectures, and students can reproduce certain model results and successfully complete in-class exercises or homework assignments. Thus, each case study is centered around student motivation that ensures their active participation in learning and understanding the case study.

I would like to thank all the authors who have contributed chapters to this book. The title page of each chapter includes the name/names, affiliation, and contact information of the corresponding author and co-authors.

Blacksburg, VA, USA   Pavel Kraikivski

Contents

Mitotic Cycle Regulation. I. Oscillations and Bistability

John J. Tyson

Introduction

In this case study we will examine mathematical models of the molecular regulatory network controlling entry into and exit from mitosis in fertilized frog eggs and in frog-egg extracts. This subject is important because mitotic cell division is at the root of all aspects of organismal growth, development and repair. Furthermore, the molecular mechanisms of mitotic regulation display intriguing commonalities across all taxa of eukaryotes that have been studied in detail [1]. Finally, the physiology and molecular biology of mitotic cycle regulation in frog eggs can be adequately addressed in much simpler terms than, say, the complex regulatory networks in mammalian somatic cells. Nonetheless, what we learn from our investigations of frog eggs reveals some basic control principles of the DNA replication-division cycle in cells of more direct relevance to human health and disease.

Physiology

The DNA replication–division cycle is the sequence of events by which a cell makes exact copies of each DNA molecule and then partitions these 'sister' chromatids to two daughter cells at division, so that each daughter receives one and only one copy of every chromosome. Because a newborn 'daughter' cell is half the size of its 'mother' cell at division, the daughter cell must grow (on average) two-fold before it becomes a dividing mother cell. This requirement of balanced growth and division is

J. J. Tyson (✉)
Department of Biological Sciences and Division of Systems Biology, Virginia Polytechnic Institute and State University, Blacksburg, VA, USA
e-mail: tyson@vt.edu

© Springer Nature Switzerland AG 2021
P. Kraikivski (ed.), *Case Studies in Systems Biology*,
https://doi.org/10.1007/978-3-030-67742-8_1

true in general, but a fertilized egg is a special case [2]. A newly fertilized frog egg is very large (~1 mm in diameter) and starts with only a single diploid nucleus. Its first order of business is to replicate this nucleus many times, to create a hollow ball (the 'blastula') of ~4000 small, undifferentiated cells, from which the zygote can begin to develop. These early division cycles proceed, *without cell growth*, by rapid, synchronous alternations between S phase (DNA replication) and M phase (mitosis) with negligible 'gaps' in between (the G1 and G2 phases that are characteristic of replication-division cycles in somatic cells). After 12 divisions, the 4096 cells of the blastula undergo a dramatic 'mid-blastula' transition: subsequent cell division cycles are much slower and asynchronous, the cells of the developing embryo exhibit waves of expression of zygotic genes, the cells also commence to grow by utilizing nutrient stores in the yolk, and they differentiate and migrate as the developing tadpole takes shape.

The early, rapid, synchronous cell division cycles are unusual in that they are not constrained by cell-cycle checkpoints that normally guard cell proliferation from errors that may compromise the integrity of the genome (for example, DNA damage induced by ionizing radiation, or mistakes in chromatid partitioning during mitosis.) The rationale for this behavior seems to be that speed is of the essence in the initial stages of development, and seriously damaged embryos can be culled after the mid-blastula transition.

For our purposes, this lack of checkpoint signaling is convenient because we can focus on the alternation between interphase (the period of DNA synthesis) and mitosis (the period of nuclear and cell division) without having to deal with the considerable mechanistic complications of checkpoint signaling.

A fundamental question confronting cell biologists of the 1970s and 80s concerned the molecular mechanisms governing this alternation between interphase and mitosis. What molecular events triggered DNA synthesis followed by entry into mitosis? What molecular events were responsible for exit from mitosis and return to interphase? These questions were attacked on two fronts: by biochemical studies and genetic analyses.

Molecular Mechanisms

On the biochemical front, Masui and Markert identified a biochemical 'factor' that promotes the maturation of frog oocytes [3], and shortly thereafter Kirschner and his colleagues showed that this factor is very active in mitotic cells of the fertilized frog egg, and inactive in interphase [4]. They called this substance—whatever it might be—MPF (M-phase promoting factor). The hunt was on to purify MPF and determine its biochemical identity, but MPF proved notoriously difficult to purify.

At about the same time, Tim Hunt was investigating protein synthesis during early mitotic cycles of sea urchin embryos [5], using the technique of polyacrylamide gel electrophoresis. To his surprise, he found that there was very little variation of protein expression during these cell cycles, except for two bands on his gels that slowly increased in intensity during interphase and then abruptly disappeared as cells exited mitosis. He called these two bands 'cyclin A' and 'cyclin B'. Presumably, the synthesis and destruction of cyclin molecules had something to do with the rise and fall of MPF activity in early embryonic cell division cycles.

Meanwhile, geneticists were seeking to identify the genes and proteins that regulate progression through the cell division cycle. These studies were undertaken not in frog eggs but in yeast cells, which were more amenable, at the time, to genetic analysis. Hartwell et al. [6, 7] made the first breakthrough, identifying a suite of 'cell division cycle' (cdc^+) genes in budding yeast. The phenotype of yeast cells carrying a cdc^- mutant allele is to become blocked uniformly at a particular stage of the cell cycle. (I am using the notation cdc^+ and cdc^- for the wild-type gene and a mutant allele, respectively.) Hartwell identified a particular cdc gene ($cdc28^+$) that seems to operate at the earliest stage of the cell cycle (called 'Start'), shortly before bud emergence and the onset of DNA synthesis.

Paul Nurse and his colleagues repeated Hartwell's screen for cdc mutations in fission yeast, with the added proviso that mutant cells should continue to grow for some time before they eventually die [8]. In addition to finding a similar suite of cdc genes in fission yeast, Nurse also identified a curious mutant cell that grew and proliferated with the same interdivision time as wild-type cells but divided at about one-half the size of wild-type cells [9]. Calling this behavior the 'wee' phenotype, Nurse realized that wee genes must be involved in the regulation of cell cycle progression rather than the nuts-and-bolts. Expanding his search for cells expressing the wee phenotype, he identified two genetic loci, $wee1$ and $wee2$. Further genetic analysis demonstrated that the $wee2^-$ gene is a rare allele of the $cdc2$ locus, and that $wee1$ closely interacts with the $cdc25$ gene in controlling the function of $cdc2$. Although $cdc2$'s point of action in fission yeast is at the G2/M transition (unlike $cdc28$'s role at Start in budding yeast), Nurse and colleagues determined that the proteins encoded by $cdc28^+$ and $cdc2^+$ are largely interchangeable in budding yeast and fission yeast, and, even more surprisingly, that human cells contain a gene that can complement $cdc2^-$ mutations in fission yeast [10].

At this point in time, a picture of the molecular controls of mitosis in frog eggs and fission yeast cells was beginning to emerge (Fig. 1). Employing the molecular cloning technologies that were newly emerging in the 1980s, Nurse and colleagues determined that $cdc2^+$ and $wee1^+$ encoded protein kinases and $cdc25^+$ encoded a protein phosphatase [11, 12].

In 1988 Lohka and Maller [13] succeeded in purifying MPF and determining that the factor is a dimer of two distinct proteins. Shortly thereafter, Maller, Hunt and Nurse showed that the two subunits of MPF are cyclin B and Cdc2 kinase [14, 15]. Apparently, the Cdc2 subunit by itself has little or no kinase activity, but is activated by binding to cyclin B. Hence, Cdc2 (and by implication Cdc28 in budding yeast) became known as 'cyclin-dependent kinase' (Cdk1).

Taken together, these discoveries led to a detailed molecular mechanism for the control of mitotic entry and exit in the early stages of frog egg development (Fig. 2).

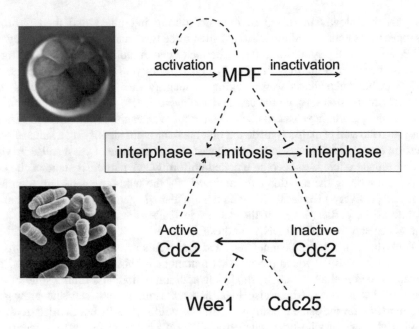

Fig. 1 Two views of cell cycle regulation circa 1980. Top: Biochemical studies of frog eggs focused on the role of M-phase promoting factor (MPF) in driving cells from interphase (DNA synthesis) into mitosis. Injection of active MPF into cells can stimulate premature entry into M phase. MPF activity disappears as cells exit mitosis and return to interphase. Bottom: Genetic studies in fission yeast focused on the roles of three genes in controlling the transitions between interphase and mitosis. The *cdc2* gene encodes a protein (Cdc2) that is a dosage-independent initiator of mitosis. The *wee1* and *cdc25* genes encode proteins that inhibit and activate Cdc2, respectively, in dosage-dependent pathways. Overexpression of *wee1* delays mitosis until cells grow larger than normal, whereas overexpression of *cdc25* induces cells to divide at a smaller size. From Tyson and Novak [22]; used by permission

Mathematical Model, Results, Predictions and Confirmations

In 1993, Novak and Tyson [16] published a mathematical model of the molecular mechanism in Fig. 2. Their model consisted of ten nonlinear ordinary differential equations (ODEs) based on standard principles of biochemical kinetics. In Table 1, I have converted the Novak-Tyson model into a simplified, two-ODE version that we shall investigate in this case study [17].

To understand the implications of this mathematical model, we must recognize that the two ODEs in Table 1 define a <u>vector field</u> in the two-dimensional <u>state space</u> (called a 'phase plane') of the variables $u(t)$, $v(t)$. A convenient tool for visualizing this vector field (and carrying out the other analytical studies and numerical simulations in this case study) is the freely available software program XPPAUT (the program description can be found in the Computational Software chapter of this textbook).

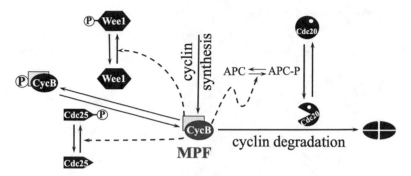

Fig. 2 Unified view of cell cycle regulation circa 1990 [1]. M-phase promoting factor (MPF) was recognized to be a heterodimer of Cdc2 (kinase subunit; gray rectangle) and cyclin B (regulatory/targeting subunit; red oval). Because kinase subunits are present in excess, MPF activity can be raised or lowered by controlling the availability of cyclin subunits. In frog eggs and fission yeast cells, this control is exercised primarily at the level of protein degradation, by controlling the activity of the APC:Cdc20 ubiquitination machinery. Alternatively, MPF activity can be inhibited by Wee1-catalyzed phosphorylation of Cdc2, and MPF activity can be restored by Cdc25-catalyzed removal of the inhibitory phosphate group. Wee1 and Cdc25 are, in turn, phosphorylated by active MPF. Phosphorylation of Wee1 decreases its kinase activity, whereas phosphorylation of Cdc25 increases its phosphatase activity. From Tyson and Novak [22]; used by permission

Table 1 The Novak-Tyson model of the MPF control system in frog egg extracts (Fig. 2)

$\dfrac{du}{dt} = k_1 - \left[A_{apc}(u) + A_{wee}(u)\right]u + A_{25}(u)[v - u]$	$u(t) =$ active MPF
$\dfrac{dv}{dt} = k_1 - A_{apc}(u)v$	$v(t) =$ total cyclin
$A_{apc}(u) = k'_{apc} + k''_{apc}G\left(u, \theta_{apc}, J_{apc}\right)$	$A_{apc}(u) =$ activity of APC:Cdc20
$A_{wee}(u) = k'_{wee} + k''_{wee}G(\theta_{wee}, u, J_{wee})$	$A_{wee}(u) =$ activity of Wee1
$A_{25}(u) = k'_{25} + k''_{25}G(u, \theta_{25}, J_{25})$	$A_{25}(u) =$ activity of Cdc25
$G(u, \theta, J) = \dfrac{2Ju}{\theta - u + J\theta + Ju + \sqrt{(\theta - u + J\theta + Ju)^2 - 4Ju(\theta - u)}}$	Goldbeter-Koshland function
$k_1 = 0.02$ min^{-1} $k'_{apc} = 0.02$ min^{-1} $k''_{apc} = 0.8$ min^{-1}	
$k'_{25} = 0.04$ min^{-1} $k''_{25} = 4$ min^{-1} $k'_{wee} = 0, k''_{wee} = 1$ min^{-1}	
$\theta_{apc} = \theta_{wee} = \theta_{25} = 0.2$ $J_{apc} = 0.01$ $J_{wee} = J_{25} = 0.03$	

These ODEs are derived by combining ideas in Novak and Tyson [16, 17]

Limit-Cycle Oscillations

To get started, we need an '.ode' file for the model in Table 1, which I provide in Table 2a. Copy it into a simple text file ('Novak_Tyson_OscExtr.ode') and store the file in a convenient folder (say, 'xpp ode files'). Next open this .ode file in XPP and select Nullcline New from the menu on the far left. You should get a 'phase plane portrait' that looks something like Fig. 3a. I have sketched in the vector field with small arrows on the phase plane, guided by the 'nullclines'. The *u*-nullcline (the

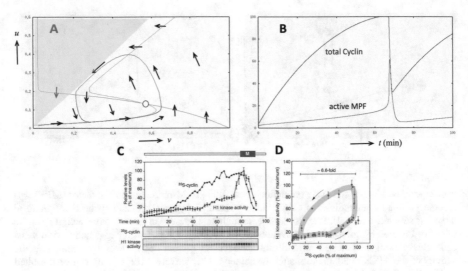

Fig. 3 Mathematical analysis and experimental confirmation of the mathematical model in Table 1: Oscillations. (**a**) Nullclines. The green curve is the u-nullcline, where $du/dt = 0$ and the vector field is horizontal; and the red curve is the v-nullcline, where $dv/dt = 0$ and the vector field is vertical. Where the nullclines intersect, both du/dt and dv/dt are 0, and the system is at a steady state (the white circle indicates that the steady state is unstable). The black curve (a 'trajectory') is a solution of the system of ODEs that starts at the initial condition specified in the .ode file and follows the arrows until it reaches a stable limit cycle oscillation (the curve that closes on itself). The gray shaded region is a "forbidden zone" where the concentration of active MPF exceeds the total concentration of cyclin molecules, which is impossible. (**b**) Time course. We plot active MPF, $u(t)$, and total cyclin, $v(t)$, as functions of time as given by the trajectory in panel A. The values of $u(t)$ and $v(t)$ have been multiplied by a scaling factor so that $v_{max} = 100\%$. (**c** and **d**) MPF oscillations in a frog egg extract; data from Pomerening et al. [21], used by permission. Murray and Kirschner [37] described how to prepare extracts of mature frog oocytes that will exhibit three or more spontaneous cycles of MPF activation and inactivation. Using such an extract, Pomerening et al. [21] made careful measurements of total cyclin concentration and MPF activity during the first mitotic cycle (panel C) and then plotted these data on a phase plane (panel D). The agreement between panels C and D and panels B and A, respectively, is remarkable and reassuring.

green curve) is the locus of points where $du/dt = 0$, i.e., where the vector field is horizontal (no change in the vertical direction); and the v-nullcline (the red curve) is the locus of points where $dv/dt = 0$, i.e., where the vector field is vertical (no change in the horizontal direction). The intersection of the nullclines (the white circle) is a steady state solution of the ODEs; in this case it is an unstable steady state. In the regions bounded by segments of the nullclines, the vector field lies along some 'compass direction' (e.g., NW, SE, etc.), as indicated by the small arrows at angles to the axes. Given any initial condition in the positive quadrant of the phase plane, the ensuing trajectory must follow the arrows, which force the trajectory to wind around the unstable steady state and settle on a closed orbit, called a 'stable limit cycle oscillation'. To find the limit cycle, select Initialconds Go and XPP will plot the trajectory in Fig. 3a. You may explore trajectories from any other initial conditions by selecting Initialconds Mouse from the menu and clicking anywhere on the phase plane. You will find that the computed solution of the system of ODEs quickly converges to the stable limit cycle in Fig. 3a. This closed orbit represents a periodically repeated alternation of active and inactive MPF, driven by the continuous synthesis and periodic degradation of the cyclin subunit (Fig. 3b). The abrupt spike in MPF activity at mitosis is driven by the phosphorylation and dephosphorylation of the kinase subunit, as described above. We will return to the experimental confirmation of the model (Fig. 3, panels C and D) after first discussing the phenomena of 'bistability' and 'hysteresis' in the control system

Bistability

In Fig. 4 we consider a simplified control system with no cyclin synthesis or degradation: total cyclin concentration is constant ($v(t) = v_{tot}$), and MPF activity is controlled solely by phosphorylation and dephosphorylation (see Table 2b). From Fig. 4a, b, we see that this system exhibits a phenomenon called 'bistability'. If v_{tot} is small ($v_{tot} < 0.23$), there is a single stable steady state with low MPF activity, because there is very little cyclin and what little dimer exists is phosphorylated and inactive. If v_{tot} is large ($v_{tot} > 0.76$), there is a single stable steady state with high MPF activity, because there is a lot of cyclin and the dimers are unphosphorylated and active. For intermediate values of v_{tot} ($0.23 < v_{tot} < 0.76$), three steady states coexist: a stable low state, a stable high state and an unstable intermediate state. Figure 4b shows the predictions of the model for an experiment in which the system is started in the low state, and v_{tot} is steadily increased (black squares): MPF activity remains low until v_{tot} exceeds 0.76, beyond which MPF abruptly activates. Conversely, if the system is started on the upper steady state, and v_{tot} is steadily decreased (white squares), then MPF activity declines, to be sure, but the dimers remain active until v_{tot} drops below 0.23, when the remaining MPF dimers are inactivated by phosphorylation of the kinase subunit.

Experimental Confirmation of Hysteresis

The simulation in Fig. 4b mimics the conditions of an experiment in frog egg extracts [18], in which there is (literally) no synthesis or degradation of cyclin molecules (see Fig. 4c). In this experiment, Solomon prepared a frog egg extract that contained all proteins of the MPF control system (Fig. 2) *except* for cyclin B, and supplemented the extract with cycloheximide, an inhibitor of protein synthesis (to prevent synthesis of endogenous cyclin B from maternal mRNA in the extract). Solomon then added a measured amount of exogenous, non-degradable cyclin B that lacked the amino-acid sequence necessary for cyclin B degradation by the APC: Cdc20 pathway. Then Solomon measured MPF activity in extracts with varying concentrations of cyclin B, to obtain a 'signal-response' relationship precisely like the black squares in Fig. 4b. The 'cyclin threshold for the activation of MPF' that

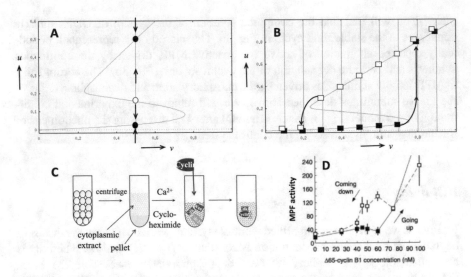

Fig. 4 Mathematical analysis and experimental confirmation of the mathematical model in Table 1: Bistability. (**a**) Phase plane diagram. The nullclines (green and red curves) are calculated from the ode file for the modified model in Table 2b. In this case, total Cyclin (v) is a constant, and only MPF activity (u) is changing in time. For $0.23 < v < 0.76$, the control system is bistable: it exhibits two stable steady states (black circles) separated by an unstable steady state (white circle). (**b**) Hysteresis. If start an extract on the lower, stable steady state and steadily increase the amount of total cyclin (black squares), the control system will stay on the lower state (interphase) until total cyclin exceeds $v = 0.76$, and thereafter it will jump abruptly to the upper steady state. If start an extract on the upper, stable steady state and steadily decrease the amount of total cyclin (white squares), the control system will stay on the upper state (mitosis) until total cyclin drops below $v = 0.23$, when it will jump abruptly to the lower state. (**c**) Solomon's protocol [18]; diagram from Tyson and Novak [22]; used by permission. Mature frog oocytes (arrested in meiosis with high MPF activity) are crushed by centrifugation and the cytoplasmic components are separated from the debris. The addition of calcium ions to the extract activates APC:Cdc20, causing degradation of cyclin B and loss of MPF activity. The addition of cycloheximide (an inhibitor of protein synthesis) prevents de novo production of cyclin B from maternal stores of mRNA. To this extract Solomon added exogenously prepared cyclin B molecules that had been genetically engineered to lack the amino acid sequence necessary for APC-induced proteolysis. In this way, Solomon created a cytoplasmic extract with a fixed and known (total) concentration of cyclin B. These cyclin B molecules combined with endogenous Cdc2 molecules to form MPF heterodimers, which could be either active (unphosphorylated) or inactive (phosphorylated on the Cdc2 subunit by the action of Wee1 in the extract). Solomon et al. observed a clear cyclin threshold for MPF activation, i.e., they observed the up-jump in panel B. The model predicts an equally clear and distinct cyclin threshold for the down-jump, which was confirmed in a different series of experiments by Sha et al. [19] and Pomerening et al. [20]. (**d**) Different cyclin thresholds for the activation and inactivation of MPF. Data from Pomerening et al. [20], used by permission

Solomon et al. observed was one-half of the hysteresis loop predicted by the model in Fig. 4b. The other half of the hysteresis loop consists of a lower cyclin threshold for the inactivation of MPF, when coming down in cyclin concentration from the upper steady state. This clear prediction of Novak and Tyson's 1993 model [16] was confirmed 10 years later, in two experimental papers, one from Jill Sible's lab at Virginia Tech [19] and the other (Fig. 4d) from Jim Ferrell's lab at Stanford [20].

Experimental Confirmation of Limit-Cycle Oscillations

Now we can return to Fig. 3c, d. These experiments [21] were carried out in an 'oscillatory extract', as in Fig. 4c but lacking cycloheximide and relying on the accumulation of endogenous (degradable) cyclin B. In this case, the extract makes its own cyclin B from maternal mRNA and degrades these cyclin B molecules abruptly at each nuclear division, when APC:Cdc20 is activated. With impressive attention to detail, Pomerening et al. measured total cyclin concentration and MPF activity at 2 min intervals and observed beautiful limit-cycle oscillation (Fig. 3c, d), exactly as predicted by the model (Fig. 3b, a, respectively).

Discussion

This case study illustrates what I call a 'dynamical paradigm for molecular cell biology' (Fig. 5) [22, 23]. The fundamental goal of cell biologists is to understand the molecular underpinnings of all the interesting and (sometimes) mysterious behaviors of living cells, as observed in laboratory experiments. The molecular machinery that implements and governs these behaviors can be ferreted out and characterized by well-developed methods of molecular genetics and biochemistry.

Molecular geneticists can identify the genes that are intimately involved in any particular aspect of cell physiology (e.g., cell cycle control); then clone the genes and characterize their protein products (e.g, protein kinases, protein phosphatases, transcription factors, etc.) and the interactions among these proteins (e.g., controlling protein synthesis and degradation, post-translational modifications, binding to activators and inhibitors, etc.). From this information molecular cell biologists can propose a 'working hypothesis' of the underlying molecular mechanism (e.g., Fig. 2). These proposals present several difficult questions: Is the working

Fig. 5 A dynamical perspective on molecular cell biology. From Tyson and Novak [22]; used by permission

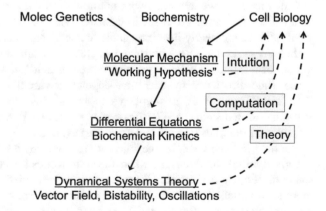

hypothesis sufficient to explain the observed cell behaviors that prompted the investigation to begin with? And, does the hypothesis make intriguing predictions that motivate new experiments?

As a first attempt to answer these questions, experimentalists might apply their biochemical intuition to the working hypothesis to try to link it up to the physiological facts in the case. For the (relatively) simple diagram in Fig. 2, this is a reasonable way to start, but it will soon be plagued by uncertainties. Why does the mechanism sometimes exhibit stable steady state behavior, as exhibited by immature frog oocytes (blocked in G2 phase of meiosis I) and mature frog oocytes (blocked in metaphase of meiosis II), and sometimes autonomous oscillations, as in the first 12 mitotic divisions of the fertilized egg? Why do these mitotic oscillations abruptly cease after 12 divisions, at the 4096-cell stage of the embryonic blastula? What is the nature of the abrupt threshold for MPF activation observed by Solomon in his experiments on frog egg extracts? It is difficult to answer these sorts of questions because of the complexity of feedback and feed-forward interactions in the mechanism, which render intuitive arguments inconclusive. The 'intuition dilemma' gets increasingly worse as we learn more and more about underlying molecular components and their interactions. Before long, the working hypothesis becomes so complicated that biochemical intuition fails calamitously.

To get around this 'curse of complexity' we, as computational cell biologists, might try to convert the molecular mechanism into a set of ordinary differential equations, based on the basic principles of biochemical kinetics. Then, in principle, we can let the computer work out the consequences of our hypothesis and compare simulations of these ODEs to the observed behavior of the cells. The sticking point of this approach is that any reasonable mathematical model of a biochemically realistic hypothesis will consist of ten or more differential equations, and each differential equation will involve a handful of kinetic 'parameters' (rate constants, binding constants, etc.), and we do not know the numerical values of these parameters *a priori*. The parameter values must be estimated from experimental data, and we may not have reliable kinetic data for most (or even a few) of these parameters. Without the 'correct' parameter values, we cannot carry out a computer simulation of the equations, so we are stuck with nothing from the model to compare with the behavior of the cells. For a realistic model with 10-100 undetermined parameter values, this is the nearly insurmountable 'curse of parameter space'.

In this case study, I have illustrated an alternative approach. We should think of a system of nonlinear ODEs as defining a 'vector field' in a high-dimensional 'state space' (one dimension for each time-dependent variable in the system of ODEs) [24, 25]. Figure 3 is a simple example of this notion, where we have a two-dimensional state space (a 'phase plane') and we can visualize the vector field by plotting nullclines (the red and green curves) in the phase plane. Once we see the vector field, we can visualize the trajectory followed by a solution of the ODEs from any given initial condition, and where the system ends up: at a stable steady state or a stable limit cycle oscillation. Of course, in a more realistic model, with a dozen or more state variables and a multitude of parameters, we cannot see the vector field with our own eyes, but mathematicians have devised a suite of powerful analytical tools (e.g., one- and two-parameter bifurcation diagrams; the topics of other 'case studies' in this book) to characterize the properties of the vector field. These theoretical tools can guide our intuition and computations to a successful account of the observed physiological characteristics of living cells. And also provide a framework for making counter-intuitive predictions about the behavior of cells under novel conditions; surprising behaviors that can be tested experimentally.

Even in the simple example given in this case study, we can begin to see the power of this dynamical perspective. The signal-response properties of frog oocytes, early frog embryos, and frog egg extracts exhibit certain characteristic traits, such as a stable G2-arrested steady state with low MPF activity (in immature oocytes), a stable M-arrested steady state with high MPF activity (in mature oocytes), sustained oscillations of MPF activity (during early embryonic divisions), and abrupt transitions between states of inactive and active MPF (in extracts). These traits can be identified with characteristic attractors of a dynamical system based on a postulated mechanism for regulating MPF activity in frog eggs: namely, stable steady states delineated by 'saddle-node' bifurcations (the turning points in the MPF activity curve—the green line—in Fig. 4a, b), and the generation of stable limit cycle oscillations around an unstable steady state, as in Fig. 3a (called a 'Hopf' bifurcation). Bifurcation theory (see the case study on 'Cell Cycle Regulation. Bifurcation Theory') assists us in designing a model that has the dynamic properties necessary to account for the observed signal-response characteristics of the cells under consideration. Bifurcation theory also points us in the direction of parameter values that exhibit these characteristics, giving us a fighting chance to simulate them in quantitative detail by computer calculations and comparison to quantitative experimental data. These comparisons, if they are successful, give us confidence in the reliability of the proposed molecular mechanism and hone our intuition on the consequences of the model. If the comparisons are unsuccessful in part, the disagreements between theory and experiment may point us in the direction of improvements to the model, to bring our mechanistic hypotheses into closer alignment with underlying molecular realities.

Once we have gained some confidence in our molecular mechanism and its mathematical model, we can use the model to make new predictions, as in Exercise #1 on 'critical slowing down' and Exercise #2 on the unreplicated DNA checkpoint. We can also extend this model of the simple MPF control system in frog oocytes and early embryonic divisions to more complex regulations of growth and division in yeast cells and mammalian cells.

Exercises

1. In this exercise, we will investigate the behavior of the mitotic control system close to the saddle-node bifurcation where MPF is abruptly activated by an increasing concentration of total cyclin. Load the .ode file for the 'interphase extract' (Table 2b) into XPPAUT and follow these steps:

   i. Select Xi vs t and hit return twice. The plot window will change to U vs t.
   ii. Select Viewaxes 2D and set Ymin to -0.05 and Ymax to 1.05. This will give you a nice plot window to view your simulations.

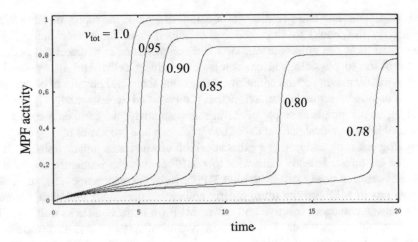

Fig. 6 Critical slowing down

iii. Open `Param` window and `IC` window. Slide both windows to the right of the plot window.

iv. Set `vtot=1.0` in the param window and click `OK`; set `v=1.0` in the `IC` window and click `OK`. Then click `Go`. XPP will plot the trajectory `u(t)` vs `t`, which starts at `u(0)=0` and ends at `u(20)=1.0`.

v. Repeat step (iv) for `vtot=v(IC)=0.95,0.90,0.85,0.80,0.78`. You will end up with the plot in Fig. 6:

What's the meaning of this plot? Each value of `vtot` is larger than the 'up-jump' point (at $v = 0.76$) on the phase plane diagram in Fig. 4b, and so u (t) (i.e., MPF activity) increases from 0 to u_{max}, the final steady-state value given by the black squares on the upper branch of the hysteresis curve. However, the time it takes for $u(t)$ to increase from 0 to $\frac{1}{2}u_{max}$, call it the 'lag' time, gets longer and longer as `vtot` approaches 0.76 from above. This observation is a general property, called 'critical slowing down', of a dynamical system close to a 'jump' point like the one at $v = 0.76$. The model's prediction of 'critical slowing down' contradicted a conclusion of Solomon's paper, that the time required for MPF activation in his protocol was independent of the amount of non-degradable cyclin B added to the extract. But Solomon reached this conclusion because he only studied cyclin B levels that were considerably larger than the up-jump threshold. Sha et al. [19] revisited this experiment and showed (see their Fig. 6 in Ref. [19]) that the lag time does indeed increase dramatically as the amount of non-degradable cyclin B gets closer to the threshold.

2. In this exercise, we will see why unreplicated DNA might delay entry of a frog egg extract into mitosis, as observed by Dasso and Newport [26]. The model suggests that this delay might be due to raising the threshold for MPF activation (the up-jump point in Fig. 4b), in which case it would take a longer time for the

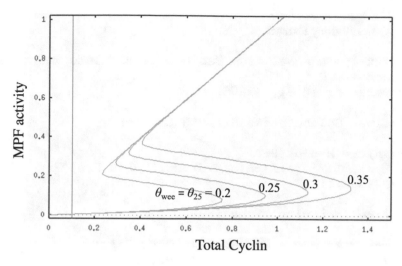

Fig. 7 Unreplicated DNA checkpoint

extract to synthesize enough cyclin B to surmount the threshold. Furthermore, the model suggests that location of the threshold is sensitively dependent on the parameters θ_{wee} and θ_{25} in Table 1. To see how this happens, load the .ode file for the 'interphase extract' (Table 2b) into XPPAUT and follow these steps:

i. Select Viewaxes 2D and set Ymin = -0.05, Xmax = 1.5 and Ymax = 1.05. This will give you a nice phase plane window to view nullclines.
ii. Open Param window and slide it to the right of the plot window.
iii. Set Hwee=0.2 and H25=0.2 in the Param window and click OK; then click Nullcline New. XPP will plot the nullclines, u vs v, for MPF activity (the green curve) and total Cyclin (the red curve). The cyclin threshold for MPF activation (the up-jump point) occurs at $v = 0.76$, as previously.
iv. Repeat step (iii) for Hwee=H25=0.25,0.30,0.35. You will end up with the phase-plane portrait in Fig. 7:

What's the meaning of this plot? If unreplicated DNA increases the activity of the phosphatase that opposes MPF kinase in the phosphorylation of Wee1 and Cdc25, then increasing amounts of unreplicated DNA in a frog egg extract will delay entry into mitosis (as observed by Dasso and Newport), because the extract must synthesize increasing amounts of cyclin B to surmount the threshold for MPF activation. This prediction of the model was also confirmed by Sha et al. [19] in their Fig. 5.

Method

Table 2 Novak_Tyson.ode files

(A) Oscillatory Extract

```
# Novak-Tyson model from Table 1 to simulate MPF oscil-
lations
# Pomerening et al. (2005)

# Define 'Goldbeter-Koshland' function

BB(u,H,J)= H-u+J*(H+u)
G(u,H,J) = (2*J*u)/(BB(u,H,J)+sqrt(BB(u,H,J)^2-
4*J*u*(H-u)))

# Define 'activity' functions

Aapc(u) = kp_apc + kpp_apc*G(u,Hapc,Japc)
Awee(u) = kp_wee + kpp_wee*G(Hwee,u,Jwee)
A25(u) = kp_25 + kpp_25*G(u,H25,J25)

# Define ODEs

du/dt = k1 - (Aapc(u)+Awee(u))*u + A25(u)*(v-u)
dv/dt = k1 - Aapc(u)*v

# Define parameter values for limit cycle oscillations

par k1=0.02
par kp_apc=0.02, kpp_apc=0.8, Hapc=0.2, Japc=0.01
par kp_wee=0, kpp_wee=1, Hwee=0.2, Jwee=0.03
par kp_25=0.04, kpp_25=4, H25=0.2, J25=0.03

# Define auxiliary variables for plotting "relative
levels"
# of total cyclin and active MPF

par sf=155
aux MPF=sf*u
aux Cyclin=sf*v

# Initial conditions

init u=0.015, v=0.025

# Default settings for XPP
```

(continued)

Table 2 (continued)

```
@ XP=v, YP=u, NMESH=100, XLO=0, XHI=1, YLO=0, YHI=0.6,
DT=0.01
@ TOTAL=100, BOUNDS=1000

Done
```

(B) Interphase Extract

```
# Novak-Tyson model to simulate bistability and hyste-
resis
# in the case of no cyclin synthesis or degradation
# Solomon et al. (1990)

# Define 'Goldbeter-Koshland' function

BB(u,H,J)= H-u+J*(H+u)
G(u,H,J) = (2*J*u)/(BB(u,H,J)+sqrt(BB(u,H,J)^2-
4*J*u*(H-u)))

# Define 'activity' functions

Awee(u) = kp_wee + kpp_wee*G(Hwee,u,Jwee)
A25(u) = kp_25 + kpp_25*G(u,H25,J25)

# Define ODEs

du/dt = A25(u)*(v-u) - Awee(u)*u
dv/dt = vtot - v

# Define parameter values for bistability

par vtot=.5
par kp_wee=0, kpp_wee=3, Hwee=0.2, Jwee=0.03
par kp_25=0.04, kpp_25=20, H25=0.2, J25=0.03

# Default settings for XPP

@ XP=v, YP=u, NMESH=100, XLO=0, XHI=1, YLO=0, YHI=0.6,
DT=0.01

done
```

Additional Reading

The earliest models of the MPF control system were presented by Hyver and Le Guyader [27], Norel and Agur [28], Goldbeter [29], and Tyson [30]. Qu, MacLellan and Weiss [31] published an excellent study of bistability and oscillations in the MPF control network. General dynamical principles of cell cycle regulation have been reviewed by Tyson, Csikasz-Nagy and Novak [32] and by Tyson and Novak [33]. Some interesting discussions of the role of mathematical modeling in molecular cell biology were published by Gunawardena [34], by Tyson and Novak [35] and by Phair [36].

References

1. Nurse P (1990) Universal control mechanism regulating onset of M-phase. Nature 344 (6266):503–508
2. Morgan D (2007) The cell cycle: principles of control. New Science Press, London
3. Masui Y, Markert CL (1971) Cytoplasmic control of nuclear behavior during meiotic maturation of frog oocytes. J Exp Zool 177(2):129–145
4. Gerhart J et al (1984) Cell cycle dynamics of an M-phase-specific cytoplasmic factor in Xenopus laevis oocytes and eggs. J Cell Biol 98(4):1247–1255
5. Evans T et al (1983) Cyclin: a protein specified by maternal mRNA in sea urchin eggs that is destroyed at each cleavage division. Cell 33(2):389–396
6. Hartwell LH et al (1974) Genetic control of the cell division cycle in yeast. Science 183 (4120):46–51
7. Hartwell LH et al (1970) Genetic control of the cell-division cycle in yeast. I Detection of mutants Proc Natl Acad Sci U S A 66(2):352–359
8. Nurse P et al (1976) Genetic control of the cell division cycle in the fission yeast Schizosaccharomyces pombe. Mol Gen Genet 146(2):167–178
9. Nurse P (1975) Genetic control of cell size at cell division in yeast. Nature 256(5518):547–551
10. Gould KL et al (1990) Complementation of the mitotic activator, p80cdc25, by a human protein-tyrosine phosphatase. Science 250(4987):1573–1576
11. Russell P, Nurse P (1987) Negative regulation of mitosis by wee1+, a gene encoding a protein kinase homolog. Cell 49(4):559–567
12. Russell P, Nurse P (1986) cdc25+ functions as an inducer in the mitotic control of fission yeast. Cell 45(1):145–153
13. Lohka MJ et al (1988) Purification of maturation-promoting factor, an intracellular regulator of early mitotic events. Proc Natl Acad Sci U S A 85(9):3009–3013
14. Gautier J et al (1990) Cyclin is a component of maturation-promoting factor from Xenopus. Cell 60(3):487–494
15. Gautier J et al (1988) Purified maturation-promoting factor contains the product of a Xenopus homolog of the fission yeast cell cycle control gene cdc2+. Cell 54(3):433–439
16. Novak B, Tyson JJ (1993) Numerical analysis of a comprehensive model of M-phase control in Xenopus oocyte extracts and intact embryos. J Cell Sci 106(Pt 4):1153–1168
17. Novak B, Tyson JJ (1993) Modeling the cell division cycle: M-phase trigger, oscillations and size control. J Theor Biol 165:101–134
18. Solomon MJ et al (1990) Cyclin activation of p34cdc2. Cell 63(5):1013–1024
19. Sha W et al (2003) Hysteresis drives cell-cycle transitions in *Xenopus laevis* egg extracts. Proc Natl Acad Sci U S A 100(3):975–980

20. Pomerening JR et al (2003) Building a cell cycle oscillator: hysteresis and bistability in the activation of Cdc2. Nat Cell Biol 5(4):346–351
21. Pomerening JR et al (2005) Systems-level dissection of the cell-cycle oscillator: bypassing positive feedback produces damped oscillations. Cell 122(4):565–578
22. Tyson JJ, Novak B (2015) Bistability, oscillations, and traveling waves in frog egg extracts. Bull Math Biol 77(5):796–816
23. Tyson JJ, Novak B (2020) A dynamical paradigm for molecular cell biology. Trends Cell Biol 30(7):504–515
24. Edelstein-Keshet L (2005) Mathematical models in biology. Society for Industrial and Applied Mathematics
25. Strogatz S (2015) Nonlinear dynamics and chaos: with applications to physics, biology, chemistry, and engineering, 2nd edn. Westview Press, Boulder
26. Dasso M, Newport JW (1990) Completion of DNA replication is monitored by a feedback system that controls the initiation of mitosis in vitro: studies in *Xenopus*. Cell 61(5):811–823
27. Hyver C, LeGuyader H (1990) MPF and cyclin: modelling of the cell cycle minimum oscillator. Biosystems 24(2):85–90
28. Norel R, Agur Z (1991) A model for the adjustment of the mitotic clock by cyclin and. MPF levels Science 251(4997):1076–1078
29. Goldbeter A (1991) A minimal cascade model for the mitotic oscillator involving cyclin and cdc2 kinase. Proc Natl Acad Sci U S A 88(20):9107–9111
30. Tyson JJ (1991) Modeling the cell division cycle: cdc2 and cyclin interactions. Proc Natl Acad Sci U S A 88(16):7328–7332
31. Qu Z et al (2003) Dynamics of the cell cycle: checkpoints, sizers, and timers. Biophys J 85 (6):3600–3611
32. Tyson JJ et al (2002) The dynamics of cell cycle regulation. BioEssays 24:1095–1109
33. Tyson JJ, Novak B (2013) Irreversible transitions, bistability and checkpoint controls in the eukaryotic cell cycle: a systems-level understanding. In: Walhout AJM, Vidal M, Dekker J (eds) Handbook of systems biology. Academic Press, Cambridge, pp 265–285
34. Gunawardena J (2014) Models in biology: 'accurate descriptions of our pathetic thinking'. BMC Biol 12:29
35. Tyson JJ, Novak B (2015) Models in biology: lessons from modeling regulation of the eukaryotic cell cycle. BMC Biol 13:46
36. Phair RD (2014) Mechanistic modeling confronts the complexity of molecular cell biology. Mol Biol Cell 25(22):3494–3496
37. Murray AW, Kirschner MW (1989) Cyclin synthesis drives the early embryonic cell cycle. Nature 339(6222):275–280

Mitotic Cycle Regulation. II. Traveling Waves

John J. Tyson

Introduction

In an earlier case study, Mitotic Cycle Regulation. I. Oscillations and Bistability, we considered the regulation of MPF activity ('M-phase Promoting Factor') in a frog-egg extract, which is (to a good approximation) a spatially homogenous, biochemical reaction mixture in a test tube. We proposed a network of biochemical reactions for the synthesis and degradation of cyclin B (one component of MPF) and for the phosphorylation and dephosphorylation of Cdk1 (the other component of MPF). We wrote a pair of nonlinear ordinary differential equations (ODEs) to express the rates of these reactions, according to standard principles of biochemical kinetics. In that previous case study, we showed that this dynamical system has some surprising properties, including multiple stable steady states ('bistability') and sustained periodic oscillations ('limit cycles'). In this case study, we consider frog-egg extract loaded into a long, thin piece of tubing, for which the dynamics of the local biochemical reactions must be supplemented with the spatial transport of chemicals by molecular diffusion along the length of the tube. We will show that, in such systems of coupled reaction and diffusion, it is possible for waves of MPF activation to propagate through the reaction vessel.

This phenomenon of traveling waves of chemical reaction is no mere mathematical curiosity. Chang and Ferrell [1] observed waves of MPF activity propagating at a speed of ~50 μm/min along thin tubes filled with frog-egg extract (Fig. 1). Similar waves of cyclic AMP production have been observed in fields of aggregating

J. J. Tyson (✉)
Department of Biological Sciences and Division of Systems Biology, Virginia Polytechnic Institute and State University, Blacksburg, VA, USA
e-mail: tyson@vt.edu

© Springer Nature Switzerland AG 2021
P. Kraikivski (ed.), *Case Studies in Systems Biology*,
https://doi.org/10.1007/978-3-030-67742-8_2

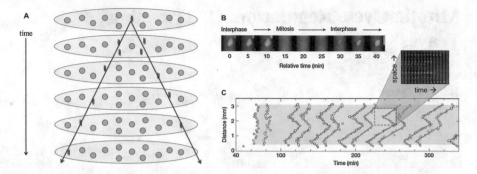

Fig. 1 Traveling waves of mitosis in a frog-egg extract. (**a**) Schematic diagram of mitotic waves propagating through a syncytium (a common cytoplasmic pool containing numerous nuclei; yellow = cytoplasm, orange circles = interphase nuclei, orange 'toothpicks' = pairs of replicated chromosomes in mitosis). In the first frame, the central nucleus is in mitosis, i.e., a region of high MPF activity. In the second frame, the central nucleus has completed mitosis and returned to interphase, while the neighboring nuclei on either side have been driven into mitosis by a wave of MPF activity that emanated from the central region. In the following frames, the wave of mitosis propagates in both directions through the syncytium. The slope of the lines (Δdistance/Δtime) is the speed of the wave. (**b**, **c**) Experimental observations of Chang and Ferrell [1]. A cytoplasmic extract of *Xenopus* eggs is loaded into a piece of Teflon tubing (0.1 mm internal diameter × 2.5 mm length), along with sperm chromatin, which assembles into discrete nuclei surrounded by nuclear envelopes, and green fluorescent protein (GFP) with a nuclear localization signal (NLS). Panel B: When the nucleus is in interphase, with an intact nuclear envelope, it accumulates GFP-NLS and glows green (lanes 0, 5, 10 min). When the nucleus enters mitosis (i.e., when it experiences an oncoming wave of MPF activity), its nuclear envelope breaks down and the green fluorescence disappears (lanes 15, 20, 25 min). When the nucleus returns to interphase, the nuclear envelope reassembles (lanes 30, 35, 40). Panel C: A space-time plot indicating when each nucleus loses green fluorescence (red circles) and regains fluorescence (blue circles). Clear waves of entry into mitosis and return to interphase propagate along the Teflon tube. The speed of the wave in the inset is ~30 μm/min. The speed of the first wave is ~45 μm/min

amoebae [2], and waves of Ca^{2+} release in fertilized eggs [3, 4]. Dramatic waves of oxidation are observed in a purely inorganic reaction mixture, the Belousov-Zhabotinsky reaction [5, 6]. Action potentials, which are waves of electrochemical activity that travel along a nerve axon [7], are probably the most well-known and best understood example of 'activity' waves driven by a coupling of local biochemical reactions with spatial fluxes generated by activity gradients. These and other examples, with representative references, are cataloged in Table 1.

The theory of traveling waves in systems of coupled reactions and diffusion has a long and distinguished history [26], which we shall illustrate in this case study. We start with a model of nerve impulse propagation, where the theory is exceptionally clear. Then we compute waves of MPF activity in a model of frog egg extract, and compare the theoretical model with the experiments of Chang and Ferrell.

Table 1 Wave propagation in biological, biochemical, biophysical and chemical systems

System	Chemical species involved	Biological function of waves	References
Early embryos	Cyclin-dependent kinase, Wee1, Cdc25, APC:Cdc20	Synchronization of DNA synthesis and nuclear divisions	[1, 8, 9]
Fertilized egg	Ca^{2+}, inositol trisphosphate (IP_3)	Initiation of early embryonic cell divisions	[3, 4, 10–12]
Aggregating amoebae	Cyclic AMP, adenylate cyclase, phosphodiesterase	Intercellular communication and synchronization	[2, 13–15]
Embryonic development	FGF, Wnt, notch signaling pathways	Somitogenesis	[16–18]
Bacterial motility	FrzE, FrzF, FrzCD	Intercellular communication and synchronization	[19, 20]
Nerve axon	Membrane potential, fast Na^+ influx, slow K^+ efflux	Propagation of nerve impulses from cell body to synapse	[7, 12, 21, 22]
Belousov-Zhabotinsky	$HBrO_3$, $HBrO_2$, Br^-, $Fe^{2+/3+}$	None. A traveling wave of oxidation of Fe^{2+} to Fe^{3+}	[5, 6, 23–25]

FitzHugh-Nagumo (FHN) Model of Nerve Impulse Propagation

We start with a simple, two-variable model of a propagating action potential [12]:

$$\frac{\partial v}{\partial t} = v(v - \theta)(1 - v) - w + I_{app} + D\frac{\partial^2 v}{\partial x^2} \tag{1a}$$

$$\frac{\partial w}{\partial t} = \varepsilon(v - \gamma w) \tag{1b}$$

This model was first proposed by FitzHugh [21] and by Nagumo et al. [22]. In Eq. (1), $v(x, t)$ = membrane potential as a function of time t and distance x along the axon, and $w(x,t)$ = ionic current transverse to the membrane. The parameters in Eq. (1) are θ = threshold for excitation, I_{app} = applied current, $D^{1/2}$ = space constant for longitudinal currents, ε = time constant for the slow transverse current ($0 < \varepsilon \ll 1$), and γ = proportionality constant connecting the slow ionic current to the local membrane potential. The term $D\partial^2 v/\partial x^2$ expresses the fact that the local buildup of charge at position x, due to longitudinal currents inside the axon, depends on the second spatial derivative of $v(x, t)$. All variables and constants have been scaled to dimensionless terms. Roughly speaking, $D \approx 1$, $\varepsilon \approx 0.1$, $v \approx (E_m - 60mV)/(120mV)$, $t \approx$ time/(0.1 ms), $x \approx$ distance/(5 mm), $0 < \theta < 1$, and $\gamma \approx 1$.

Action potential. First, let's investigate the dynamics of the 'local' ODEs, i.e., no diffusion, $D = 0$ in Eq. (1). Write an ode file to implement these ODEs in XPP, with $I_{app} = 0$, $\varepsilon = 0.01$, $\theta = 0.2$, and $\gamma = 2.5$. Load your ode file into XPP and draw the phase plane portrait illustrated in Fig. 2a. The ODEs have a single stable steady state at $v = w = 0$. Small perturbations away from the steady state in any direction return directly to the resting membrane potential, $v = 0$ ($E_m \approx -60$ mV). But a sufficiently

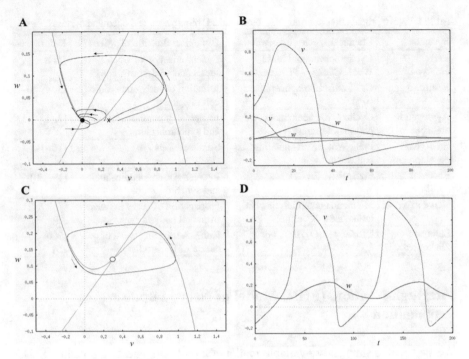

Fig. 2 FitzHugh-Nagumo model of an action potential. XPP screen shots of Eq. (1) for $D = 0$, $\varepsilon = 0.01$, $\theta = 0.2$, and $\gamma = 2.5$. In the phase plane portraits, the v-nullcline is red, and the w-nullcline is green. In the time courses, *real time* $= t/(10$ ms$)$. (**a, b**) Excitability; $I_{app} = 0$. The steady state (black circle) is stable but 'excitable;' i.e., small perturbations come directly back to the steady state, but large perturbations (x) go off on a long transient before returning to the steady state. The transient excitation (depolarization) of the membrane is illustrated in panel B. Small depolarizations return directly to the resting potential, $v = 0$, and $w(t)$ never rises above 0.015. For depolarizations above $v = 0.3$, the membrane becomes hyperpolarized to a maximum potential of $v = 0.88$ within 1.5 ms, then hypopolarized to $v = -0.22$ at 4 ms, and finally recovers to the resting potential at ~10 ms. Meanwhile, $w(t)$ rises to ~0.15 during the phase of membrane depolarization, and falls back to zero as the membrane potential returns to rest. (**c, d**) Oscillations; $I_{app} = 0.1$. If a sufficiently large, constant current is applied across the membrane, the axon will generate periodic action potentials. In this case, the period is ~9 ms. During the up-jump, $w \approx 0.08$, and during the down-jump $w \approx 0.21$

large depolarization of the membrane, say to $v = 0.3$ ($E_m \approx -25$ mV), triggers a reversal of membrane polarity, i.e., the membrane becomes hyperpolarized to a maximum potential of $E_m \approx +45$ mV within 1.5 ms (see the time course in Fig. 2b). Subsequently the slow K$^+$ channel opens (w increases), and the membrane potential drops to $v = -0.22$ ($E_m \approx -85$ mV) at $t \approx 4$ ms. Finally, as $w(t)$ decreases back to 0, the membrane returns to its resting potential at ~10 ms. These numbers agree reasonably well with the action potential observed in a squid giant axon [7, 12].

Next, change I_{app} to 0.1 and plot the phase plane portrait and time course in Fig. 2c, d. The applied current induces sustained oscillations of membrane potential, with a period of ~9 ms. The oscillations agree reasonably well with experiments at an applied current of 20 μA/cm².

Traveling wave front. Now imagine a long, slender nerve axon, where the membrane potential $v(x, t)$ is governed by Eqs. (1a,b), which couples local interactions between v and w that are driving transmembrane currents, as just described, and a longitudinal current driven by the second derivative of $v(x, t)$ in the spatial direction. In due course, we will solve this partial differential equation (PDE) numerically, but first we look at a simplified version of Eq. (1). Notice, in Fig. 2, that $v(t)$ changes more rapidly than $w(t)$, which is a consequence of our assumption that $\varepsilon \ll 1$, i.e., that $w(t)$ is a 'slow' variable. This observation suggests that, to a first approximation, we set $w = 0$ in Eq. (1), and look for a solution of the single PDE.

$$\frac{\partial v}{\partial t} = kv(v - \theta)(1 - v) + D\frac{\partial^2 v}{\partial x^2} \tag{2}$$

In this equation we have introduced a new parameter, k, which governs the local rate of change of $v(x, t)$. Because v is a dimensionless variable, k has units time^{-1} and D has units length^2 time^{-1}.

In Fig. 3, we compute a solution to Eq. (2), using XPP and the ode file in the Methods section M1. The figure shows a wave of excitation, v increasing from 0 (blue) ahead of the wave to 1 (red) behind the wave, moving from left-to-right across the domain at a speed of 0.6 space units per time unit. The shape of the wave front is nearly time invariant as it moves across the domain.

This computation suggests that we should look for a 'traveling wave' solution to Eq. (2) of the form

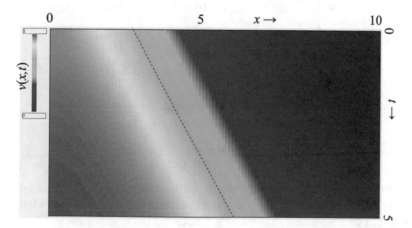

Fig. 3 Traveling wave front solution to PDE (2). Using XPP, we compute the solution $v(x,t)$ to Eq. (2) from the initial condition $v(x,0) = 1/(1+e^{(x-3.3)})$. Parameter values: $k = 2$, $\theta = 0.2$, $D = 1$. The dashed line tracks the position of the 'wave front' (where $v(x,t) = 0.5$) across the domain as time proceeds. The slope of this line is the wave speed, $c = 0.6$

$$v(x, t) = p(z), \text{where } z = x - ct. \tag{3}$$

A solution of this form represents a fixed waveform, $p(z)$, that is propagating left-to-right at speed $c > 0$. If we make this change of variables, then the PDE (2) becomes

$$-c\frac{dp}{dz} = kp(p - \theta)(1 - p) + D\frac{d^2p}{dz^2}, \quad p(-\infty) = 1, p(+\infty) = 0, \tag{4}$$

which is a second order, nonlinear ODE for $p(z)$, which must satisfy the given boundary conditions at $z = \pm\infty$.

Analytical solution. Quite surprisingly, this ODE has an analytic solution, which we can find by making a 'lucky' guess:

$$p(z) = \frac{1}{1 + e^{Az}}. \tag{5}$$

This waveform connects the stable state $p = 1$, as $z \to -\infty$, to the stable steady state $p = 0$, as $z \to +\infty$. It represents a sigmoidally shaped wave front with an inflection point at $z = 0$, $p = \frac{1}{2}$, where the slope of the wave front is $-A/4$. Differentiating the function $p(z)$ in Eq. (5), we obtain

$$\frac{dp}{dz} = -Ae^{Az}p^2, \text{and} \quad \frac{d^2p}{dz^2} = -A^2 e^{Az}(1 - e^{Az})p^3; \tag{6}$$

and substituting these expressions in Eq. (4), we get

$$cAe^{Az}p^2 = kp(p - \theta)(1 - p) - DA^2 e^{Az}(1 - e^{Az})p^3. \tag{7}$$

After a little algebra, Eq. (7) becomes

$$0 = (k(1 - \theta) - cA - DA^2) - (k\theta + cA - DA^2)e^{Az}. \tag{8}$$

For this identity to be true for all $-\infty < z < \infty$, we must insist that

$$A = \sqrt{\frac{k}{2D}}, \text{and} \quad c = \sqrt{\frac{kD}{2}}(1 - 2\theta). \tag{9}$$

These two equations give us the speed, c, of the wave front and a measure of the width of the wave front, $A^{-1} = \sqrt{2D/k}$. (Show that, in the wave front, v falls from 0.88 to 0.12 over a distance $\Delta x = 4A^{-1}$). Notice that the speed of the wave front is > 0 (i.e., the wave moves to the right) for $\theta < 0.5$, and $c < 0$ (i.e., the wave moves to the left) for $\theta > 0.5$, and the wave front is stationary (unmoving) for $\theta = 0.5$.

Graphical solution. By direct verification, we have shown that PDE (2) has a solution $v(x, t) = p(x - ct) = 1/(1 + e^{A(x - ct)})$, which is a wave front traveling left-to-right at a characteristic speed, c. It is also possible to prove the existence of a traveling wave front of speed c by a graphical method that is applicable to PDEs of the form of Eq. (2), with the cubic nonlinearity, $u(u - \theta)(1 - u)$, replaced by a more general nonlinear function, $f(u)$, with a 'cubic' shape. We proceed by defining $q(z) = dp/dz$, and writing Eq. (4) as a pair of first order, nonlinear ODEs

$$\frac{dp}{dz} = q \tag{10a}$$

$$D\frac{dq}{dz} = -cq - kp(p - \theta)(1 - p) \tag{10b}$$

This pair of ODEs defines a vector field in the phase plane, which we shall view (as usual) by XPP. Write an ode file to implement the ODEs in Eq. (10) and plot the phase plane portrait given in Fig. 4.

The middle one of the three mustard-colored curves in Fig. 4 is the solution of Eq. (1) that we are looking for. It 'starts' at $p(-\infty) = 1$ and 'ends' at $p(+\infty) = 0$;

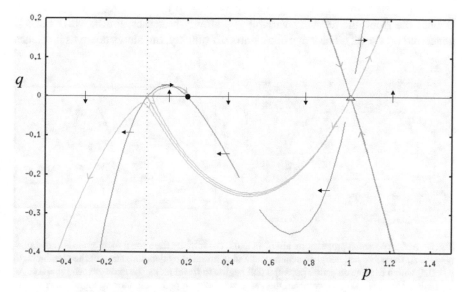

Fig. 4 Phase plane portrait for ODEs (10a, b). Parameter values: $k = 2$, $D = 1$, $\theta = 0.2$, and $c = 0.6$. The p- and q-nullclines (red and green curves, respectively) intersect at three steady states: $p,q = 0,0$ (saddle point), $= 0.2,0$ (stable focus), and $= 1,0$ (saddle point). The turquoise curves are the two trajectories that proceed directly into the saddle point at $p = 1$ as $z \to \infty$, and the mustard colored curves are the trajectories that come directly out of the saddle point (i.e., they fall into the saddle point as $z \to -\infty$). The middle mustard curve, for $c = 0.6$, connects the two saddle points, starting at $p = 1$ (as $z \to -\infty$) and ending at $p = 0$ (as $z \to +\infty$). For $c = 0.58$, the mustard curve passes to the left of the saddle point at $p = 0$ and sails off to the 'southwest'. For $c = 0.62$, the mustard curve passes to the right of the saddle point at $p = 0$ and proceeds to the stable focus at $p = 0.2$

along the way, $p(z)$ is monotone decreasing, i.e., $q(z) = dp/dz < 0$, and the traveling wave, $v(x,t) = p(x - ct)$, is moving left-to-right at speed $c = 0.6$. There is a unique speed c for the traveling wave: if $c < 0.6$, then $p(z)$ shoots off to $-\infty$ as $z \to \infty$, and if $c > 0.6$, then $p(z)$ goes to θ as $z \to \infty$. Neither case satisfies the two end-point conditions in Eq. (4). It should be obvious that this graphical argument holds if the nonlinear function $kp(p-\theta)(1-p)$ is replaced by any nonlinear function $f(p)$ with a cubic shape. For instance, in Exercise 1 you are asked to find the speed of the traveling wave solution of Eq. (10) when $f(p) = p(1-p) - \gamma\frac{p-\delta}{p+\delta}$.

Traveling pulse. Now we add back the slow variable, $w(x,t)$, retaining the assumption that $w \approx$ constant in the wave front. The result is a wave front of depolarization ('excitation') propagating along the axon at a speed $c = 0.6$, followed by a wave of repolarization ('recovery') moving behind the front at the same speed. Behind the wave back, the axonal membrane recovers its resting state, ready to propagate another action potential when triggered by a super-threshold membrane depolarization at the axon hillock (Fig. 5).

To check this theory, we will compute a traveling pulse solution to the full FHN model, Eq. (1). The ode file needed for this calculation is provided in Methods section M2. Running this file as directed, one gets the results in Fig. 6. A traveling pulse (the yellow band) runs across the domain, from left to right. First, a wave front kicks the system from its resting state ($v = w = 0$) to an excited state ($v \approx 0.8$). Behind the front, the recovery variable (w) slowly increases, until a wave back is generated (at $t \approx 22$). The wave back starts off quickly, but slows down as it catches

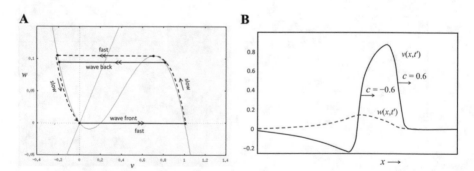

Fig. 5 Action potential propagation along an axon. (**a**) Initially the axonal membrane is at rest ($v = w = 0$). At the nerve cell body, an action potential is initiated by depolarization of the membrane to $v = 0.3$, which triggers an action potential that begins to travel along the axon at speed $c = 0.6$. As the action potential moves away from the nerve cell body, it pulls neighboring regions away from the resting potential $v = 0$ to the fully depolarized state $v = 1$; see panel B. (We are assuming that w changes little over the narrow width of the wave front.) During the depolarization phase, w increases slowly from 0 to ~0.1, as w reaches the top of the v-nullcline. At this point there develops a wave back, initially with speed $c = -1$ (because $\theta = 1$). This wave of repolarization carries v to -0.2. Behind the wave back, v and w both recover slowly to the rest state, $v = w = 0$. Initially the wave back (the dashed horizontal line) travels faster than the wave front and, consequently, moves into regions of lower values of w, until it eventually stabilizes at a speed $c = -0.6$ (when $\theta = 0.8$). (**b**) The action potential $v(x,t')$ as a function of space x at a fixed time t'. The wave front and back are both moving left-to-right at speed 0.6

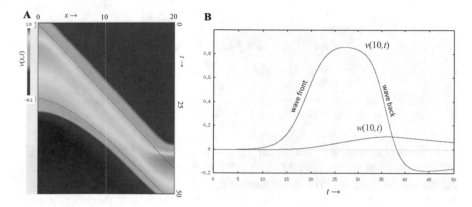

Fig. 6 Traveling pulse solution to the FHN equations (1). (**a**) Plot of membrane potential (v) versus space (x) and time (t). The wave front is traveling at a speed of ~0.47. The wave back travels quickly at first, then slows down to a speed of ~0.51 by the end of the record. Both waves are annihilated when they hit the no-flux boundary at $x = 20$. (**b**) Plot of membrane potential (v) and slow ionic current (w) in the middle of the domain ($x=10$), as functions of time. As the wave front passes this point, the membrane rapidly depolarizes; then recovers slowly back to rest

up to the wave front. By the end of the record, the wave front and back are both moving at a speed of ~0.5 su/tu. For the FHN equations, 1 tu ≈ 0.1 ms and 1 su ≈ 5 mm, so the predicted wave speed is

$$c \approx 0.5 \; \frac{\text{su}}{\text{tu}} \; \frac{1 \text{ tu}}{0.1 \text{ ms}} \; \frac{5 \text{ mm}}{1 \text{ su}} = 25 \; \frac{\text{mm}}{\text{ms}}. \tag{11}$$

The speed of an action potential on the squid giant axon is ~20 mm/ms [7].

Novak-Tyson Model of MPF Waves in Frog-Egg Extracts

Finally, we will compute a traveling pulse of MPF activation in a model of frog-egg extracts and compare the model results with the experimental observations of Chang & Ferrell [1], illustrated in Fig. 1. To this end, we use the Novak-Tyson model in Table 1 of the case study on 'Mitotic Cycle Regulation. I. Oscillations and Bistability':

$$\frac{\partial u}{\partial t} = k_1 - A_{\text{apc}}(u)u - A_{\text{wee}}(u)u + A_{25}(u)(v - u) + D\frac{\partial^2 u}{\partial x^2} \tag{12a}$$

$$\frac{\partial v}{\partial t} = k_1 - A_{\text{apc}}(u)v + D\frac{\partial^2 v}{\partial x^2} \tag{12b}$$

where

$$A_{\mathrm{apc}}(u) = k'_{\mathrm{apc}} + k''_{\mathrm{apc}} G(u, \theta_{\mathrm{apc}}, J_{\mathrm{apc}}) \tag{12c}$$

$$A_{\mathrm{wee}}(u) = k'_{\mathrm{wee}} + k''_{\mathrm{wee}} G(\theta_{\mathrm{wee}}, u, J_{\mathrm{wee}}) \tag{12d}$$

$$A_{25}(u) = k'_{25} + k''_{25} G(u, \theta_{25}, J_{25}) \tag{12e}$$

$$G(u, \theta, J) = \frac{2Ju}{\theta - u + J\theta + Ju + \sqrt{(\theta - u + J\theta + Ju)^2 - 4Ju(\theta - u)}} \tag{12f}$$

In these equations, $u = $ [active MPF], $v - u = $ [inactive MPF], and $v = $ [active MPF] + [inactive MPF] = [total CycB]. All concentrations are scaled appropriately. As a reminder, 'active' MPF is the mitosis-specific cyclin-dependent kinase, Cdk1: CycB, and 'inactive' MPF is the tyrosine-phosphorylated form of the dimer. Wee1 is the protein kinase that phosphorylates and inactivates Cdk1, and Cdc25 is the protein phosphatase that removes the inhibitory phosphate group. Furthermore, as is appropriate for frog egg extracts, CycB is synthesized at a constant rate, k_1, and degraded by the APC/proteasome pathway. Novak and Tyson assume that Cdc25 and APC are activated by and Wee1 is inhibited by active MPF. The activities of these enzymes, A_{apc}, etc., are described by 'Goldbeter-Koshland' functions, $G(u, \theta, J)$ [27]. D is the diffusion constant of the Cdk1:CycB dimer, $D \approx 400 \; \mu\mathrm{m}^2 \; \mathrm{min}^{-1}$. The rate parameters we use are:

$k_1 = 0.025$	$D = 1$		
$k_{\mathrm{apc}}' = 0.025$	$k_{\mathrm{apc}}'' = 0.8$	$\theta_{\mathrm{apc}} = 0.2$	$J_{\mathrm{apc}} = 0.01$
$k_{\mathrm{wee}}' = 0$	$k_{\mathrm{wee}}'' = 1$	$\theta_{\mathrm{wee}} = 0.2$	$J_{\mathrm{wee}} = 0.03$
$k_{25}' = 0.04$	$k_{25}'' = 4$	$\theta_{25} = 0.2$	$J_{25} = 0.03$

where all k's have units min^{-1}. By choosing $x = $ distance/(20 μm), we may set $D = 1 \; \mathrm{min}^{-1}$ in Eq. (12).

As we did for the FHN model, we solve Eq. (12) for a traveling pulse solution, using the ode file in Methods section M3. The result is displayed in Fig. 7. The pulse of MPF activation (i.e., sperm nuclei in mitosis, nuclear envelope broken down, and NLS-GFP dissipated) travels with a speed

$$c \approx 2.5 \frac{\mathrm{su}}{\mathrm{tu}} \frac{1 \; \mathrm{tu}}{\mathrm{min}} \frac{20 \; \mu\mathrm{m}}{1 \; \mathrm{su}} = 50 \frac{\mu\mathrm{m}}{\mathrm{min}} \quad . \tag{13}$$

This result is in good agreement with the observations of Chang & Ferrell [1]; see Fig. 1 above.

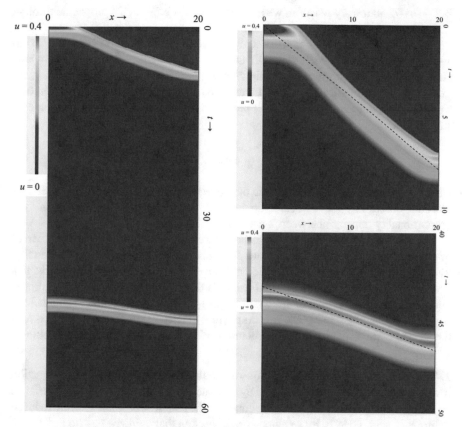

Fig. 7 Traveling pulse in the Novak-Tyson model of MPF activation in an oscillating frog egg extract. Plot of MPF activity (u) as a function of space ($0 \leq x \leq 20$) and time ($0 \leq t \leq 60$). (Left) Within a few minutes, a sharp wave front develops from the initial conditions and travels across the domain at a speed of ~2.5 su/tu. It is followed shortly after by a diffuse wave back traveling at the same speed. Forty-three minutes later a second pulse is generated at $x = 0$, and it travels across the domain at a faster speed of ~6 su/tu. The first wave is the traveling pulse driven by diffusion; the second wave is a 'phase wave' induced in the oscillatory medium by the first wave Compare this plot to Fig. 1c, taking into account that the computed domain is 200 μm (0.2 mm) by 60 min duration. (Right) Close-ups of the traveling pulse and the phase wave. Their speeds are estimated from the slopes of the dashed lines

Discussion

Traveling waves of 'excitation' are commonly employed by living cells to communicate information over large distances in short times (Table 1). The most well-known example is the action potential, which is a wave of membrane depolarization that travels along the squid giant axon at a speed of 20 m/s. Traveling waves of cAMP coordinate the aggregation of cellular slime molds, and waves of MPF activation can be observed in multinucleated cells. MPF waves travel at a speed of ~50 μm/min. Excitation waves are also observed in chemical systems. In the

Belousov-Zhabotinsky reaction, waves of oxidation ($Fe^{2+} \rightarrow Fe^{3+}$) propagate through an aqueous solution at speed 1—10 mm/min.

In all cases, the propagation of these waves arises from the coupling of an autocatalytic process, which kicks off an excitation locally, with diffusion, which allows the excitation to spread to neighboring regions, one place after another. By examining several examples of such waves, by analytical and numerical methods, we found that, in general, the speed of propagation of excitation waves is given by a simple formula

$$c = a\sqrt{kD}, \tag{14}$$

where D is the diffusion coefficient of the active species, k is a first-order rate constant describing the autocatalytic process that drives the wave, and a is a proportionality constant between 2 and 10.

As early as 1906, R. Luther [28, 29] discussed the possibility of traveling waves of excitation in homogeneous, aqueous solutions, and he proposed, rather casually, that such waves would move at a speed given by Eq. (14). Luther did not derive this equation; most likely he had in mind a simple dimensional analysis [30]. A proper derivation is quite demanding, as illustrated in this case study. Nonetheless, 'Luther's equation', $c = a\sqrt{kD}$, is a simple rule to keep in mind in order to estimate the speed of waves of excitation in chemical, biochemical and biophysical settings.

Exercises

1. Traveling wave front in the Belousov-Zhabotinsky reaction. The BZ reaction involves the oxidation of malonic acid by bromate ions (BrO_3^-) in acidic solution, with an Fe^{2+}/Fe^{3+} redox catalyst [31]. In the course of the reaction, bromate ions are reduced to bromous acid, $HBrO_2$, by an autocatalytic reaction that oxidizes Fe^{2+} to Fe^{3+}:

$$BrO_3^- + HBrO_2 + 3H^+ + 2Fe^{2+} \rightarrow 2HBrO_2 + H_2O + 2Fe^{3+} \tag{R5}$$

The resulting 'explosion' in $HBrO_2$ concentration is limited by the disproportionation reaction:

$$2HBrO_2 \rightarrow BrO_3^- + HOBr + H^+ \tag{R4}$$

The rate law for Reaction (R5) is [32]:

$$V_{R5} = \frac{d[HBrO_2]}{dt} = (k_{R5}[H^+][BrO_3^-])[HBrO_2],$$

$$k_{R5} = 42 \ M^{-2} \ s^{-1}.$$

In addition, $HBrO_2$ is removed by a complex set of reactions that involve the reduction of Fe^{3+} back to Fe^{2+}. Hence, as $HBrO_2$ increases during the autocatalytic phase, the solution changes color from red (Fe^{2+}) to blue (Fe^{3+}), and during the subsequent $HBrO_2$ clearance phase, the solution changes color from blue (Fe^{3+}) to red (Fe^{2+}). Including diffusion of $HBrO_2$ in the solution, we can write the following reaction-diffusion equation for the scaled concentration, $v(x,t)$, of $HBrO_2$,

$$\frac{\partial v}{\partial t} = k\left(v(1-v) - \gamma\frac{v-\delta}{v+\delta}\right) + D\frac{\partial^2 v}{\partial x^2} \tag{15}$$

where $k = k_{R5}[H^+][BrO_3^-]$ is the rate constant for the autocatalytic explosion of $HBrO_2$, γ is a parameter for the concentration of Fe^{2+} in the reaction front, and D is the diffusion coefficient of $HBrO_2$ ($D \approx 2 \cdot 10^{-5}$ cm^2 s^{-1}). We can scale the space and time coordinates, $\hat{t} = kt$ and $\hat{x} = \sqrt{k/D}x$, so that $k = D = 1$ in Eq. (15). We want to find a traveling wave solution to this equation of the form $v(x,t) = p(x-ct)$. As in the main text, this change of coordinate system converts Eq. (15) into the system of nonlinear ODEs

$$\frac{dp}{dz} = q \tag{16a}$$

$$\frac{dq}{dz} = -cq - \left(p(1-p) - \gamma\frac{p-\delta}{p+\delta}\right) \tag{16b}$$

(a) Show that the function $f(p) = p(1-p) - \gamma\frac{p-\delta}{p+\delta}$ is 'cubic' shaped. Let $\gamma = 0.1$, $\delta = 0.005$.

(b) Write an ode file to implement ODEs (16) and compute the phase plane portrait in Fig. 8 for $\gamma = 0.1$, $\delta = 0.005$ and $c = 0.7$. Clearly, the traveling wave solution has a speed a little less than 0.7. Show that the wave speed that connects the saddle point at $p = 0.886$ to the saddle point at $p = 0.00557$ is between 0.68 and 0.69.

(c) Translating this wave speed back to dimensions of cm/s, we get $c = 0.685\sqrt{kD}$. For the values of k and D given above, show that

$$c = 0.02 \text{ cm s}^{-1}\text{M}^{-1}\sqrt{[H^+][BrO_3^-]}.$$

How does this estimate agree with observations of Field & Noyes [33] and Showalter [34]?

2. Traveling wave front in the Novak-Tyson model. Let's look for a traveling front solution to Eq. (12a) when $v(x,t) = $ constant. Letting $u(x,t) = p(z)$, $z = x-ct$, we transform Eq. (12a) into

$$-c\frac{dp}{dz} = k_1 - A_{apc}(p)p - A_{wee}(p)p + A_{25}(p)(v-p) + D\frac{d^2p}{dz^2} \tag{17}$$

Then, defining $q(z) = dp/dz$, we convert (17) into a pair of nonlinear, first-order ODEs

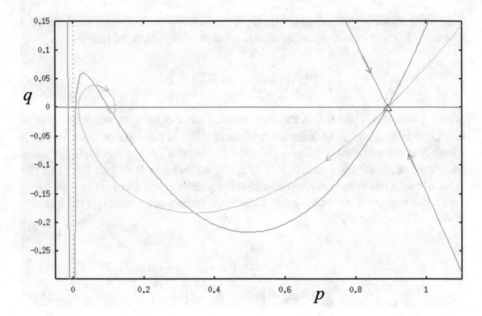

Fig. 8 Phase plane portrait for the system of ODEs (16). To get this portrait, open the nUmerics tab and set Method to (S)tiff (accept the default settings) and Ncline ctrl to 100. Use Sing pts to find the saddle point at $p = 0.886$, and Draw Invariant Sets. The blue curves are the invariant sets that approach the saddle point, and the mustard-colored curves are the invariant sets that leave the saddle point

$$\frac{dp}{dz} = q \tag{18}$$

$$D\frac{dq}{dz} = -cq - \left[k_1 - A_{\text{apc}}(p)p - A_{\text{wee}}(p)p + A_{25}(p)(v - p)\right].$$

(a) Show that the function $f(p) = k_1 - A_{\text{apc}}(p)p - A_{\text{wee}}(p)p + A_{25}(p)(v - p)$ is 'cubic' shaped. The $A..(p)$ functions are defined in Eq. (12c–f), and the parameter values are given in the table below Eq. (12).

(b) Write an ode file to implement ODEs (18) and compute the phase plane portrait for $v = 0.5, D = 1$ and $c = 2$. You should get a result similar to Fig. 8, if you plot the nullclines on the window $0 \le p \le 0.6, -0.4 \le q \le 0.1$. Show that the wave speed that connects the saddle point at $p = 0.416$ to the saddle point at $p = 0.0576$ is between 1.9 and 2.0. Notice that this value is close to the speed of the first wave front in Fig. 7.

3. Traveling wave front in Fisher's equation. In 1937, R.A. Fisher, the renowned English statistician and mathematical biologist, published a model of the spread of an advantageous gene through a population of interbreeding organisms distributed along a linear habitat [35]:

$$\frac{\partial v}{\partial t} = kv(1-v) + D\frac{\partial^2 v}{\partial x^2} \tag{19}$$

where $v(x,t)$ is the local gene frequency ($0 \le v \le 1$), k is the intensity of selection in favor of the mutant gene, and D is the diffusion coefficient for the random movements of organisms along the habitat. In traveling wave coordinates, $v(x,t) = p(x-ct)$, this PDE becomes

$$\frac{dp}{dz} = q$$

$$D\frac{dq}{dz} = -cq - kp(1-p). \tag{20}$$

We are looking for a traveling wave solution of Eq. (20) that satisfies the end-conditions:

$$p(-\infty) = 1, \quad p(+\infty) = 0, \quad \text{and} \quad q = \frac{dp}{dz} < 0. \tag{21}$$

(a) Write an ode file to implement ODEs (20). As usual, you may assume that space and time are scaled so that $k = D = 1$. For $c = 2$, plot the nullclines on the window $-0.1 \le p \le 1.1$, $-0.15 \le q \le 0.05$. Show that there is a trajectory that connects the saddle point at $p = 1$ to the steady state at $p = 0$, with $q < 0$. This is the traveling wave we are looking for. Find the eigenvalues of the steady states at the two end points:

$$(p,q) = (1,0) : \quad \text{eigenvalues} = 0.41457 \text{ and} - 2.41457 \text{ (saddle point)}$$

$$(p,q) = (0,0) : \quad \text{eigenvalues} = -1 \text{ and} - 1 \text{ (stable node)}$$

(b) Repeat the previous step for $c = 3$ and $c = 1$. Explain why the solution at $c = 3$ is a perfectly acceptable traveling wave for this problem, but the solution at $c = 1$ is not. (Hint: p is a gene frequency, i.e., $0 \le p \le 1$.)

(c) Using linear stability analysis, show that the steady state at $(p,q) = (1,0)$ is always a saddle point, whereas the steady state at $(p,q) = (0,0)$ is a stable node for $c \ge 2$ and a stable focus for $0 < c < 2$. Explain that, as a consequence, Fisher's equation (19) exhibits traveling wave solution for all speeds $c \ge 2$, but not for speeds $0 < c < 2$.

4. Bistability and traveling waves in a model of MAP kinase activation. In 2004, Markevich et al. [36] made a surprising discovery about the activation of mitogen-activate kinase (MAPK), which is governed by the phosphorylation and dephosphorylation cycles shown in Fig. 9.

The reaction rates, $v_1 \dots v_4$, given in Fig. 9 are Michaelis-Menten rate laws for distributive-ordered phosphorylations and dephosphorylations of the MAPK

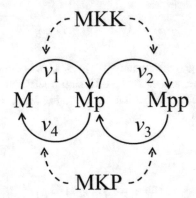

Fig. 9 MAP kinase signaling motif. $M =$ unphosphorylated MAPK; Mp $=$ mono-phosphorylated MAPK; Mpp $=$ doubly-phosphorylated MAPK; MKK $=$ MAP kinase kinase; MKP $=$ MAP kinase phosphatase. The reaction rates, $v_1 \ldots v_4$, are given by $v_1 = \frac{k_1 * MKK * M}{K_{m1} + M + K_{m1} Mp / K_{m2}}$, $v_2 = \frac{k_2 * MKK * Mp}{K_{m2} + Mp + K_{m2} M / K_{m1}}$, $v_3 = \frac{k_3 * MKP * Mpp}{K_{m3} + Mpp + K_{m3} Mp / K_{m4} + K_{m3} M / K_{m5}}$, $v_4 = \frac{k_4 * MKP * Mp}{K_{m4} + Mp + K_{m4} Mpp / K_{m3} + K_{m4} M / K_{m5}}$

species. *MKK* and *MKP* are the fixed concentrations of the kinase and phosphatase enzymes, and the only tricky thing here is the authors' assumption that the product, M, of the second step of the dephosphorylation reaction does not readily dissociate from MKP. As a consequence, M (i.e., unphosphorylated MAPK) is an inhibitor of MKP, with a dissociation constant K_{m5}.

(a) Explore this reaction motif with the following ode file:

```
# MAPK bistability
# Conservation condition, MT = total MAPK
Mp = MT - M - Mpp
#
dM/dt = (k4*MKP*Mp)/(Km4+MP+Km4*Mpp/Km3+Km4*M/Km5) - (k1*MKK*M)/
(Km1+M+Km1*Mp/Km2)
#
dMpp/dt = (k2*MKK*Mp)/(Km2+MP+Km2*M/Km1) - (k3*MKP*Mpp)/(Km3+Mpp
+Km3*Mp/Km4+Km3*M/Km5)
#
par MKK=20, MKP=1, MT = 360
par k1=1, Km1=50, k2=20, Km2=50
par k3=40, Km3=2, k4=360, Km4=18, Km5=100
@ XP=M, YP=Mpp, XLO=0, YLO=0, XHI=400, YHI=400
done
```

Draw the phase plane portrait and show that the system has three steady states at

$$(M, Mp, Mpp) = (1.9, 5.7, 352.4), (263.8, 29.6, 66.6), (354, 4.7, 1.3).$$

Which of these steady states are stable? Unstable?

(b) In 2006, Markevich et al. [37] pointed out that this bistable switch can generate a wave of MAPK phosphorylation that might propagate signals over long distances in a cell. To verify this claim, use **M3** as a template to write an ode file that simulates this model as a reaction-diffusion system in one spatial dimension. Use the same parameter values as in part (a), with $D = 1$, $N = 200$, $h = 0.1$ (i.e., $L = 20$), total $= 100$, dt $= 0.001$, and Bounds $= 1000$. For initial conditions, use

$$M[1..200] = 354/(1 + \exp((40 - [j])/10));$$

$$Mpp[1..200] = 360/(1 + \exp(([j] - 40)/10)).$$

Plot the result using `Viewaxes Array`, and fill in the dialog box as follows:

```
Column 1:Mpp1; NCols:200; Row 1:0; Nrows:1001; RowSkip:100; Zmin:0;
Zmax:400, Autoplot:0; Colskip:1.
```

You should get Fig. 10. What is the speed of the wave front?

(c) Bistability in MAP kinase signaling cascades has been observed in dozens of circumstances. Although bistability is often attributed to positive feedback from the end of the cascade to an early step [38], there are cases where such positive feedback is definitely not operative [39]. In these cases, the distributive phosphorylation mechanism of Markevich et al. [36] is definitely a prime candidate for the source of bistability [40]. Read these papers and provide a brief summary of the arguments of the evidence the 'positive feedback' and 'distributive phosphorylation' mechanisms.

Methods
M1. Traveling front solution to Eq. (2).
To compute a traveling wave solution to Eq. (2) use the following ode file:

```
# FitzHugh-Nagumo traveling wave
!del = D/h^2
v0 = v1
v[1..100]' = del*(v[j+1]-2*v[j]+v[j-1]) + k*v[j]*(v[j]-the)*(1-v[j])
v101 = v100
par D=1, k=2, the=0.2, h=0.1
@ total=5, dt=.005
done
```

The line `v[1..100]'` defines a set of 100 ODEs that simulate a solution of Eq. (2) by the 'method of lines'. Given $N=100$ and $h=0.1$, the length of the spatial domain is $L=10$. The statements $v0 = v1$ and $v101 = v100$ enforce 'no flux' boundary conditions at the two ends of the domain, $x=0$ and $x=10$. Load this ode file into XPP. Before we can run the simulator, we must prescribe an initial condition

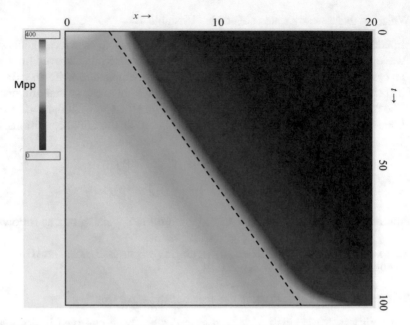

Fig. 10 Traveling wave front of MAPK activation. A front of MAPK phosphorylation propagates into a field of unphosphorylated MAPK (blue)

for $v(x,0)$. To do this, click on `Initialconds formUla`. At the prompt for the 'variable', type `v[1..100]` and hit return. At the 'formula' prompt, type `1/(1+exp((([j]-33)/10))`, and hit return. When the 'variable' prompt reappears, hit return. XPP will then solve Eq. (2) over $0 \leq x \leq 10, 0 \leq t \leq 5$. To plot the result, click on `Viewaxes Array`, and fill in the dialog box as in Table 2(1).

Click `OK` and a new window appears, like the one in Fig. 3, only blank. Click on the `Redraw` tab and you will see Fig. 3.

M2. Traveling pulse solution to Eq. (1).

To compute a traveling pulse solution to Eq. (1) use the following `ode` file:

```
# FitzHugh-Nagumo traveling wave
!del = D/h^2
v0 = v1
v[1..200]' = del*(v[j+1]-2*v[j]+v[j-1]) + k*(v[j]*(v[j]-the)*(1-v
[j])-w[j])
v201 = v200
w0 = w1
w[1..200]' = eps*(v[j]-gam*w[j])
w201 = w200
par D=1, k=2, the=0.2, eps=0.01, gam=2.5, h=0.1
@ total=50, dt=.005
done
```

Table 2 (1, 2, 3) Dialog boxes for traveling wave array plots in M1, M2 and M3, respectively

Column 1: v1	Column 1: v1	Column 1: u1
NCols: 100	NCols: 200	NCols: 200
Row 1: 0	Row 1: 0	Row 1: 0
Nrows: 201	Nrows: 501	Nrows: 601
RowSkip: 5	RowSkip: 20	RowSkip: 20
Zmin: 0	Zmin: -0.2	Zmin: 0
Zmax: 1	Zmax: 1	Zmax: .4
Autoplot(0/1): 0	Autoplot(0/1): 0	Autoplot(0/1): 0
Colskip: 1	Colskip: 1	Colskip: 1
(1)	(2)	(3)

Load this file into XPP and prescribe initial conditions for $v(x,0)$ as in section M1. Click on Initialconds formUla, and at the prompt for the 'variable', type v [1..200]; return. At the 'formula' prompt, type 1/(1+exp(([j]-10)/10)); return. The 'variable' prompt will reappear type w[1..200]; return. At the 'formula' prompt, type 0; return. The 'variable' prompt will reappear; hit return. XPP will then solve Eq. (2) over $0 \leq x \leq 20$, $0 \leq t \leq 50$. To plot the result, click on Viewaxes Array, and fill in the dialog box as in Table 2(2). Click OK and a new window appears, like the one in Fig. 6, only blank. Click on the Redraw tab and you will see Fig. 6.

M3. Traveling pulse solution to Eq. (12).

To compute a traveling pulse solution to Eq. (12) use the following ode file:

```
# MPF traveling wave
# Define 'Goldbeter-Koshland' function
BB(u,H,J) = H-u+J*(H+u)
G(u,H,J) = (2*J*u)/(BB(u,H,J)+sqrt(BB(u,H,J)^2-4*J*u*(H-u)))
# Define 'activity' functions
Aapc(u) = kp_apc + kpp_apc*G(u,Hapc,Japc)
Awee(u) = kp_wee + kpp_wee*G(Hwee,u,Jwee)
A25(u) = kp_25 + kpp_25*G(u,H25,J25)
# Define PDEs
!del = D/h^2
u0 = u1
u[1..200]'=del*(u[j+1]-2*u[j]+u[j-1])+k1-Aapc(u[j])*u[j]-Awee(u
[j])*u[j]+A25(u[j])*(v[j]-u[j])
u201 = u200
v0 = v1
v[1..200]' = del*(v[j+1]-2*v[j]+v[j-1])+k1-Aapc(u[j])*v[j]
v201 = v200
par k1=0.025
par kp_apc=0.025, kpp_apc=0.8, Hapc=0.2, Japc=0.01
par kp_wee=0, kpp_wee=1, Hwee=0.2, Jwee=0.03
par kp_25=0.04, kpp_25=4, H25=0.2, J25=0.03
par D=1, h=0.1
@total=60, dt=.005
done
```

Load this file into XPP and go through the routine: Initialconds formUla
| u[1..200] | 0.5/(1+exp((([j]-30)/10)) | v[1..200] | 0.6 | |;
where each | is a return. To plot the result, select Viewaxes Array, and fill in the
dialog box as in Table 2(3). Click OK and, after the new window appears, click
Redraw and you will see Fig. 7.

Additional Reading

Notable review articles on chemical waves in cell and developmental biology have
been written by Goldbeter [41, 42], Gelens [43], and Di Talia [44].

Waves of oxidation (target patterns and spiral waves) in shallow layers of
Belousov-Zhabotinsky solutions were first observed by Zaikin and Zhabotinsky
[5, 45] and by Winfree [6, 24]. In deeper layers, Winfree first reported rotating
'scroll waves' of oxidation [23]. The theory of traveling waves in excitable media
(like Fig. 2a) was reviewed by Tyson and Keener [26] for waves in one- and two-
spatial dimensions, and by Keener and Tyson [46] for scroll waves in three spatial
dimensions. For an introduction to scroll wave geometry, see Tyson and Strogatz
[47]. For a masterful treatment of electrochemical waves in three dimensions, the
must-read is Winfree's book, *When Time Breaks Down* [48].

References

1. Chang JB, Ferrell JE Jr (2013) Mitotic trigger waves and the spatial coordination of the
   Xenopus cell cycle. Nature 500(7464):603–607
2. Tomchik KJ, Devreotes PN (1981) Adenosine 3',5'-monophosphate waves in *Dictyostelium
   discoideum*: a demonstration by isotope dilution--fluorography. Science 212(4493):443–446
3. Gilkey JC et al (1978) A free calcium wave traverses the activating egg of the medaka, *Oryzias
   latipes*. J Cell Biol 76(2):448–466
4. Busa WB, Nuccitelli R (1985) An elevated free cytosolic Ca2+ wave follows fertilization in
   eggs of the frog, *Xenopus laevis*. J Cell Biol 100(4):1325–1329
5. Zaikin AN, Zhabotinsky AM (1970) Concentration wave propagation in two-dimensional
   liquid-phase self-oscillating system. Nature 225(5232):535–537
6. Winfree AT (1972) Spiral waves of chemical activity. Science 175(4022):634–636
7. Hodgkin AL, Huxley AF (1952) A quantitative description of membrane current and its
   application to conduction and excitation in nerve. J Physiol 117(4):500–544
8. Deneke VE et al (2016) Waves of Cdk1 activity in S phase synchronize the cell cycle in
   Drosophila embryos. Dev Cell 38(4):399–412
9. Novak B, Tyson JJ (1993) Modeling the cell division cycle: M-phase trigger, oscillations and
   size control. J Theor Biol 165:101–134
10. Jaffe LF (1993) Classes and mechanisms of calcium waves. Cell Calcium 14(10):736–745
11. Dupont G, Goldbeter A (1994) Properties of intracellular Ca2+ waves generated by a model
    based on Ca(2+)-induced Ca2+ release. Biophys J 67(6):2191–2204
12. Keener J, Sneyd J (2009) Mathematical physiology. I: cellular physiology, 2nd edn. Springer,
    Cham

13. Devreotes PN et al (1983) Quantitative analysis of cyclic AMP waves mediating aggregation in *Dictyostelium discoideum*. Dev Biol 96(2):405–415
14. Tyson JJ et al (1989) Sprial waves of cyclic AMP in a model of slime mold aggregation. Physica D 34:193–207
15. Halloy J et al (1998) Modeling oscillations and waves of cAMP in *Dictyostelium discoideum* cells. Biophys Chem 72(1-2):9–19
16. Pourquie O (2003) The segmentation clock: converting embryonic time into spatial pattern. Science 301(5631):328–330
17. Goldbeter A, Pourquie O (2008) Modeling the segmentation clock as a network of coupled oscillations in the Notch, Wnt and FGF signaling pathways. J Theor Biol 252(3):574–585
18. Oates AC et al (2012) Patterning embryos with oscillations: structure, function and dynamics of the vertebrate segmentation clock. Development 139(4):625–639
19. Sliusarenko O et al (2006) From biochemistry to morphogenesis in myxobacteria. Bull Math Biol 68(5):1039–1051
20. Welch R, Kaiser D (2001) Cell behavior in traveling wave patterns of myxobacteria. Proc Natl Acad Sci U S A 98(26):14907–14912
21. Fitzhugh R (1961) Impulses and physiological states in theoretical models of nerve membrane. Biophys J 1(6):445–466
22. Nagumo J et al (1964) An active pulse transmission line simulating nerve axon. Proc Inst Radio Engin 50:2061–2070
23. Winfree AT (1973) Scroll-shaped waves of chemical activity in three dimensions. Science 181 (4103):937–939
24. Winfree AT (1974) Rotating chemical reactions. Sci Amer 230(6):82–95
25. Tyson JJ, Fife PC (1980) Target patterns in a realistic model of the Belousov-Zhabotinskii reaction. J Chem Phys 73:2224–2237
26. Tyson JJ, Keener JP (1988) Singular perturbation theory of traveling waves in excitable media (a review). Physica D 32:327–361
27. Goldbeter A, Koshland DE Jr (1981) An amplified sensitivity arising from covalent modification in biological systems. Proc Natl Acad Sci U S A 78(11):6840–6844
28. Luther R (1906) Propagation of chemical reactions in space. Z Elektrochemie 12(32):596
29. Luther R et al (1987) Propagation of chemical reactions in space. J Chem Educ 64:740–742
30. Showalter K, Tyson JJ (1987) Luther's 1906 discovery and analysis of chemical waves. J Chem Educ 64:742–744
31. Tyson JJ What everyone should know about the Belousov-Zhabotinsky reaction. In: Levin SA (ed) Frontiers in mathematical biology. Springer-Verlag, Cham, pp 569–587
32. Field RJ, Foersterling H-D (1986) On the oxybromine chemistry rate constants with cerium ions in the Field-Koros-Noyes mechanism of the Belousov-Zhabotinskii reaction. J Phys Chem 90:5400–5407
33. Field RJ, Noyes RM (1974) Oscillations in chemical systems, V: quantitative explanation of band migration in the Belousov-Zhabotinskii reaction. J Am Chem Soc 96:2001–2006
34. Showalter K (1981) Trigger waves in the acidic bromate oxidation of ferroin. J Phys Chem 85:440–447
35. Fisher RA (1937) The wave of advance of advantageous genes. Ann Eugenics 7:353–369
36. Markevich NI et al (2004) Signaling switches and bistability arising from multisite phosphorylation in protein kinase cascades. J Cell Biol 164(3):353–359
37. Markevich NI et al (2006) Long-range signaling by phosphoprotein waves arising from bistability in protein kinase cascades. Mol Syst Biol 2:61
38. Ferrell JE Jr, Machleder EM (1998) The biochemical basis of an all-or-none cell fate switch in Xenopus oocytes. Science 280(5365):895–898
39. Harding A et al (2005) Subcellular localization determines MAP kinase signal output. Curr Biol 15(9):869–873
40. Kholodenko BN (2006) Cell-signalling dynamics in time and space. Nat Rev Mol Cell Biol 7 (3):165–176

41. Dupont G, Goldbeter A (1992) Oscillations and waves of cytosolic calcium: insights from theoretical models. BioEssays 14(7):485–493
42. Goldbeter A (2006) Oscillations and waves of cyclic AMP in *Dictyostelium*: a prototype for spatio-temporal organization and pulsatile intercellular communication. Bull Math Biol 68 (5):1095–1109
43. Gelens L et al (2014) Spatial trigger waves: positive feedback gets you a long way. Mol Biol Cell 25(22):3486–3493
44. Deneke VE, Di Talia S (2018) Chemical waves in cell and developmental biology. J Cell Biol 217(4):1193–1204
45. Zhabotinsky AM, Zaikin AN (1973) Autowave processes in a distributed chemical system. J Theor Biol 40(1):45–61
46. Keener JP, Tyson JJ (1992) The dynamics of scroll waves in excitable media. SIAM Rev 34 (1):1–39
47. Tyson JJ, Strogatz SH (1991) The differential geometry of scroll waves. Int J Bifur Chaos 1 (4):723–744
48. Winfree AT (1987) When time breaks down: the three-dimensional dynamics of electrochemical waves and cardiac arrhythmias. Princeton Univ Press

Cell Cycle Regulation. Bifurcation Theory

John J. Tyson

Introduction

In earlier case studies, we have seen how a molecular control system can be described by a system of nonlinear differential equations (a 'dynamical system'), and that solutions of these equations can be identified with laboratory observations and can predict the outcome of novel experiments. We called this approach the 'dynamical perspective on molecular cell biology'. In this case study we elaborate on this notion by showing that the behavior of a dynamical system can be characterized by a 'bifurcation diagram', which indicates how the dynamical variables of the system respond to changes in parameter values. A bifurcation diagram is a mathematician's depiction of a physiologist's signal-response curve, which measures how some adaptable characteristic of a cell (i.e., a variable in the mathematical model) responds to changes in the cell's external environment or internal conditions (i.e., a parameter in the model). A summary of the generic bifurcations of dynamical systems is, in this context, a catalog of the basic types of signal-response curves that can be exhibited by living cells. We apply this idea to a model of cell cycle regulation in fission yeast.

J. J. Tyson (✉)
Department of Biological Sciences and Division of Systems Biology, Virginia Polytechnic
Institute and State University, Blacksburg, VA, USA
e-mail: tyson@vt.edu

© Springer Nature Switzerland AG 2021
P. Kraikivski (ed.), *Case Studies in Systems Biology*,
https://doi.org/10.1007/978-3-030-67742-8_3

The Growth-Controlled Fission-Yeast Cell Cycle

The earlier case study focused on bistability and oscillations in frog embryos and frog egg extracts, where DNA synthesis (S phase) and mitosis (M phase) are triggered by MPF, a heterodimer of Cdk1 kinase and its binding partner, cyclin B [4]. Figure 2 in that case study described the underlying molecular mechanism (cyclin synthesis and degradation and Cdk1 phosphorylation and dephosphorylation) that controls MPF activity in both frog eggs and fission yeast cells. The DNA replication-division cycles during early embryonic development of a frog egg proceed without cell growth, but fission yeast cells progress through S and M phases only if they are growing. Indeed, the division of yeast cells (and most types of cells, both eukaryotes and prokaryotes) is coordinated with cell growth so that the average interdivision time is equal to the population mass doubling time, and the cells maintain a stable size distribution around some optimal DNA-to-mass ratio [5]. This phenomenon of 'size regulation' is built into the MPF control system that regulates DNA synthesis and division in fission yeast cells. It can be illustrated by the schematic diagram of MPF-control of the fission yeast cell cycle in Fig. 1.

Figure 1 shows how the MPF control system in fission yeast responds to changes in cell size. Let's follow the red 'cell cycle trajectory,' starting in the lower left corner

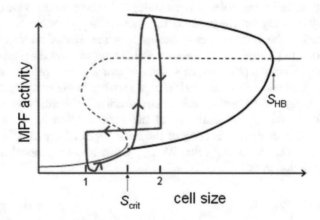

Fig. 1 A schematic diagram of progression through the fission yeast cell cycle [2]. The characteristic behavior of the MPF control system in fission yeast depends on the size of the cell, as illustrated in this one-parameter bifurcation diagram. In drawing this diagram, we assume that cell size is fixed at a certain value (i.e., the cell is not growing), and we compute the asymptotic state of the nonlinear ODEs describing the MPF control system. The faint solid lines (left and right) denote stable steady states of low and high MPF activity (G1/S/G2 phase and M phase of the cell cycle, respectively). The dashed line indicates unstable steady states. The heavy solid lines indicate the maximum and minimum levels of MPF activity over the course of a limit cycle oscillation at some fixed value of cell size between S_{crit} and S_{HB}. S_{HB} is a Hopf bifurcation point, where MPF oscillations disappear as the steady state regains stability. S_{crit} is a SNIPER bifurcation point ('saddle-node infinite-period'), where large amplitude limit cycle oscillations abruptly give way to the stable steady state of low MPF activity. The red path indicates the trajectory of a fission yeast cell during its cycle of growth and division, as described in the main text

of the diagram. Small newborn cells ($S \approx 1$) are captured by a stable steady state of low MPF activity (G1 phase). As the cell grows, MPF rises to an activity sufficient to initiate DNA synthesis but not mitosis. During this growth phase ($S < S_{crit}$), the cell proceeds through S phase into G2. When the cell finally reaches the critical size (S_{crit}) where the G1/S/G2 steady state disappears, it hops onto a large amplitude limit cycle, indicated by the solid black lines. MPF activity shoots high enough to trigger mitosis, and as the cell proceeds out of mitosis, MPF activity drops precipitously. This drop is the signal for cell division, which creates two newborn cells of half the size of their mother, and they are captured by the low-MPF state, in order to repeat the cycle. Notice that the cell cycle time (the time required to progress around the red loop) is identical to the mass-doubling time (the time necessary to grow from birth size =1 to division size = 2).

In the next sections, we provide an introduction to the basic concepts of bifurcation theory that are necessary to understand Fig. 1. Much of the following material is reproduced from my article in the *Encyclopedia of Systems Biology* [6] (used by permission).

What Is Bifurcation Theory?

Bifurcation theory provides a classification of the expected ways in which the number and/or stability of invariant solutions ('attractors' and 'repellors') of nonlinear ODEs may change as parameter values are changed. The most common qualitative changes are saddle-node bifurcations, Hopf bifurcations, and SNIPER bifurcations. At a saddle-node bifurcation, a pair of steady states, usually a stable node and an unstable saddle point, coalesce and disappear. At a Hopf bifurcation, a stable focus changes to an unstable focus and makes way for small amplitude periodic solutions (limit cycles). At a SNIPER bifurcation, the coalescence of a saddle point and a stable node creates an infinite-period limit cycle oscillation. These bifurcations have clear physiological correlates in the regulation of DNA replication, mitosis and cell division. Saddle-node bifurcations are related to checkpoints in the cell cycle: the establishment and removal of checkpoints correspond to the creation and annihilation of stable steady states at saddle-node bifurcations. The repetitive nature of the cell cycle (G1-S-G2-M-G1-etc.) is related to limit cycle solutions of the underlying kinetic equations: the ability to oscillate spontaneously arises at either a Hopf or a SNIPER bifurcation.

Historical Background

Since the days of Isaac Newton, ordinary differential equations (ODEs) have been used through-out the physical and life sciences to describe the temporal development of dynamical systems: from the solar system, to the clock radio, to the regulation of

DNA replication and cell division. Initially the focus was on ODEs that could be solved exactly in terms of the elementary functions of high-school algebra and trigonometry or the special functions of mathematical physics. But in the 1890's Poincaré [7] introduced a qualitative theory of dynamical systems, which are systems of n nonlinear ODEs of the general form

$$\frac{d\mathbf{u}}{dt} = f(\mathbf{u}, \mathbf{p}); \quad \mathbf{u}(t) = (u_1, \ldots, u_n), \quad \mathbf{p} = (p_1, \ldots, p_m), \tag{1}$$

where $u_i(t)$ are the variables that describe the behavior of the dynamical system, and $\mathbf{p}$ is a set of parameter values that characterize the rates of change of the u_i's. Poincaré proposed to interpret these equations as a vector field in n-dimensional state space, $\mathbf{u}$, and to characterize this vector field by its invariant solutions, which can be either attractors or repellors. The crucial question for Poincaré was not 'what is the exact solution of the ODE?' but 'how do the qualitative features of the attractors and repellors depend on the values of the parameters?' This latter question is the subject of bifurcation theory, which was developed in the mid-20th Century by Andronov's school of Russian physicists and engineers [8], and later by a host of mathematicians, as summarized by Kuznetsov [9]. An excellent introduction to bifurcation theory has been provided by Strogatz [10].

A one-parameter bifurcation diagram is a plot of the steady state value of a chosen dynamical variable, u_i, as a function of a chosen parameter, p_j, the 'bifurcation parameter'. In Fig. 2a, we plot a typical bifurcation diagram for a bistable system. Between the two thresholds, $\theta_{\text{inact}} < p_j < \theta_{\text{act}}$, the system can persist in either of two stable steady states (u_i small or u_i large). Precisely at the thresholds, $p_j = \theta_{\text{inact}}$ and $p_j = \theta_{\text{act}}$, the dynamical system undergoes a bifurcation from one type of behavior (a single stable steady state) to a qualitatively different type of behavior (bistability). This type of bifurcation is called a saddle-node. In Fig. 2b, we illustrate a Hopf

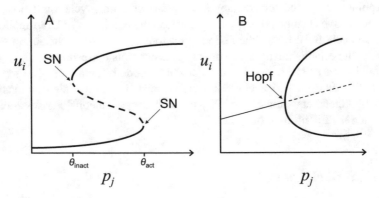

Fig. 2 One-parameter bifurcation diagrams. (**a**) Saddle-node bifurcation. Solid line: stable steady state; dashed line: unstable steady state. (**b**) Hopf bifurcation. Thin solid line: stable steady state; thin dashed line: unstable steady state; thick solid line: maximum and minimum values attained by a stable limit cycle oscillation

bifurcation, in which a stable steady state loses stability and gives rise to stable limit-cycle oscillations. The limit cycles are born with small amplitude and grow in size as the parameter value pulls away from the bifurcation point.

One-Parameter Bifurcation Diagrams and Signal-Response Curves

The connection between bifurcation theory and cell physiology is the signal-response curve [1, 3]. In a typical experiment, a molecular cell biologist might challenge cells with increasing amounts of an extracellular signal molecule and measure whether certain downstream genes are expressed or not. And a typical result is that, for low signal levels there is no expression, but for signal levels above a certain threshold there is strong expression of the gene (Fig. 3). In this circumstance, it is natural to ask what happens if the signal level is steadily decreased in cells that are expressing protein R? Do they turn off at the same signal strength where they turned on? Or at a much lower signal strength? Or not at all?

In the first case, the signal-response curve is perfectly smooth and reversible; there are no qualitative changes in the behavior of the control system as the signal

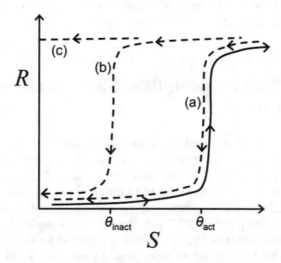

Fig. 3 Signal-response curve. An experimentalist can vary the signal strength S (say, the concentration of an extracellular ligand) and observe the response R (say, the expression level of a gene induced by the signal). As S is slowly increased (solid curve), the expression of R turns on abruptly; a typical threshold-type response. What happens as S is slowly decreased? There are three possibilities. (**a**) The gene expression turns on and off at the same threshold signal strength: the signal-response curve is smooth and reversible; e.g., a Hill function. (**b**) The threshold for gene inactivation (θ_{inact}) is lower than the threshold for gene activation (θ_{act}): the signal-response curve has a region of bistability and functions like a toggle switch. (**c**) The gene cannot be inactivated by lowering the signal strength even to zero: the control system functions as a one-way switch

varies up and down. In the second case, there is a region of bistability between the two thresholds, and the behavior of the control system is qualitatively different over three ranges of signal strength: for $S < \theta_{inact}$ there is a single stable steady state with R small; for $\theta_{inact} < S < \theta_{act}$ the control system can persist in either of two attractors (R small or R large) that are separated by an unstable steady state; and for $S > \theta_{act}$ there is a single stable steady state with R large. This is exactly the case of a one-parameter bifurcation diagram with saddle-node bifurcations bounding a zone of bistability (Fig. 2a).

It is possible that $\theta_{inact} < 0$. In this case the signal S, which is the concentration of a signaling molecule (i.e., a positive number), cannot be made small enough to flip the switch off. Hence, by increasing S the switch can be turned on, but it can't be turned off by decreasing S. The control system is said to exhibit a one-way switch.

There are many convincing examples of toggle switches in molecular and cell biology generally [3] and in cell cycle regulation particularly (see the case study on 'Mitotic Cycle Regulation').

In another common physiological situation, a cellular process begins to oscillate when a stimulating signal gets large enough [11]. In this case, the signal-response curve exhibits a Hopf bifurcation, as in Fig. 2b.

These examples suggest that the signal-response curves often measured by cell physiologists are none other than one-parameter bifurcation diagrams in the parlance of applied mathematicians. If we may associate abrupt, qualitative changes in signal-response characteristics of living cells with bifurcations in vector fields of nonlinear dynamical systems, then it is natural to ask:

How Many Types of Generic Bifurcations Are Exhibited by Dynamical Systems?

Are there hundreds of different types of bifurcations to match the seemingly boundless variety of cellular behaviors? Or are all the peculiarities of cellular signaling simply variations on a few common themes? The answer is the latter [9]. In addition to the saddle-node and Hopf bifurcations illustrated in Fig. 2, there are only a few other common, generic, one-parameter bifurcations: subcritical Hopf, cyclic fold, saddle-loop and SNIPER bifurcations (Fig. 4 and Table 1).

Of this we can be certain: Cell physiology is governed by underlying regulatory networks involving biochemical reactions among genes, RNAs and proteins. These networks are dynamical systems; their dynamics are governed by nonlinear ODEs (biochemical kinetic equations). The solutions of these equations determine the time-dependent behavior of the cell, and the nature of these solutions are determined by the attractors and repellors of the vector field in state space. Qualitative changes in the behavior of cells must be reflections of qualitative changes in the nature of these attractors and repellors, i.e., reflections of the generic bifurcations of nonlinear vector fields. Hence, the six types of bifurcations we have introduced must be the basic building blocks of all signal-response curves exhibited by cells.

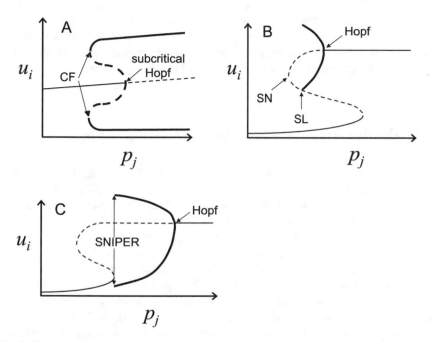

Fig. 4 The other common types of bifurcation points. (**a**) Subcritical Hopf bifurcation and cyclic fold (CF) bifurcation. (**b**) Saddle-loop (SL) bifurcation. (**c**) Saddle-node infinite-period (SNIPER) bifurcation

Table 1 Generic bifurcations of dynamical systems [9]

Name	Characteristics	Cell cycle correlate
Saddle-node	Creation and annihilation of pairs of steady states	Irreversible transitions; checkpoints
Hopf, supercritical	Birth of stable limit cycles of small amplitude and finite frequency	Spontaneous MPF oscillations in embryos
Hopf, subcritical	Birth of unstable limit cycles of small amplitude and finite frequency	Subcrit Hopf and cyclic fold bifurcations occur in pairs and may correlate with embryonic MPF oscillations
Cyclic fold	Creation and annihilation of pairs of limit cycles	
Saddle-loop	Annihilation of a limit cycle by a homoclinic orbit at a saddle point; finite amplitude and small frequency	SL and SNIPER bifurcations are closely related; they are involved in irreversible transitions in the yeast cell cycle (budding yeast and fission yeast)
SNIPER[a]	Annihilation of a limit cycle by a homoclinic orbit at a saddle-node; finite amplitude and small frequency	

[a]Also known as a SNIC bifurcation ('saddle-node invariant-circle')

An important caveat to this interpretation of bifurcation theory is the fact that single cells are very small, with limited numbers of molecules (10s, 100s, 1000s) of each of the interacting species. Hence, continuous ODEs are only a first approximation to the dynamics of intracellular molecular control systems. The effects of stochastic variations of small numbers of molecules can have significant effects on the qualitative features of dynamical systems. Stochastic effects must be given due consideration, but only after the modeler has made a thorough study of the continuous dynamic system (the nonlinear ODEs) by bifurcation theory.

A Model Illustrating the Bifurcation Diagram in Fig. 1

To illustrate some of these generic bifurcations, we start with the same model of MPF control introduced in Table 1 of the case study on 'Mitotic Cycle Regulation.' We will examine this dynamical system using XPP-AUTO, starting with the ode file in Table 2a in the Methods section. Copy this code into a simple text file (call it, say, mpf.ode) and store it in a convenient folder. Next open mpf.ode in XPP and select Nullcline New from the menu on the far left. You should get a 'phase plane portrait' that looks like Fig. 5a. Find the singular point at the intersection of the nullclines and determine that it is unstable. What are the eigenvalues at this unstable steady state? Next, find the limit cycle oscillation around the unstable steady state, and plot the time-courses $u(t)$ and $v(t)$ for $0 \leq t \leq 300$, as in Fig. 5b. These oscillations are reminiscent of the growth-independent mitotic cycles in an early frog embryo, but, of course, fission yeast cells do not behave this way. Nonetheless, let us continue to investigate the dynamical properties of this MPF control system at a variety of fixed values of cell size, S. In Fig. 5c, we superimpose the phase plane portraits for $S = 0.8$, 0.9 and 1.0. For $S = 0.8$, the MPF control system has a single steady state (confirm that it is stable) on the lower branch of the u-nullcline (i.e., low MPF activity characteristic of G1/S/G2). For $S = 0.9$, the nullclines intersect in three places; confirm that only one steady state (G1/S/G2) is stable. At $S = 1.0$, the stable G1/S/G2 steady state (a 'node') coalesces with the unstable saddle point (i.e., a SN bifurcation), and the stable steady state is replaced by a stable limit cycle oscillation of large amplitude and long period; this situation is called a SNIPER bifurcation ('saddle-node bifurcating to an infinite-period oscillation'). You should explore the dynamical system in the neighborhood of $S = 1$ to confirm these properties.

Next, let's calculate the one-parameter bifurcation diagram that summarizes the results in Fig. 5. First step: set $S = 0.2$ and integrate the ODEs (multiple repeats of Initialconds|Last) until the stable steady state is loaded into the Initial Data file. Then click on File|Auto, and step down the menu on the left side. (1) Parameter: verify that S is the primary bifurcation parameter. (2) Axes|hi-lo: verify that the one-parameter bifurcation diagram will plot U as a function of S; set Xmin=0, Ymin=0, Xmax=2.5, Ymax=0.3. (3) Numerics: set Ntst=30, Nmax=2000, NPr=0, Ds=0.0002, Dsmin=1e-05, EPSL=1e-05, Dsmax=0.005, Par Min=0, Par Max=2.5. Leave other numerical parameters unchanged. (4) Click on Run|Steady state, and you should get the bifurcation diagram in Fig. 6a. Auto finds three bifurcation points: two SN bifurcations, at $S = 0.8542$ and 0.9969, and a HB at $S = 1.946$.

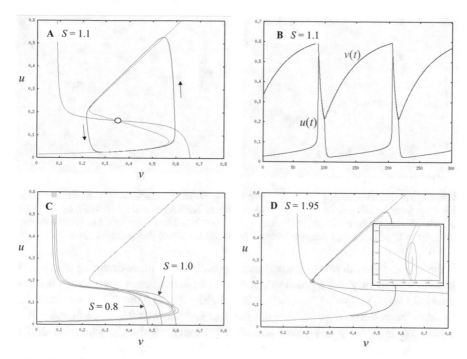

Fig. 5 Phase plane portraits of the MPF model in Table 2a. (**a**) The u-nullcline (green) and v-nullcline (red) for $S = 1.1$. At the intersection of the nullclines (o) there is a single, unstable steady state surrounded by a limit cycle oscillation (solid black line). (**b**) The limit cycle oscillation is plotted as a time course. The period is 116 min. (**c**) Superimposition of phase plane portraits for $S = 1.0, 0.9$ and 0.8. As S increases, the red nullcline shifts to the left, and the green nullcline shifts to the right. $S = 1.0$ is very close to the SNIPER bifurcation point. $S = 0.9$ is within the region of three steady states (one stable, two unstable), and $S = 0.8$ is just below the SN bifurcation, where there is a single, stable steady state on the lower branch of the u-nullcline. (**d**) $S = 1.95$ is just beyond the HB point, where the steady state on the upper branch of the u-nullcline is a stable focus (i.e., damped oscillations, see inset)

Our last step with this ode file is to follow the stable limit cycles that bifurcate from HB at $S = 1.946$. First, go to Axes | hi-lo and set Ymax=0.6. Then click on Grab, to get a listing of the labelled points on the bifurcation diagram. 'Tab' through the points until you get the HB point (label 4) at $S = 1.946$. Hit 'Enter,' and choose Run | Periodic. You should get the bifurcation diagram in Fig. 6b. The loci of green and blue circles trace out the maximum and minimum values of $u(t)$ on the limit cycle oscillation at each fixed value of S. The Hopf bifurcation is 'subcritical,' as indicated in the inset to Fig. 6b. There is a 'cyclic fold' bifurcation at label 6, where loci of unstable and stable limit cycles converge. This type of bifurcation is also known as a 'saddle-node of periodic orbits.'

Our next undertaking is to add growth and division (S changing) to our model of the MPF control system at constant S. To this end, we convert S from a fixed parameter to a dynamic variable that increases exponentially, $dS/dt = \mu S$, where μ

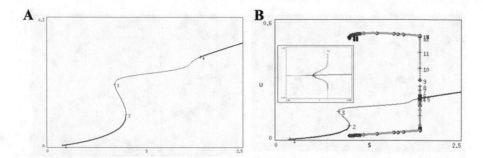

Fig. 6 One-parameter bifurcation diagram for the model in Table 2a. (**a**) First pass through Auto, to find the SN and HB bifurcation points (labels 2, 3, 4). (**b**) Calculation of periodic orbits that bifurcate from the HB point at $S = 1.946$ and end at the SNIPER point at $S = 0.9969$. Inset: the HB is 'subcritical,' i.e., the small amplitude limit cycles (open blue circles) are unstable. Label 6 denotes a 'cyclic fold' bifurcation, where the unstable limit cycles become stable (green circles)

is the specific growth rate of the cell and $\tau = \ln 2/\mu$ is the mass-doubling time of cell growth. We specify that a growing cell divides in half whenever $u(t)$ drops below a threshold level, $u(t) = 0.1$; i.e., for a cell to divide, first $u(t)$ must increase above 0.3 to drive the cell into mitosis, then it must decrease through 0.1 to allow the cell to divide. We implement this 'growth-controlled' model in Table 2b. Run this ode file in XPP to reproduce the time courses in Fig. 7a. Next, plot $u(t)$ versus $S(t)$, and superimpose the curve on top of the bifurcation diagram in Fig. 6b to get Fig. 7b, which is to be compared with the 'schematic diagram' in Fig. 1. In these calculation, $\mu = 0.007$ min^{-1} and the mass-doubling time $\tau = 99$ min. The time between cell divisions is also 99 min. Try varying the value of μ and show that the interdivision time is always equal to the mass-doubling time.

Discussion

The 'dynamical perspective on molecular cell biology' [1] maintains that molecular regulatory networks are dynamical systems, governed, to a first approximation, by systems of nonlinear ordinary differential equations (ODEs) derived from the basic principles of biochemical reaction kinetics. In general terms, these ODEs, $\frac{d\mathbf{u}}{dt} = f(\mathbf{u}, \mathbf{p})$, describe the rates of change of biochemical concentrations, $u_i(t)$, in dependence on reaction rate constants, p_j ('parameter values') in the network. Dynamical Systems Theory (DST) proposes that the best way to understand such systems of ODEs is in terms of a vector field on a multidimensional state space, $\mathbf{u}$, and, in particular, on the attractors and repellors of the vector field (steady states, periodic orbits, and more complex 'strange' attractors). Furthermore, bifurcation theory is the most appropriate way to understand how attractors and repellors arise and disappear in dependence on the parameter values in the ODEs. A one-parameter bifurcation diagram depicts how the attractors and repellors of a chosen variable, say u_1, depend

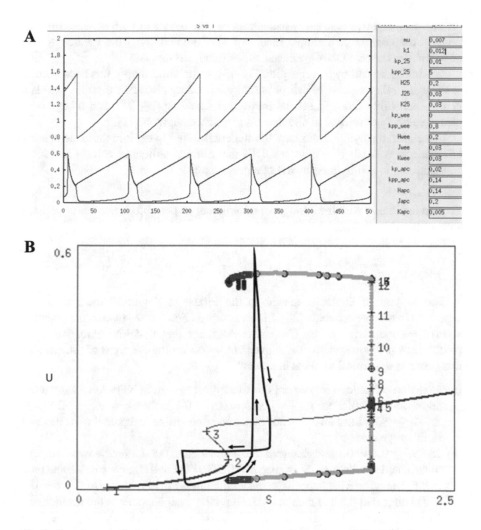

Fig. 7 Growth-controlled cell cycle for fission yeast, according to the model in Table 2b. (**a**) Time courses of $S(t)$ (black), $u(t)$ (red) and $v(t)$ (blue). (**b**) The 'cell cycle trajectory' (black curve: $u(t)$ vs. $S(t)$) is superimposed on the bifurcation diagram in Fig. 6b. Compare this computed diagram with the schematic diagram in Fig. 1

on a particular parameter, say p_1. DST associates a one-parameter bifurcation diagram with a cell physiologist's 'signal-response' curve, which depicts how a biochemical response of a regulatory network, say the activity of a protein, depends on a parameter value under experimental control, say the concentration of a hormone in the cellular growth medium.

If this perspective is a productive way to view molecular regulatory networks, then it is imperative that molecular cell biologists understand the basic principles of bifurcation theory; in particular, the classification of elementary bifurcations exhibited by nonlinear ODEs of the type in Eq. (1). Table 1 lists these elementary bifurcations, and this case study has introduced the reader to each of these bifurcations in the context of a simple model of MPF regulation in the control system governing eukaryotic DNA synthesis, mitosis and cell division.

In addition to cell cycle regulation [12–15], bifurcation theory has been used to study many other aspects of cell physiology, including circadian rhythms [16–19], neural excitability [20–23], second-messenger signaling [24, 25], and other examples, splendidly described in Goldbeter's classic monograph [11].

In the accompanying 'Exercises' the student is asked to explore these elementary bifurcations in a pair of nonlinear ODEs that mimic membrane voltage and ionic conductance, in a simplified model of a nerve axon.

Exercises

To gain familiarity with the basic types of bifurcations possible in nonlinear ODEs, a simple model to practice on is

$$\frac{du}{dt} = u(1-u)(1+u) - v, \quad \frac{dv}{dt} = (u-a)(b-y) - c$$

This system of ODEs is related to the FitzHugh-Nagumo model of action potentials in a nerve cell axon [26]. In this interpretation, $u(t) =$ membrane potential and $v(t) =$ conductance of the K^+ channel. Although this model is not realistic for a biochemical reaction system, it is convenient for observing the types of bifurcations that occur in dynamical systems in general.

(a) **Write** an ode file to implement this dynamical system in XPP-Auto. Start it up, and **draw** the nullclines for $a = 0.9, b = 0, c = 0.1$ in the phase plane $-2 \leq u \leq 2, -1 \leq v \leq 1$. **Find** all the steady states and **determine** their stability. Is there an oscillatory solution?

(b) Keep $b = 0, c = 0.1$ and change a to 0.7 and then to 0.5. **Describe** what happens to the u and v nullclines. Now keep $a = 0.5, c = 0.1$ and change b to 0.3 and then to 0.6. **Describe** what happens to the v nullclines. Finally, keep $a = 0.5, b = 0.3$ and change c to 0.2 and then to 0.05. **Describe** what happens to the v nullclines. Now you know how to change a, b and c to make any of the possible phase plane portraits for this dynamical system.

(c) Let's start with $a = -0.5, b = 0.7, c = 0.1$. **Draw** the phase plane portrait. **Find** the two unstable steady states and the limit cycle oscillation. Now, keeping b and c fixed, decrease a to $-0.6, -0.7, -0.8$. **What** qualitative change to the dynamical system happens between $a = -0.6$ and -0.7?

(d) Set $a = -1, b = 0.7, c = 0.1$, and integrate the ODEs (Initialconds|Last) multiple times until the stable steady state is loaded into the Initial Data window. Click on File|Auto, and on the Axes|hi-lo tab set Xmin=-1.1, Ymin=-1.1, Xmax=1.1, Ymax=1.1. On the Numerics tab, set Par Min=-2. Then Run|Steady state. You should get the one-parameter bifurcation diagram in Fig. 8a, with Hopf bifurcation (HB) points at $a = -0.6423$ and 0.179. Grab the point label 2 and Run|Periodic. You should get the bifurcation diagram in Fig. 8b.

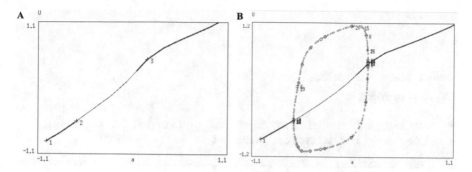

Fig. 8 One-parameter bifurcation diagrams for Exercise (d). (**a**) Locus of steady states, with HB at $a = -0.6423$ (label 2) and $+0.179$ (label 3). (**b**) Periodic solutions (green curve) bifurcating from the HB points

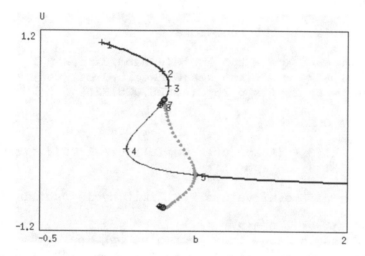

Fig. 9 One-parameter bifurcation diagram for Exercise (f). SN bifurcation points at $b = 0.5438$ (label 3) and 0.2073 (label 4). HB at $b = 0.7576$ (label 5). The stable periodic orbits that arise at HB are extinguished at a 'saddle loop' bifurcation at $b \approx 0.5$ (label 8)

(e) This time, start at $a = -0.7$, $b = 0$, $c = 0.2$. **Plot** the phase plane portrait; then **repeat** for $b = 0.15, 0.2, 0.25$. How do the qualitative features of the dynamical system change as b passes through the value 0.2? Next, **plot** phase plane portraits for $b = 0.5, 0.75, 1.0$. How do the qualitative features of the dynamical system change as b passes through the value 0.75?

(f) **Challenge**: use Auto to calculate the one-parameter bifurcation diagram in Fig. 9, for $a = -0.7$, $c = 0.2$, and b varying.

Methods

Table 2 Fission yeast ode files

(A) Non-growing cell

```
# Novak-Tyson model for fission yeast cell cycle
# Size, S = bifurcation parameter

# Define 'Goldbeter-Koshland' function

BB(u,v,J,K)= v-u+v*J+u*K

G(u,v,J,K) = (2*K*u)/(BB(u,v,J,K)+sqrt(BB(u,v,J,K)^2-
4*K*u*(v-u)))

# Define 'activity' functions

A_apc(u) = kp_apc + kpp_apc*G(u,Hapc,Japc,Kapc)
A_wee(u) = kp_wee + kpp_wee*G(Hwee,u,Jwee,Kwee)
A_25(u) = kp_25 + kpp_25*G(u,H25,J25,K25)

# Define ODEs

du/dt = k1*S - (A_apc(u)+A_wee(u))*u + A_25(u)*(v-u)
dv/dt = k1*S - A_apc(u)*v

# Define parameter values for limit cycle oscillations

par S=1.1, k1 = 0.012
par kp_25=0.01, kpp_25=5, H25=0.2, J25=.03, K25=.03
par kp_wee=0, kpp_wee=.8, Hwee=0.2, Jwee=.03, Kwee=.03
par kp_apc=0.02, kpp_apc=0.14, Hapc=0.14, Japc=0.2,
Kapc=.005

# Default settings for XPP

@ XP=v, YP=u, NMESH=100, XLO=0, XHI=1, YLO=0, YHI=0.6

Done
```

(continued)

Table 2 (continued)

(B) Growth-controlled model

```
# Novak-Tyson model for fission yeast cell cycle
# Size S = dynamic variable

# Define 'Goldbeter-Koshland' function

BB(u,v,J,K)= v-u+v*J+u*K
G(u,v,J,K) = (2*K*u)/(BB(u,v,J,K)+sqrt(BB(u,v,J,K)^2-
4*K*u*(v-u)))

# Define 'activity' functions

A_apc(u) = kp_apc + kpp_apc*G(u,Hapc,Japc,Kapc)
A_wee(u) = kp_wee + kpp_wee*G(Hwee,u,Jwee,Kwee)
A_25(u) = kp_25 + kpp_25*G(u,H25,J25,K25)

# Define ODEs

dS/dt = mu*S
du/dt = k1*S - (A_apc(u)+A_wee(u))*u + A_25(u)*(v-u)
dv/dt = k1*S - A_apc(u)*v

# Define parameter values for limit cycle oscillations

par mu=.007, k1 = 0.012
par kp_25=0.01, kpp_25=5, H25=0.2, J25=.03, K25=.03
par kp_wee=0, kpp_wee=.8, Hwee=0.2, Jwee=.03, Kwee=.03
par kp_apc=0.02, kpp_apc=0.14, Hapc=0.14, Japc=.2,
Kapc=.005

# Use 'global' command to divide the cell in half when

# u(t) decreases through the value 0.1

global -1 u-0.1 [S=S/2]

# Default settings for XPP

@ XP=v, YP=u, NMESH=100, XLO=0, XHI=1, YLO=0, YHI=0.6

Done
```

Additional Reading

For an introduction to dynamical systems and bifurcation theory, consult the text-books by Strogatz [10] and Edelstein-Keshet [27], or the brief survey by Odell [28]. For a more advanced treatment of bifurcation theory, a good source is Kuznetsov's textbook [9].

For a thorough guide to XPP-AUT, Ermentrout's book [29] is indispensable; for a quick guide, read Appendix C of Ingalls' textbook [30].

For more details on modeling the eukaryotic cell cycle control system, see review articles by Tyson & Novak [2, 31, 32] and by Ferrell [33, 34].

References

1. Tyson JJ, Novak B (2020) A dynamical paradigm for molecular cell biology. Trends Cell Biol 30(7):504–515
2. Tyson JJ, Csikasz-Nagy A, Novak B (2002) The dynamics of cell cycle regulation. BioEssays 24:1095–1109
3. Tyson J et al (2003) Sniffers, buzzers, toggles and blinkiers: dynamics of regulatory and signaling pathways in the cell. Curr Opin Cell Biol 15:221–231
4. Tyson JJ, Novak B (2015) Bistability, oscillations, and traveling waves in frog egg extracts. Bull Math Biol 77(5):796–816
5. Tyson JJ (1985) The coordination of cell growth and division – intentional or incidental? BioEssays 2(2):72–77
6. Tyson JJ (2013) Cell cycle model analysis, bifurcation theory. In: Dubitzky W et al (eds) Encyclopedia of systems biology. Springer, New York, pp 274–278
7. Poincaré, H. (1899) Les méthodes nouvelles de la mécanique céleste. Gauthiers-Villars
8. Andronov A et al (1966) Theory of oscillators. Pergamon
9. Kuznetsov YA (2004) Elements of applied bifurcation theory, 3rd edn. Springer-Verlag
10. Strogatz S (2015) Nonlinear dynamics and chaos: with applications to physics, biology, chemistry, and engineering, 2nd edn. Westview Press
11. Goldbeter A (1996) Biochemical oscillations and cellular rhythms. Cambridge Univ Press
12. Borisuk MT, Tyson JJ (1998) Bifurcation analysis of a model of mitotic control in frog eggs. J Theor Biol 195(1):69–85
13. Qu Z et al (2003) Dynamics of the cell cycle: checkpoints, sizers, and timers. Biophys J 85 (6):3600–3611
14. Battogtokh D, Tyson JJ (2004) Bifurcation analysis of a model of the budding yeast cell cycle. Chaos 14(3):653–661
15. Csikasz-Nagy A et al (2006) Analysis of a generic model of eukaryotic cell-cycle regulation. Biophys J 90(12):4361–4379
16. Goldbeter A (1995) A model for circadian oscillations in the Drosophila period protein (PER). Proc Biol Sci 261(1362):319–324
17. Leloup JC et al (1999) Limit cycle models for circadian rhythms based on transcriptional regulation in Drosophila and Neurospora. J Biol Rhythms 14(6):433–448
18. Kim JK, Forger DB (2012) A mechanism for robust circadian timekeeping via stoichiometric balance. Mol Syst Biol 8:630
19. Battogtokh D, Tyson JJ (2018) Deciphering the dynamics of interlocked feedback loops in a model of the mammalian circadian clock. Biophys J 115(10):2055–2066

20. Chay TR, Keizer J (1983) Minimal model for membrane oscillations in the pancreatic beta-cell. Biophys J 42(2):181–190
21. Keizer J, Levine L (1996) Ryanodine receptor adaptation and Ca^{2+}-induced Ca^{2+} release-dependent Ca^{2+} oscillations. Biophys J 71(6):3477–3487
22. Bertram R et al (1995) Topological and phenomenological classification of bursting oscillations. Bull Math Biol 57:413–439
23. Izhikevich EM (2000) Neural excitability, spiking and bursting. Int J Bifur Chaos 10 (6):1171–1266
24. Goldbeter A, Segel LA (1977) Unified mechanism for relay and oscillation of cyclic AMP in *Dictyostelium discoideum*. Proc Natl Acad Sci U S A 74(4):1543–1547
25. Dupont G, Goldbeter A (1993) One-pool model for Ca^{2+} oscillations involving Ca2+ and inositol 1,4,5-trisphosphate as co-agonists for Ca^{2+} release. Cell Calcium 14(4):311–322
26. Keener J, Sneyd J (2009) Mathematical physiology. I: cellular physiology, 2nd edn. Springer
27. Edelstein-Keshet L (2005) Mathematical models in biology. Society for Industrial and Applied Mathematics
28. Odell GM (1980) Qualitative theory of systems of ordinary differential equations, including phase plane analysis and the Hopf bifurcation theorem. In: Segel LA (ed) Mathematical models in molecular and cellular biology. Cambridge Univ. Press, pp 649–727
29. Ermentrout B (2002) Simulating, analyzing, and animating dynamical systems. SIAM
30. Ingalls BP (2013) Mathematical modeling in systems biology: an introduction. MIT Press
31. Tyson JJ, Novak B (2013) Irreversible transitions, bistability and checkpoint controls in the eukaryotic cell cycle: a systems-level understanding. In: Walhout AJM, Vidal M, Dekker J (eds) Handbook of systems biology. Academic Press, pp 265–285
32. Tyson JJ, Novak B (2015) Models in biology: lessons from modeling regulation of the eukaryotic cell cycle. BMC Biol 13:46
33. Ferrell JJ (2013) Feedback loops and reciprocal regulation: recurring motifs in the systems biology of the cell cycle. Curr Opin Cell Biol 25(6):676–686
34. Ferrell JJ et al (2011) Modeling the cell cycle: why do certain circuits oscillate? Cell 144 (6):874–885

Glycolytic Oscillations

Pavel Kraikivski

Introduction

Glycolysis occurs in the cells of almost all organisms, including animals, plants, and unicellular microorganisms. Living cells use the process of glycolysis to generate energy by splitting a six-carbon glucose molecule into two three-carbon molecules called pyruvates. This process results in the net production of two molecules of NADH that play a very critical role in cell metabolism, and two molecules of ATP that is the primary carrier of energy in living cells. Therefore, glycolysis allows a living cell to use glucose as its main energy source. Interestingly, energy production in glycolysis is not always a steady process, but, under certain conditions, it can exhibit oscillatory dynamics.

Glycolytic oscillations were one of the earliest observations indicating that biochemical systems can exhibit oscillatory behavior. The first observations of spontaneous oscillations in glycolysis were reported in 1957 by Duysens and Amesz [1]. Specifically, they observed that fluorescence of the glycolytic intermediate, NADH, undergoes damped oscillations in a suspension of yeast cells. Also, in 1964, Chance and coworkers [2, 3], demonstrated that glycolytic oscillations could be sustained in yeast suspensions for relatively long periods of time; see Fig. 1a. Moreover, a sustained oscillation with constant amplitude and frequency can be induced in yeast cells undergoing cyanide poisoning, subsequent to glucose feeding after starvation [4]. Figure 1b shows such a sustained glycolytic oscillation observed by Bier and coworkers in Ref. [5]. Even though the reaction pathway of glycolysis in yeast was well-established before 1957, oscillatory behavior in glycolysis could not be explained without a deeper understanding of glycolytic regulatory kinetics. To

P. Kraikivski (✉)
Academy of Integrated Science, Division of Systems Biology, Virginia Polytechnic Institute and State University, Blacksburg, VA, USA
e-mail: pavelkr@vt.edu

© Springer Nature Switzerland AG 2021
P. Kraikivski (ed.), *Case Studies in Systems Biology*,
https://doi.org/10.1007/978-3-030-67742-8_4

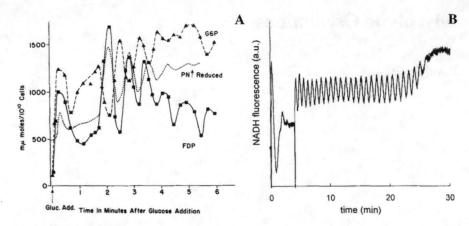

Fig. 1 Experimental glycolytic oscillations. (**a**) A damped sinusoidal oscillation of G-6-P and FDP is observed in a suspension of *S. carlsbergensis*. From Ghosh & Chance [3]; used by permission. (**b**) Oscillations of NAD(P)H fluorescence are observed in yeast cells. At t = 0 min, 20 mM glucose was added, followed by 4 mM KCN at t = 4 min. Upon cyanide addition, after a few cycles of transient oscillations, a sustained oscillation with constant amplitude and frequency is observed. From Bier et al. [5]; used by permission

test different hypothetical mechanisms that might lead to oscillatory behavior in glycolysis, many researchers applied mathematical modeling and systems biology approaches [5–7]. The successful explanation of glycolytic oscillations by mathematical models demonstrated that computational modeling can be an effective and crucial tool to analyze and understand the complex dynamic behavior of biological systems.

Physiology and Molecular Mechanism

Glycolysis is a biochemical pathway that breaks down glucose into intermediates, so as to generate and store energy in the form of the high-energy molecules, ATP and NADH. The glycolysis process occurs in intracellular fluids of bacteria, yeast, and animal and plant cells. Glycolysis consists of two main phases: the energy-requiring phase and the energy-releasing phase. In the energy-requiring phase, glucose is modified into an unstable sugar that has two additional phosphate groups attached to it. The resulting two phosphate-bearing, six-carbon sugar is called fructose-1,6-bisphosphate. Two ATP molecules are used to attach phosphate groups for this step of glycolysis. Because fructose-1,6-bisphosphate is unstable, it can be split in half, resulting in two three-carbon sugars. Then, in the energy-releasing phase, each three-carbon sugar is further modified into another three-carbon molecule, pyruvate. Overall, the reactions in the energy-releasing phase also result in the production of four ATP, and two NADH, molecules. Considering that two ATP molecules were consumed in the energy-requiring phase, the net products of the two phases are two

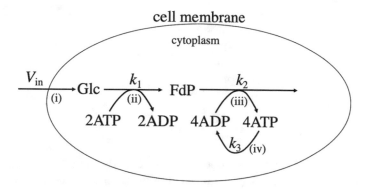

Fig. 2 Schematic illustration of the glycolytic pathway taking place in the cytoplasm. The model shows four steps of the glycolytic pathway: (i) an influx of glucose Glc into cytoplasm, (ii) conversion of Glc to FdP with dephosphorylation of two ATP molecules, (iii) the disappearance of one FdP molecule coupled to production of four ATP molecules and (iv) a reaction hydrolyzing ATP

molecules of ATP and two of NADH. These steps in glycolysis do not require oxygen, but each reaction in glycolysis requires its own enzyme to proceed. In the presence of oxygen, the three-carbon pyruvate can be broken down to carbon dioxide. This process produces additional energy in the form of ATP molecules.

We will consider a simplified scheme of the glycolytic pathway that can still exhibit oscillations under certain conditions; see Fig. 2. This scheme only includes four steps: (i) an influx of glucose, Glc, (ii) the conversion of molecules, i.e., glucose modification to fructose-1,6-bisphosphate, FdP, with the consumption of two ATP molecules, (iii) the production of four ATP molecules in the process of breaking down one FdP molecule, and (iv) ATP hydrolysis. Although, this scheme does not include all processes involved in the glycolytic pathway, it is sufficient to explain oscillatory dynamics in the glycolytic system.

The scheme in Fig. 2 does not include NADH; however, we can monitor oscillations in the glycolytic pathway by measuring temporal evolution of ATP concentration. This assumption is supported by experimental observations that other intermediates in the glycolytic pathway exhibit oscillatory dynamics, similar to those of NADH [3] (see Fig. 1a).

Mathematical Model

We will use a mathematical model of the glycolytic pathway to demonstrate that the molecular components involved in the glycolytic mechanism may exhibit oscillatory dynamics. A simple mathematical model of the glycolytic mechanism, shown in Fig. 2, was proposed by Cortassa and coworkers in 1991 [6], which is described by the following system of ordinary differential equations:

$$\frac{d[\text{Glc}]}{dt} = V_{in} - V_1$$

$$\frac{d[\text{ATP}]}{dt} = -2V_1 + 4V_2 - V_3$$

$$\frac{d[\text{FdP}]}{dt} = V_1 - V_2. \tag{1}$$

The first equation in (1) describes glucose, Glc, dynamics that occur due to glucose uptake at the rate V_{in} (influx), and conversion of glucose into unstable, two phosphate-bearing, six-carbon intermediate, FdP, with the rate V_1 (outflux) described by the law of mass action as $V_1 = k_1[\text{Glc}][\text{ATP}]$, where k_1 is a rate constant. The influx of glucose V_{in} is assumed to be at a constant rate. The second equation in system (1) describes the rate of change of the ATP concentration due to the following three processes: (i) the consumption of two ATP molecules in the energy-requiring phase of glycolysis (the conversion of glucose into FdP) with the rate $2V_1$, (ii) the production of four ATP molecules in the energy-releasing phase of glycolysis with the rate $4V_2$ where $V_2 = k_2[\text{FdP}]([\text{A}] - [\text{ATP}])[\text{P}]$ and k_2 is a rate constant, and (iii) the non-glycolytic ATP-consuming processes described by Michaelis-Menten kinetics as $V_3 = k_3 \frac{[\text{ATP}]}{K+[\text{ATP}]}$, where K is the Michaelis constant of the ATPase and k_3 represents the maximum rate achieved by the non-glycolytic ATP-consuming processes. The concentration of the phosphate [P] is assumed to be constant. Also, [A]=[ATP] + [ADP] is constant, so as to satisfy a conservation relationship for the total adenine nucleotides. The third equation in system (1) describes the FdP dynamics governed by the process of FdP production occurring with the rate V_1, and the process of conversion of FdP into pyruvates occurring with the rate V_2.

Note that there are many mathematical models that have been proposed to explain glycolytic oscillations. For example, the mathematical model describing a very general metabolic scheme of the glycolytic pathway was developed by Goldbeter and Lefever in 1972 [7]. Their model includes additional components in the glycolytic pathway, such as phosphofructokinase and adenosine monophosphate, that play an important role in their explanation of glycolytic oscillations. By contrast, Cortassa's mathematical model (1) does not include these molecular components and can be further simplified. As was shown by Bier and coworkers in Ref. [5], a "core model" consisting of only two differential equations (see Exercise 1) that describe the dynamics of glucose and ATP, can produce oscillations similar to Cortassa's model. In this chapter, we will analyze model (1), as it describes four well-established processes of glycolysis, i.e., (i) import of glucose, (ii) conversion of glucose into fructose-1,6-bisphosphate (FdP), (iii) the process of breaking down one molecule of FdP coupled to the production of four ATP molecules, and (iv) ATP hydrolysis.

Results

The mathematical model (1) can be simulated using any computational tool that is equipped to solve a system of ordinary differential equations. Here, we will use XPPAUT to analyze the dynamic behavior of system (1). The XPP code that was used to obtain all results in this section, is provided in the Method section.

The system of Eq. (1) has seven parameters; however, Cortassa and coworkers showed that two parameters: (i) the influx of glucose V_{in}, and (ii) the maximum rate achieved by non-glycolytic ATP-consuming processes, k_3, control transition between the steady energy production regime and the oscillatory energy production regime [4]. Thus, the alteration of these two parameters can qualitatively change dynamic behavior of the system. For example, the solution of system (1) becomes oscillatory at some values of V_{in} and k_3, which indicates the existence of a Hopf bifurcation point (the point at which a steady state solution switches to an unstable steady state solution with a periodic solution or limit cycle). The qualitatively different dynamic behavior of Glc, ATP, and FdP components, depending on V_{in} and k_3 parameter values, is demonstrated in Fig. 3a–c. Importantly, the influx of glucose V_{in} is the parameter that can be experimentally controlled in the lab. Thus, the predicted dynamic behavior of system (1), depending on the V_{in} parameter, can be experimentally validated.

At a relatively high rate of glucose influx V_{in}, the intracellular level of glucose concentration Glc and the energy production attain steady state values, as shown in Fig. 3a. The oscillations appear when either the rate of glucose influx V_{in} is significantly decreased, or the maximum rate of ATP breakdown reaction k_3 is increased. Other parameters in model (1) are not significant for the appearance of glycolytic oscillations. At low rates of glucose influx V_{in}, the glycolytic system exhibits damped oscillations, as shown in Fig. 3b. At high rates of the ATP

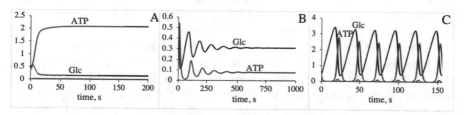

Fig. 3 The dynamic behavior of metabolites in the glycolytic pathway depending on the values of parameters V_{in} and k_3. The blue curve is ATP, the black curve is Glc and the red curve is FdP. (a) Metabolite concentrations quickly approach steady-state values for the following parameter values: $V_{in} = 0.1$ mM s^{-1} and $k_3 = 0.3905$ mM s^{-1}. (b) The damped oscillations of metabolite concentrations appear for $V_{in} = 0.0075$ mM s^{-1} and $k_3 = 0.3905$ mM s^{-1}. (c) The glycolytic system exhibits sustained oscillations when $V_{in} = 0.2$ mM s^{-1} and $k_3 = 2.5$ mM s^{-1}. The initial values for concentrations are the following: [Glc] = 0.5 mM, [ATP] = 0.4 mM and [FdP] = 0 mM. Other parameter values used for these simulation results are: [P] = [A] = 10 mM, K = 2 mM, $k_1 = 0.3$ mM s^{-1}, $k_2 = 0.1$ mM s^{-1}. XPP code that was used to simulate the temporal evolution of metabolites in the glycolytic pathway is provided in the Method section

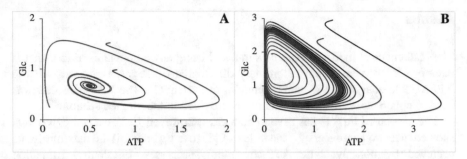

Fig. 4 Numerical solutions of model equations (1) in the phase plane. (**a**) Three solution trajectories (black, red and blue curves) are converging to a stable fixed point. The system of differential equations (1) has been integrated starting at three different initial points. These solutions are obtained for $k_3 = 1$ mM s^{-1}. (**b**) Three solution trajectories (black, red, and blue curves) are converging to a limit cycle. The blue trajectory is started inside the limit cycle and diverges away from the unstable fixed point in the direction of the limit cycle, whereas the red and black trajectories are started outside of the limit cycle and converge toward it. These solutions are obtained for $k_3 = 2$ mM s^{-1}. Thus, there is a Hopf bifurcation point that occurs when the k_3 parameter has some value between 1 mM s^{-1} and 2 mM s^{-1}. In these simulations, the initial value of FdP is zero, and other parameters are: $V_{in} = 0.1$ mM s^{-1}, [P] = [A] = 10 mM, K = 2 mM, $k_1 = 0.3$ mM s^{-1}, $k_2 = 0.1$ mM s^{-1}

breakdown reaction, sustained oscillations are observed for all modeled variables (metabolite concentrations), as shown in Fig. 3c.

As has been shown by Bier and coworkers in Ref. [5], the dynamic behavior of system (1) is independent of the FdP component. Model (1) can be simplified to the two differential equations model (see Exercise 1). In this case, we can analyze model (1) in the phase plane (Glc vs. ATP) and also observe damped and sustained oscillations exhibited by the system (see Fig. 4). For example, simulation results in Fig. 4a, b show how the solution of system (1) changes from a stable steady-state solution to a limit cycle, when parameter k_3 is changed from 1 mM s^{-1} to 2 mM s^{-1}. Therefore, there should be a k_3 parameter value at which the stable steady-state solution switches to the unstable solution with the limit cycle. This value of the k_3 parameter corresponds to the Hopf bifurcation point.

At the glucose flux rate of $V_{in} = 0.1$ mM s^{-1}, the Hopf bifurcation occurs when parameter $k_3 = 1.59$ mM s^{-1}, as shown in the one-parameter bifurcation diagram in Fig. 5a. The two-parameter bifurcation diagram in Fig. 5b shows values of both V_{in} and k_3 parameters at which Hopf bifurcation occurs. The diagram shows the parameter values for which the sustained glycolytic oscillations can be observed. The blue line in Fig. 5b separates the parameter region that corresponds to the stable steady-state solution (damped oscillations), and the parameter region corresponding to the unstable steady-state solution where the limit cycle exists.

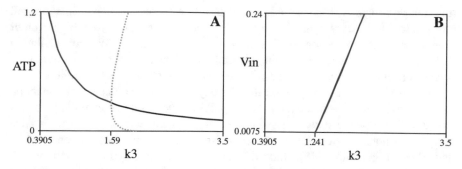

Fig. 5 One- and two-parameter bifurcation diagrams. (**a**) One-parameter bifurcation diagram for ATP concentration as a function of k_3 parameter that is varied from 0.3905 mM s^{-1} to 3.5 mM s^{-1} and $V_{in} = 0.1$ mM s^{-1}. The steady state solution is shown along the red line until the parameter k_3 reaches the Hopf bifurcation value at $k_3 = 1.59$ mM s^{-1}; then ATP concentration exhibits periodic behavior. The green dotted line shows the maximum and minimum values of ATP for the oscillatory solution. For k_3 parameter values between 0.3905 mM s^{-1} and 1.59 mM s^{-1} (where the red line is obtained), the system would exhibit the dynamics shown in Fig. 3a, b, and for k_3 parameter values larger than 1.59 mM s^{-1} (where the green dotted line is obtained), the system would exhibit the dynamics shown in Fig. 3c. (**b**) Two-parameter bifurcation diagram. The line shows values of parameters V_{in} and k_3 at which the Hopf bifurcation is observed. This line separates the parameter regions with the stable steady-state solution for parameter values on the left side and above the blue line and the unstable solution with a limit cycle for parameter values on the right side and below the blue line. Other model parameters have the same values as were used to simulate results in Fig. 3

Discussion

We learned that the biochemical system that is responsible for the breakdown of glucose to extract energy can exhibit an interesting dynamic behavior. In particular, the simple molecular mechanism shown in Fig. 2, involving only four processes of glycolysis, can produce oscillatory dynamics. However, it was not immediately intuitive that this glycolytic pathway can produce oscillations. Thus, we used mathematical modeling to analyze this biochemical system and confirmed that the proposed scheme of the glycolytic pathway can indeed exhibit oscillatory dynamics.

For the glycolytic oscillations, the modeling results reveal that the positive feedback of ATP on its own production rate and the non-glycolytic ATP-consuming process described by the nonlinear Michaelis-Menten reaction, play a crucial role in appearance of oscillations [5, 6]. The positive feedback loop through ATP creates an autocatalytic process in which the number of ATP molecules produced in the energy-releasing phase exceeds the number of consumed ATP molecules in the energy-requiring phase. As was computationally proven by Cortassa and coworkers in Ref. [4], this positive feedback loop through ATP was required for the appearance of oscillations. Also, the appearance of oscillations can be controlled by varying either the influx of glucose V_{in} or the rate constant k_3 describing non-glycolytic ATP-consuming processes, but not by varying other model parameters. Bier and coworkers demonstrated that replacing the Michaelis-Menten type of saturability of the non-glycolytic ATP-consuming reaction with a power law approximation, did not produce oscillatory dynamics in the glycolytic system. Therefore, higher order nonlinear dependencies on ATP concentration are essential for the oscillatory behavior of the glycolytic system. Overall, the dynamic behavior of nonlinear systems is often not intuitive but can be revealed by mathematical modeling.

It is also important to emphasize that the physiological significance of metabolic oscillations is not well-understood. The physiologically normal metabolic state of a single cell is a steady state. It has been postulated that metabolic oscillations arise only when homeostatic regulatory mechanisms are compromised in abnormal or pathological conditions [5]. Indeed, the sustained glycolytic oscillations are observed only under some specific conditions, such as cyanide poisoning (Fig. 1b). Also, glycolytic oscillations are sensitive to changes in the activity of some enzymes. Thus, in the case of metabolic processes, the tendency of living cells to oscillate is not a robust behavior.

Exercises
Exercise 1. Simulate the following "core model" of glycolysis

$$\frac{d[\text{Glc}]}{dt} = V_{in} - k_1[\text{Glc}][\text{ATP}]$$

$$\frac{d[\text{ATP}]}{dt} = 2k_1[\text{Glc}][\text{ATP}] - k_3\frac{[\text{ATP}]}{K + [\text{ATP}]} - k_4[\text{ATP}]$$

proposed by Bier and coworkers [5]. Show that this model can produce oscillatory dynamic behavior, similar to that seen in Fig. 3a–c.

Exercise 2. Use XPPAUT to analyze the model from Exercise 1 in the phase plane. Plot nullclines and trajectories, similar to that shown in Fig. 4.

Exercise 3. Use AUTO to show that the core model from Exercise 1 produces one- and two-parameter bifurcation diagrams, similar to that shown in Fig. 5.

Method
Model (1) was simulated using XPPAUT computational software. The XPP code below can be used to obtain all results shown in Figs. 3, 4, and 5. This code has to be written in a text file with the file extension, .ode, as:

```
#XPP code-Glycolytic Oscillations
dGlc/dt = Vin-k1*Glc*ATP
dATP/dt = -2*k1*Glc*ATP+4*k2*FdP*(A-ATP)*P-k3*ATP/(K+ATP)
dFdP/dt = k1*Glc*ATP-k2*FdP*(A-ATP)*P
par Vin=0.1, k3=0.3905
par P=10, A=10, k1=0.3, k2=0.1, K=2
init Glc=0.5, ATP=0.4, FdP=0
# final time
@ total=1000
# set the plotting window size:
@ xlo=0, xhi=200, ylo=0, yhi=4
done
```

To perform simulations in XPP, open the .ode file in XPP. The function of each XPP menu item is explained in the Computational Software chapter of this book.

To produce results in Fig. 3a, open the Initial Data window by using the ICs button and click GO. This should plot Glc as a function of time (the black curve in Fig. 3a). Open Graphic stuff menu and (A)dd curve window, then set Y-axis: FdP and Color:1 and click OK to add the FdP curve (the red line in Fig. 3a). Repeat Graphic stuff->(A)dd curve, set Y-axis:ATP, Color:9 and click OK to add the ATP curve (the blue line in Fig. 3a).

To simulate results in Fig. 3b, set Xmax:1000 and Ymax:0.6 in the Viewaxes->2D window and click OK. Click Erase to clean the graphics area. Open the Parameters window by clicking on Param button and change parameter Vin value to 0.0075, click OK and then GO. This should produce three curves that are identical to those in Fig. 3b.

To produce results in Fig. 3c, set Xmax:150, Ymax:4 in the Viewaxes->2D window and click OK. Click Erase to clean the graphics area. In the Parameters window, change Vin value to 0.2 and k3 value to 2.5, click OK and then GO. Glc, ATP, and FdP curves should appear in the graphics area.

To plot trajectories in the phase plane, set X-axis:ATP, Yaxis:Glc, Xmin:0, Ymin:0, Xmax:3.5, Ymax:3 in the Viewaxes->2D window and click OK. In the Parameters window, set Vin=0.1 and k3=1 then click OK. Go to Initialconds and click m(I)ce and use the computer mouse cursor to click on the graphic area to indicate the starting point for the trajectory. This should produce trajectories as in Fig. 4a, but all trajectories will be plotted in black color. You may need to use Erase and Graphic stuff ->(R)emove all commands or to restart XPP to clean the graphic area from all previous curves. To obtain the limit cycle, as in Fig. 4b, set k3=2 in the Parameters window and click OK. Go to Initialconds and click m(I)ce and use the computer mouse cursor to click on the graphic area in several locations.

To produce one- and two-parameter bifurcation diagrams in Fig. 5, set Vin to 0.1 and k3 to 0.3905 in the Parameters window and click OK and GO. Click Initialconds->(L)ast and repeat this command five times. Now the solution of the system of equations should be at the non-oscillating steady state.

Open File->Auto. In the opened AUTO window go to Parameter menu and set Par1:k3 and Par2:Vin and click OK. Open Axes and then hI-lo and in the opened window set Y-axis:ATP, Xmin:0.3905, Ymin:0, Xmax:3.5, Ymax:1.2 and click OK. Open Numerics and in the opened window, set Par Max:3.5 and click OK. Go to Run and click Steady state, the red line (stable steady state solution) and the black line (unstable steady state solution) as in Fig. 5a should appear in the graphics area. Click Grab and grab (using the computer mouse cursor) a Hopf bifurcation point which corresponds to k3=1.59 (should be indicated by a cross sign where the red line ends and the black line begins), then click Enter. Go to Run and choose the Periodic option. Now, the green dotted curve should appear. This completes the one-parameter bifurcation diagram in Fig. 5a.

To obtain the two-parameter bifurcation diagram (Fig. 5b), click Grab and grab the Hopf bifurcation point again by using the mouse cursor and click Enter. Next, open Axes and Two par and in the opened window set Ymin:0.0075 and Ymax:0.24 and click OK. Then, do Run and Two Param, and the blue line should appear. To complete the blue line down, click Grab and grab the end of the line, open Numerics and change Ds:-0.02 (negative), and click OK. Do Run and Two Param again. Now the blue curve should be identical to the one in Fig. 5b. This blue line separates the region with stable and unstable steady state solutions.

Additional Reading

The earliest mathematical models of Glycolytic Oscillations were presented by Higgins [8], Sel'kov [9] and Goldbeter and Lefever [7].

References

1. Duysens LN, Amesz J (1957) Fluorescence spectrophotometry of reduced phosphopyridine nucleotide in intact cells in the near-ultraviolet and visible region. Biochim Biophys Acta 24 (1):19–26
2. Chance B, Schoener B, Elsaesser S (1964) Control of the waveform of oscillations of the reduced pyridine nucleotide level in a cell-free extract. Proc Natl Acad Sci U S A 52:337–341
3. Ghosh A, Chance B (1964) Oscillations of glycolytic intermediates in yeast cells. Biochem Biophys Res Commun 16(2):174–181
4. Richard P et al (1994) Yeast cells with a specific cellular make-up and an environment that removes acetaldehyde are prone to sustained glycolytic oscillations. FEBS Lett 341 (2-3):223–226
5. Bier M et al (1996) Control analysis of glycolytic oscillations. Biophys Chem 62(1-3):15–24
6. Cortassa S, Aon MA, Westerhoff HV (1991) Linear nonequilibrium thermodynamics describes the dynamics of an autocatalytic system. Biophys J 60(4):794–803
7. Goldbeter A, Lefever R (1972) Dissipative structures for an allosteric model. Application to glycolytic oscillations. Biophys J 12(10):1302–1315
8. Higgins J (1967) The theory of oscillating reactions - kinetics symposium. Industrial & Engineering Chemistry 59(5):18–62. https://doi.org/10.1021/ie50689a006
9. Sel'kov EE (1968) Self-oscillations in glycolysis 1. A simple kinetic model. Eur J Biochem 4 (1):79–86

NF-κB Spiky Oscillations

Pavel Kraikivski

Introduction

NF-κB is a family of transcription factors that are involved in the regulation of important cellular processes, such as innate and adaptive immunity, inflammation, stress responses, and apoptosis. NF-κB is activated by many external stimuli, including bacteria, viruses, and various stresses and proteins, after which it initiates proper cell responses by regulating expression of various genes. The NF-κB signaling system is conserved to operate in many different cell types and species, and it has been suggested that the NF-κB family has evolutionarily conserved mediators of immune responses [1]. In addition to its important biological roles, NF-κB regulation produces a very interesting dynamic behavior that may play an important role in the temporal encoding of signals. Mathematical models that were successful in explaining the dynamic behavior of NF-κB are now used to demonstrate the power of mathematical modeling approaches in Biological Sciences.

The transcription factor NF-κB and its regulation remain an exciting and active area of study. Active NF-κB participates in the control of transcription of >500 target genes [2]. The differential transcription regulation of situation-specific genes by NF-κB determines the fate of cells. Despite the fact that the activation steps of NF-κB are well-established, little is known about how the information is processed by the NF-κB system in order to regulate differential transcription of situation-specific genes. Thus, it would be of interest to learn how the signal-processing elements of the NF-κB system differentiate and select to transcribe specific genes.

In 2002, Hoffman and coworkers developed a computational model that explains observed oscillations in the temporal response of NF-κB activity [3]. Then, in 2004,

P. Kraikivski (✉)
Academy of Integrated Science, Division of Systems Biology, Virginia Polytechnic Institute and State University, Blacksburg, VA, USA
e-mail: pavelkr@vt.edu

© Springer Nature Switzerland AG 2021
P. Kraikivski (ed.), *Case Studies in Systems Biology*,
https://doi.org/10.1007/978-3-030-67742-8_5

Nelson and coworkers used single-cell time-lapse imaging experiments to reveal spiky oscillations in the nuclear-cytoplasmic translocation of the NF-κB transcription factor [4]. They proposed that NF-κB oscillations may be involved in temporal encoding of the signal, which would allow the NF-κB system to regulate its interaction with other cellular signaling pathways [4]. It was known that NF-κB cooperates with multiple other signaling molecules and pathways [5]. Thus, in combination with other oscillatory transcription factor pathways, the NF-κB oscillatory system may be involved in differential regulation of cell fate in response to different stimuli. In 2006, Krishna and coworkers developed a simple core model of NF-κB regulation that successfully explains the spiky oscillations in the nuclear-cytoplasmic translocation of NF-κB [6]. In this chapter, we will study the oscillatory dynamic behavior of the NF-κB system by analyzing their mathematical model.

Physiology and Molecular Mechanism

The transcription factor NF-κB regulates expression of numerous cytokines and adhesion molecules in response to infection, stress, and injury [1]. Cytokines are necessary to protect uninfected cells from viral infection, but they also can induce the acute phase response in inflammation. Because NF-κB is an important component in the inducible expression of many proteins, including cytokines, NF-κB is a major element in the regulation of innate immune response and inflammation. In certain situations, NF-κB is also an important player involved in the regulation of cell division and apoptosis. One of the interesting and enigmatic properties of the NF-κB system is that it produces spiky asynchronous oscillations in the nuclear-cytoplasmic translocation of the NF-κB system components in response to a stimulation that activates NF-κB. For example, spiky NF-κB oscillations appear in cells stimulated by tumor necrosis factor alpha (TNFα), which induces activation of the NF-κB system. The nuclear translocation of activated NF-κB transcription factor can be observed using single-cell time-lapse imaging experiments [4].

The molecular mechanism that includes key interactions of the NF-κB signaling system, is shown in Fig. 1. In the cytoplasm of unstimulated cells, NF-κB is sequestered by binding to its inhibitor, IκB. NF-κB-activating stimuli activate the inhibitor kappa B kinase (IKK) that phosphorylates IκB which is then degraded (note that the degradation of IκB is not shown in Fig. 1). Also, IKK targets IκB for phosphorylation when it is in a complex with NF-κB. Once IκB is degraded, the freed NF-κB molecules translocate to the nucleus, where they bind to DNA consensus sequences of target genes. Also, NF-κB transcription factor regulates IκB transcription, increasing the production of its own inhibitor. Generally, only free nuclear NF-κB is imported into the nucleus. The import reaction of free NF-κB molecules is marked as (i) in Fig. 1. The inhibitor IκB can freely shuttle between the cytoplasm and the nucleus. Thus, in the nucleus, free NF-κB can be resequestered by IκB and inhibited again. Once NF-κB is bound by IκB, this complex is exported

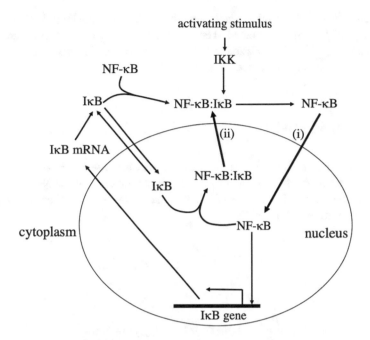

Fig. 1 Molecular mechanism that shows key interactions in the NF-κB signaling system. The key variables of the NF-κB signaling system are the concentrations of NF-κB inside and outside the cell nucleus, IκB inside and outside, the complex NF-κB:IκB inside and outside, and the concentration of the mRNA of IκB. Two compartments (cell cytoplasm and nucleus), and the activating stimulus (e.g., TNF-α) that acts on the inhibitor kappa B kinase (IKK) and the IκB gene are also shown to support the demonstration of how the information from external stimulus is processed towards gene regulation in the cell. (i) and (ii) are used to indicate transport in and out of the cell nucleus, respectively

from the nucleus back into the cytoplasm. The export reaction of NF-κB: IκB complex is marked as (ii) in Fig. 1.

The nuclear-cytoplasmic shuttling of NF-κB is observed to be oscillatory. Moreover, the oscillatory pattern is "spiky" considering that the periodic sharp peaks are observed in the nuclear localization of NF-κB following cell stimulation with the NF-κB-activating stimulus. The pattern of peak timing and amplitudes is also observed to be different in different cell types [4]. The purpose and function of differential temporal behavior of NF-κB in gene regulation remain unknown. One of the hypotheses is that temporal encoding of the signal allows the NF-κB signaling system to regulate its interaction with other cellular signaling pathways [4]. Also, Zambrano and coworkers discovered that NF-κB oscillations drive the expression of distinct sets of genes whose expression dynamics and functions are correlated [7]. Thus, NF-κB oscillations are translated into functionally related patterns of gene expression.

The negative feedback loop in the NF-κB signaling system is a key mechanism that produces oscillations. There are two key negative regulations shown in Fig. 1: (i) IκB negatively regulates nuclear NF-κB localization, and (ii) NF-κB promotes the degradation of IκB by forming a complex with IκB, which can be phosphorylated by IKK, and subsequently degraded. The mechanism (ii) turns out to be crucial for oscillations [6]. We will analyze the dynamic behavior of NF-κB regulatory mechanisms using mathematical modeling.

Mathematical Model

To study oscillatory behavior of NF-κB, we will analyze a simple mathematical model that describes the dynamics of three key components of the NF-κB signaling system: nuclear NF-κB denoted as N_{nuc}, mRNA of the IκB inhibitor denoted as I_m, and cytoplasmic IκB denoted as I_{cyt}.

$$\frac{dN_{nuc}}{dt} = a\frac{1 - N_{nuc}}{K_i + I_{cyt}} - b\frac{I_{cyt}N_{nuc}}{K_n + N_{nuc}}$$

$$\frac{dI_m}{dt} = N_{nuc}{}^2 - I_m \tag{1}$$

$$\frac{dI_{cyt}}{dt} = I_m - C_{ikk}(1 - N_{nuc})\frac{I_{cyt}}{K_i + I_{cyt}}$$

This mathematical model (1) was proposed by Krishna, Jensen and Sneppen in 2006 [6], and is successful in producing the spiky oscillatory dynamics of NF-κB. This core model was constructed based on a model developed by Hoffman and coworkers [3], which includes many chemical reactions that describe the dynamics of 26 different molecules in the NF-κB system. However, this core model (1) is sufficient to study oscillatory dynamics of the NF-κB system. In this model, all variables in the system of equations (1) are rescaled to be dimensionless. Also, the total (nuclear and cytoplasmic) NF-κB is normalized to one; thus, the cytoplasmic NF-κB is given by $1 - N_{nuc}$. Importantly, the activating external stimuli act on the inhibitor, kappa B kinase (IKK), and this action is controlled by the input parameter C_{ikk} in model (1). Therefore, the dynamic behavior of any NF-κB component in (1), measured as a function of the C_{ikk} parameter, can represent a signal response curve for the NF-κB signaling system. This signal response curve will be obtained below by solving the system of equations (1). Model (1) has also four other dimensionless parameters: a, b, K_n, and K_i, that can alter dynamic behavior of the NF-κB signaling system.

The first term in the first equation of system (1) represents the import of cytoplasmic NF-κB, $(1 - N_{nuc})$, into the nucleus. As mentioned previously, only free nuclear NF-κB is imported into the nucleus. This import is inhibited by cytoplasmic IκB (I_{cyt}), that forms the complex with NF-κB. The complex formation is described by Michaelis-Menten type kinetics, where K_i sets the concentration of IκB at which half of the cytoplasmic NF-κB is in the complex with its inhibitor. And the parameter a is proportional to the NF-κB nuclear import rate. The second term in the first equation of system (1) represents the export of NF-κB from the nucleus via the NF-κB:IκB complex, and this depends on both N_{nuc} and I_{cyt}. Parameter b is proportional to the IκB nuclear import rate. We also note here that only the complex NF-κB:IκB can be exported from inside the nucleus, not the free NF-κB. The parameter K_n sets the concentration of nuclear NF-κB at which half the nuclear IκB is in complex with NF-κB. K_n depends on the rates of association and dissociation of the complex, as well as on the export rate of the complex from the cell nucleus to the cytoplasm.

The first term in the second equation of system (1) describes the rate of production of IκB mRNA. Transcription of the IκB gene is activated by the NF-κB dimer and, thus, the rate of mRNA production depends on the square of N_{nuc}. The second term in the second equation of system (1) describes the degradation of IκB mRNA. In this model, the degradation rate sets the overall timescale for dynamics of all components in the NF-κB system.

The first term in the third equation of system (1) describes the production of cytoplasmic IκB from its mRNA. The second term describes degradation of cytoplasmic IκB due to the presence of IKK, which is proportional to C_{ikk}. This process also depends on the concentration of the NF-κB:IκB complex. This complex is formed in the cytoplasm and, therefore, its formation depends on both cytoplasmic IκB (I_{cyt}) and cytoplasmic NF-κB ($1 - N_{nuc}$). In the denominator of this term, the K_i parameter sets the concentration of IκB at which half of the cytoplasmic NF-κB is in the complex with its inhibitor.

Results

The mathematical model (1) can be conveniently solved and analyzed using XPPAUT computational tool (see Computational Software chapter in this book). The XPP code is provided in the Method section.

First, we want to confirm that model (1) can produce oscillatory behavior, as observed in experiments. The experimental data in Fig. 2a–c show three types of nuclear NF-κB oscillations that are observed in cells under different doses and durations of TNF-α: (i) damped oscillations (Fig. 2a), (ii) spiky oscillations (Fig. 2b), and (iii) soft (non-spiky) oscillations (Fig. 2c). In these experiments, TNF-α is an activating stimulus that activates IKK kinase and, thus, promotes nuclear localization of NF-κB. Figure 2 also shows an example of each type of oscillation, generated from model (1) using different C_{ikk} parameter values: (i) damped oscillation (Fig. 2d), (ii) spiky oscillation (Fig. 2e), and (iii) soft oscillation (Fig. 2f). Thus, qualitatively, model (1) is successful in explaining different types of NF-κB oscillations observed in living cells.

As can be observed in Fig. 2d–f, the time duration that NF-κB is present inside the nucleus, above its average value, depends on IKK activity, C_{ikk}. Thus, the NF-κB system can control transcription of specific genes by tuning oscillatory behavior in

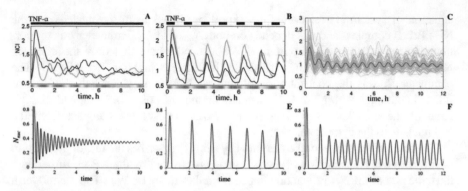

Fig. 2 Dynamics of NF-κB nuclear localization. (**a–c**) Experimentally detected oscillations of nuclear factor NF-κB p65 subunit in several cells. These three figures are adapted from the manuscript in Ref. [7]. (**a**) Cells are exposed by constant flow of 10 ng/mL TNF-α (indicated by the black bar above the plot). Blue, red and green curves are obtained by measuring the NF-κB p65 subunit in three different cells. (**b**) Alternating TNF-α doses $D_1 = 10$ ng/mL, $D_2 = 0$ ng/mL, of 45 min duration each (indicated by black/white alternating segments above the plot) were applied. (**c**) Nuclear factor NF-κB p65 subunit time series measured in single cells (green curves) and then averaged (black curve). Alternating TNF-α doses $D_1 = 10$ ng/mL, $D_2 = 0$ ng/mL, of 22.5 min duration each were applied. (**d–f**) Numerical simulations of model (1) for $C_{ikk} = 0.2$, $C_{ikk} = 0.045$ and $C_{ikk} = 0.08$, respectively. Other parameters in these simulations have the following values: $a = 0.007$, $b = 1000$, $K_n = 0.03$ and $K_i = 0.00002$. Initial conditions: $N_{nuc} = 0.005$, $I_m = 0.016$ and $I_{cyt} = 0.007$

response to an external stimulus and adjusting the amount of time that NF-κB is active inside the nucleus.

Figure 3a shows that only the concentration of nuclear NF-κB shows sharp peaks, whereas IκB oscillations are not spiky. This is observed both in experiments [4] and in the model simulation results.

All different oscillatory regimes can be demonstrated by a one-parameter bifurcation diagram that represents the signal-response curve. The diagram in Fig. 3b shows dynamic behavior of nuclear NF-κB, N_{nuc}, as the function of C_{ikk} parameter (signal). The green loop corresponds to a periodic solution with spikes for C_{ikk} parameter values in [0.0078, 0.039] and soft oscillations for $0.039 \leq C_{ikk} \leq 0.161$; the red line, $C_{ikk} \geq 0.161$, corresponds to a damped oscillation solution converging to a stable steady state.

Discussion

This case study shows how mathematical modeling can help us explain and understand the origin of spiky oscillations observed in the NF-κB system. The mathematical model (1) is described only by three equations and five parameters. Thus, we can easily explore and understand the range of dynamic behavior that the model can

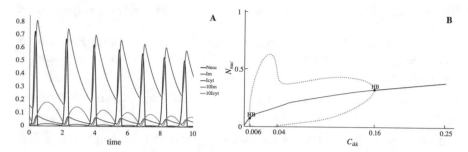

Fig. 3 NF-κB model simulation results. (**a**) Periodic behavior of nuclear NF-κB, N_{nuc}, mRNA of IκB inhibitor, I_m, and cytoplasmic IκB, I_{cyt}, components when parameter $C_{ikk} = 0.045$. Also, I_m and I_{cyt} components are shown magnified by a factor of ten. (**b**) One-parameter bifurcation diagram that shows the nuclear NF-κB depending on the activity of IKK kinase (C_{ikk}). Hopf bifurcation occurs at $C_{ikk} = 0.0058$ and $C_{ikk} = 0.161$. Spiky oscillations appear for C_{ikk} parameter values in the following range [0.0078, 0.039]. Non-spiky (soft) oscillations occur when $0.039 \leq C_{ikk} \leq 0.161$. Damped oscillations occur for $C_{ikk} \geq 0.161$. Other parameters in these simulations have the following values: $a = 0.007$, $b = 1000$, $K_n = 0.03$ and $K_i = 0.00002$

exhibit. We learned that the model can produce both spiky and soft oscillations, and that spiky oscillations can be obtained for a wide range of the C_{ikk} parameter values, as shown in the one-parameter bifurcation diagram of Fig. 3b. Therefore, the spiky oscillations are a robust dynamic property of the NF-κB system.

The time duration that NF-κB is present inside the cell nucleus in sufficiently large concentrations is sensitive to the stimulus signal (TNF-α) that activates IKK. Because NF-κB needs time to dimerize and bind to gene promoters, the tuning mechanism for spike duration can be used by the NF-κB system to control transcription of specific genes.

The negative feedback loop in the NF-κB signaling system is a key mechanism that allows the system to produce oscillatory behavior. Models that include a negative feedback with time delay are often used to explain oscillations. For example, a similar mechanism has been used to model the oscillations in a p53 regulatory system [8] and in Hes1 protein oscillations [9]. Functional roles of pulsing behavior in various genetic circuits have been extensively reviewed in Ref. [10]. It has been shown that negative feedback loops appear commonly in many pulsatile circuits. The potential functions that pulsatile systems might provide in the cell include a) co-regulation of targets of pulsing transcription factors with other regulators, b) synchronization between multiple pulsing signaling pathways or several pulsing factors, enabling them to more productively activate target expression, and c) efficiently distributing limited resources by alternating between conflicting regulatory programs and separating them in a timely manner. Also, more pulsatile systems and corresponding functions are expected to be discovered.

Exercises

Exercise 1. Simulate model (1) using XPPAUT, and

(a) show that spiky and soft oscillations can be obtained for different values of the K_i parameter
(b) use AUTO to produce one-parameter bifurcation diagram for the K_i parameter
(c) investigate how dynamic behavior of the model depends on a, b and K_n parameters.

Exercise 2. Simulate model (1) using XPPAUT and answer the following questions:

(a) What are the parameters that control the negative feedback loops in the NF-κB signaling system?
(b) What will happen if these parameters are set to zero? Demonstrate it using XPP plots showing the change in dynamic behavior of the NF-κB signaling system.

Method
The corresponding description of the model (1) for XPPAUT computational software has to be written in a text file with the file extension .ode and has the following form:

```
# NF-kB Oscillator

dNnuc/dt = a*(1-Nnuc)/(Ki+Icyt) - b*Icyt*Nnuc/(Kn+Nnuc)
dIm/dt = Nnuc^2 - Im
dIcyt/dt = Im - Cikk*(1-Nnuc)*Icyt/(Ki+Icyt)

aux plotIm=10*Im
aux plotIcyt=10*Icyt

par Cikk=0.045, a=0.007, b=1000
par Ki=0.00002, Kn=0.03

init Nnuc=0.005, Im=0.016, Icyt=0.007

done
```

To run this code in XXPAUT, use the following steps:

(1) In the XPP main window, set a stiff solver in nUmerics, then Method then (S)tiff by accepting default Tolerance =0.001, minimum step =1e-12, maximum step =1 by Enter keyboard key. Exit nUmerics by clicking on [Esc] key.
(2) On the top of the main XPP Window click ICs and also Param buttons. Initial Data and Parameters windows will appear.
(3) In Initial Data window, click the Go button, and the plot of Nnuc variable as a function of time should appear in the Graphics area. Axes ranges and variables can be changed in Viewaxes of the main XPP Window, in 2D by writing variable names next to X-axis: and Y-axis: lines and by setting Xmin, Ymin, Xmax, Ymax values.

To compute the one-parameter bifurcation diagram in AUTO, use the following steps:

(1) In the `Parameters` window set the parameter `Cikk` $=0.25$ and then press `OK` and `GO` buttons in this window. The plot of the Nnuc variable as a function of time will appear in the Graphics area.
(2) In `Initialconds` click `(L)ast` and repeat it at least three times to establish non-oscillating conditions.
(3) Open `File` and click `Auto`, a new window will appear.
(4) In `Axes` click `hI-lo`, and set `Xmin=0`, `Ymin=0`, `Xmax=0.26`, `Ymax=1`.
(5) Select `Numerics` and set `Ds:-0.02`, `Par Min:0` and `Par Max:0.26`. `Par Min` and `Par Max` set the range in which parameter `Cikk` will be varied. The negative `Ds` means that the parameter change starts at maximum parameter value and goes towards the minimum parameter value.
(6) In the `Run` click `Steady state`; this command will plot steady state solution (red lines).
(7) Click `Grab` and on the right, grab the point at the end of the red line where the black line begins with the computer mouse cursor selecting the point and by pressing Enter key.
(8) In the `Run` click `Periodic`; this command will plot the periodic solution (a green line). The one-parameter bifurcation diagram should look like in Fig. 3b.

References

1. Ghosh S, May MJ, Kopp EB (1998) NF-kappa B and Rel proteins: evolutionarily conserved mediators of immune responses. Annu Rev Immunol 16:225–260
2. http://www.bu.edu/nf-kb/gene-resources/target-genes/
3. Hoffmann A et al (2002) The IkappaB-NF-kappaB signaling module: temporal control and selective gene activation. Science 298(5596):1241–1245
4. Nelson DE et al (2004) Oscillations in NF-kappaB signaling control the dynamics of gene expression. Science 306(5696):704–708
5. Hoesel B, Schmid JA (2013) The complexity of NF-kappaB signaling in inflammation and cancer. Mol Cancer 12:86
6. Krishna S, Jensen MH, Sneppen K (2006) Minimal model of spiky oscillations in NF-kappaB signaling. Proc Natl Acad Sci U S A 103(29):10840–10845
7. Zambrano S et al (2016) NF-kappaB oscillations translate into functionally related patterns of gene expression. Elife 5:e09100
8. Tiana G, Jensen MH, Sneppen K (2002) Time delay as a key to apoptosis induction in the p53 network. Eur Phys J B 29(1):135–140
9. Jensen MH, Sneppen K, Tiana G (2003) Sustained oscillations and time delays in gene expression of protein Hes1. FEBS Lett 541(1-3):176–177
10. Levine JH, Lin Y, Elowitz MB (2013) Functional roles of pulsing in genetic circuits. Science 342(6163):1193–1200

Tick, Tock, Circadian Clocks

Jae Kyoung Kim

Introduction

The circadian (~24 h) clock is a self-sustained endogenous oscillator, which times diverse behavioral and physiological processes such as the sleep-wake cycle, body temperature, blood pressure and hormone secretion [1]. The disruption of circadian rhythms increases the risk of getting various chronic diseases such as cancer, diabetes, mood disorders and sleep disorders [2]. The key oscillatory mechanism of the circadian clock is a transcriptional-translational negative feedback loop [3, 4]. However, the majority of biological systems with negative feedback loops reach steady homeostasis rather than generating sustained oscillations. It turns out that strong non-linearity and a time delay need to be incorporated into the negative feedback loop to generate sustained circadian rhythms [5, 6]. By studying this, we can learn how these elements function together to generate sustained rhythms, which are also commonly observed in other biological oscillators such as the cell cycle and segmentation clock [7]. Furthermore, this common principle of biological oscillators can be used to design synthetic oscillators.

Physiology and Molecular Mechanisms

Nearly all organisms exposed to day-night cycles have an innate time keeping system, the circadian clock, to keep track of time and optimize daily life [1, 2]. As a result, in mammals, ~50% of the protein-coding genome is transcribed in a

J. K. Kim (✉)
Department of Mathematical Sciences, Korea Advanced Institute of Science and Technology, Daejeon, South Korea

Biomedical Mathematics Group, Institute for Basic Science, Daejeon, Republic of Korea
e-mail: jaekkim@kaist.ac.kr

© Springer Nature Switzerland AG 2021
P. Kraikivski (ed.), *Case Studies in Systems Biology*,
https://doi.org/10.1007/978-3-030-67742-8_6

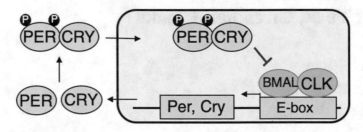

Fig. 1 The core transcriptional-translational negative feedback loop of the mammalian circadian clock. Transcriptional activator complex BMAL1:CLOCK promotes the transcription of period (*Per*) 1/2 and cryptochrome (*Cry*) 1/2. The translated proteins PER1/2, which are phosphorylated by the CK1δ/ε, bind to the CRY1/2. Then, PER1/2:CRY1/2 translocate to the nucleus, where they bind to BMAL1:CLOCK and repress the transcriptional activity of BMAL1:CLOCK. This transcriptional repression is released by the degradation of PER1/2, which is mainly triggered by further phosphorylation by CK1δ/ε

circadian rhythmic manner, and >80% of currently approved drug targets show rhythmicity, leading to the dependence of therapeutic effects on circadian rhythms [8–11]. For instance, the efficacy and toxicity of various anticancer drugs differ depending on dosing time (e.g. morning vs. night) [12]. Furthermore, heart surgery in the afternoon is less likely to trigger tissue damage than that in the morning because Rev-erbα, which inhibits protection of heart cells, oscillates with a peak in the morning [13]. Importantly, when circadian rhythms are disrupted (e.g., due to shift work and electric illumination after sunset), the risk of getting various chronic diseases, including metabolic diseases and cancer, is known to increase [2].

The mammalian circadian clock consists of about ten transcription regulators forming interlocked transcriptional feedback loops to generate sustained rhythms [3, 4]. The core clock mechanism is a transcriptional-translational negative feedback loop, which is conserved within the animal kingdom and was initially deciphered in the fruit fly [8]. This led to the 2017 Nobel Prize in Physiology or Medicine awarded to Michael Young, Michael Rosbash, and Jeffrey Hall. In the mammalian circadian clock, the core negative feedback loop consists of repressor complexes PER1/2: CRY1/2, which repress their own transcriptional activator complex BMAL1: CLOCK in the nucleus (Fig. 1). In particular, the repression of transcription mainly occurs via *protein sequestration*: the repressor complexes sequester the activator complex in an inactive 1:1 stoichiometric complex rather than directly binding with their own genes to repress the transcription [14–16].

Mathematical Model

Various mathematical models describing the mammalian circadian clock have been developed [17–20]. Recently, Kim and Forger developed detailed and simple versions of a mathematical model describing the mammalian circadian clock [21]. The detailed model consists of hundreds of variables, including the majority of core clock molecules such as PER1/2, CRY1/2, BMAL1, CLOCK, NPAS2, Rev-erbs and CK1δ/ε. Due to the accuracy of the model, it has been used to understand how circadian rhythms are altered by genetic mutations [22, 23], temperature change [24, 25] and clock-modulating molecules, including a new drug candidate [26–29]. On the other hand, the simple Kim-Forger model consists of only three variables. Although the model is simple, it accurately captures the key dynamics of the core transcriptional-translational negative feedback loop based on protein sequestration of the mammalian circadian clock. Thus, it has been widely used to investigate various properties of the mammalian circadian clock, such as molecular mechanisms for robust circadian rhythms against temperature change [24, 30], intrinsic noise [31], and a coupling signal [32], and circadian regulation of a cell cycle component [33, 34]. With this simple Kim-Forger model, in this section, we will illustrate how to simplify mathematical models with important tools of mathematical modeling: quasi-steady state approximation (QSSA) and non-dimensionalization. Then, using the simplified model, we will investigate key molecular mechanisms that generate sustained circadian rhythms in the results section.

In the simple Kim-Forger model (Fig. 2a, b), repressor mRNA (M) is translated to repressor protein (C) in the cytoplasm. After its translocation to the nucleus, the repressor protein (F) binds with a free active activator (A) to form an inactive 1:1 stoichiometric complex (A:F) and thus represses its own transcription. To describe this transcriptional repression via protein sequestration, the transcription of mRNA is assumed to be proportional to the fraction of free activator (A) among the total activator in the nucleus ($A_T = A + A : F$). All species are subject to degradation (β_i) except for the activator.

The model (Fig. 1b) can be simplified by using the QSSA under the assumption of rapid binding between repressors and activators [17, 21, 35]. For this, we first introduce a new variable $R \equiv F + A : F$, the concentration of total repressor in the

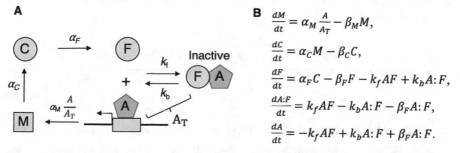

Fig. 2 Kim-Forger model for the mammalian circadian clock without model reduction. (**a, b**) Model diagram (**a**) and ordinary differential equations describing the model (**b**). M, concentration of repressor mRNA; C, concentration of repressor protein in the cytoplasm; F, nuclear concentration of free repressor protein; A, nuclear concentration of free activator; $A{:}F$, nuclear concentration of stoichiometric complex of repressor and activator; A_T, nuclear concentration of total activator (i.e. $A + A{:}F$), which is conserved

nucleus. Then, using $\frac{dR}{dt} = \frac{dF}{dt} + \frac{dA:F}{dt}$ and $F = R - A : F$, we can replace F with R in the model as follows:

$$\frac{dM}{dt} = \alpha_M \frac{A}{A_T} - \beta_M M,$$

$$\frac{dC}{dt} = \alpha_C M - \beta_C C,$$

$$\frac{dR}{dt} = \alpha_F C - \beta_F R,$$

$$\frac{dA : F}{dt} = k_f A(R - A : F) - k_b A : F - \beta_F A : F,$$

$$\frac{dA}{dt} = -k_f A(R - A : F) + k_b A : F + \beta_F A : F.$$

As the total activator concentration ($A_T = A + A : F$) is conserved (i.e., $\frac{dA:F}{dt} + \frac{dA}{dt}$ =0), we can replace $A : F = A_T - A$ and further simplify the model.

$$\frac{dM}{dt} = \alpha_M \frac{A}{A_T} - \beta_M M,$$

$$\frac{dC}{dt} = \alpha_C M - \beta_C C,$$

$$\frac{dR}{dt} = \alpha_F C - \beta_F R,$$

$$\frac{dA}{dt} = -k_f A(R - (A_T - A)) + k_b(A_T - A) + \beta_F(A_T - A).$$

Due to the rapid binding between repressors and activators (i.e., large k_f and k_b), the concentration of free activator (A) rapidly reaches the quasi-steady state (QSS), which can be derived by solving the QSS equation for the A (i.e., $\frac{dA}{dt} = 0$ or $k_f A$ $(R - (A_T - A)) - k_b(A_T - A) - \beta_F(A_T - A) = 0$) (Fig. 3a) [17, 21, 35]. By dividing the QSS of A with A_T, we can derive the QSS of $\frac{A}{A_T}$, which determines the transcription in the model as follows:

$$\frac{A(R)}{A_T} = \frac{1 - R/A_T - K_d/A_T + \sqrt{(1 - R/A_T - K_d/A_T)^2 + 4K_d/A_T}}{2} \xrightarrow[K_d \to 0]{} \left| 1 - \frac{R}{A_T} \right|,$$

(1)

where $K_d = \frac{k_b + \beta_F}{k_f} \approx \frac{k_b}{k_f}$ as $k_b \gg \beta_F$ is an approximate dissociation constant (Fig. 3b).

The QSS of A/A_T (Eq. 1) decreases as the molar ratio R/A_T increases. In particular, if K_d is small (solid line; Fig. 3b), the fraction of free activator approximately follows a piecewise linear function of the molar ratio R/A_T (i.e., $\left| 1 - \frac{R}{A_T} \right|$). With the QSS of A/A_T (Eq. 1), the fraction of free activator (A/A_T) can be determined solely by the molar ratio between total repressor and activator (R/A_T) without solving the

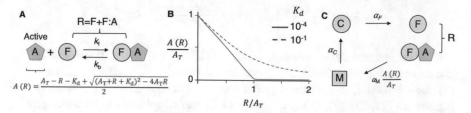

Fig. 3 Model reduction with QSSA. (**a**) Due to the rapid binding and unbinding between the free repressor (F) and free activator (A), the free activator can be assumed in the QSS. $K_d = \frac{k_b + \beta_F}{k_f} \approx \frac{k_b}{k_f}$ is an approximate dissociation constant. (**b**) As the molar ratio between total repressor and total activator (R/A_T) increases, the QSS of the fraction of free activator ($A(R)/A_T$) decreases. As the dissociation constant (K_d) becomes smaller, $A(R)/A_T$ shows a sharper transition at $R/A_T \approx 1$ and becomes more similar to a piece-wise linear function $\left|1 - \frac{R}{A_T}\right|$ (Eq. 1). (**c**) With the substitution of the QSS of free activator, the simplified Kim-Forger model can be derived (Eq. 2)

differential equation to track the concentration of free activator (A). Thus, we can simplify the model by replacing A with its QSS as follows (Fig. 3c):

$$\frac{dM}{dt} = \alpha_M \frac{A(R)}{A_T} - \beta_M M,$$
$$\frac{dC}{dt} = \alpha_C M - \beta_C C, \qquad (2)$$
$$\frac{dR}{dt} = \alpha_F C - \beta_F R.$$

With the reduced model, we can more easily understand the mechanisms underlying the transcriptional negative feedback loop. Specifically, when the repressor is absent (i.e., $R/A_T = 0$), all of the activator is free (Fig. 3b), and thus transcription occurs at the maximal rate (α_M). As the repressor level increases, the fraction of free activator and thus the transcription rate decrease approximately linearly. Furthermore, for the case of tight binding (solid line; Fig. 3b), when the repressor level becomes similar to that of the total activator (i.e., their molar ratio is 1:1), most activators are sequestered by repressors and thus transcription is almost completely repressed.

Although the model is reduced (Eq. 2), it still has a large number of parameters. Thus, we need to reduce the number of parameters to facilitate the investigation of the parameter region leading to sustained oscillation [17, 21, 32]. To this end, we assume that the degradation rates of all molecules are the same (*i.e.* $\beta_M = \beta_c = \beta_F = \beta$), which increases the chance of oscillations [36, 37]. Then, the number of parameters can be further reduced by scaling parameters, variables and time as

$$M = \frac{\alpha_M}{\beta} \overline{M}, \quad C = \frac{\alpha_M \alpha_C}{\beta^2} \overline{C}, \quad R = \frac{\alpha_M \alpha_C \alpha_R}{\beta^3} \overline{R},$$

$$K_d = \frac{\alpha_M \alpha_C \alpha_R}{\beta^3} \overline{K}_d, \quad A_T = \frac{\alpha_M \alpha_C \alpha_R}{\beta^3} \overline{A}_T, \quad t = \frac{1}{\beta} \overline{t}. \tag{3}$$

See supplementary information in Ref. [21] for a detailed description of the choice of scaling. Substitution of these scaled parameters, variables and time into the model (Eq. 2) non-dimensionalizes the model and normalizes most of the parameters [17, 21, 32]. Thus, we can get a further simplified model with two free dimensionless parameters, $\overline{K}_d$ and $\overline{A}_T$, as follows:

$$\frac{d\overline{M}}{d\overline{t}} = \frac{1 - \overline{R}/\overline{A}_T - \overline{K}_d/\overline{A}_T + \sqrt{\left(1 - \overline{R}/\overline{A}_T - \overline{K}_d/\overline{A}_T\right)^2 + 4\overline{K}_d/\overline{A}_T}}{2} - \overline{M}$$

$$\frac{d\overline{C}}{d\overline{t}} = \overline{M} - \overline{C}, \tag{4}$$

$$\frac{d\overline{R}}{d\overline{t}} = \overline{C} - \overline{R}.$$

For simplicity, we will keep using the notation of the original variables (e.g. M instead of $\overline{M}$) and parameters (e.g. K_d instead of $\overline{K}_d$) throughout this chapter.

Results

The dynamics of the nondimensionalized model (Eq. 4) are fully determined by the two dimensionless parameters A_T and K_d. Thus, we just need to adjust the two parameters to identify the parameter region generating sustained oscillation. The relationship between the choice for the values of these two parameters and the resulting amplitude of oscillations can be effectively illustrated with a heat map of relative amplitude whose X-axis and Y-axis represent the value of A_T and K_d, respectively. This will allow us to identify the region of the parameter space where the oscillation occurs (Fig. 4a). However, unfortunately, it is difficult to interpret the biological meaning of the identified region.

In fact, we need to choose the axis of the heat map more cleverly. Let's look at the function describing the transcription in Eq. (4) where A_T and K_d appear in the model. The function is determined by two ratios, R/A_T and K_d/A_T. Thus, these two ratios are a more natural choice for the X-axis and Y-axis of the heat map than A_T and K_d. If we choose certain values of A_T and K_d, then K_d/A_T will be a single number. However, unlike K_d/A_T, R/A_T involves a variable R, which would oscillate and keep changing. In this case, one can use the average of R over a cycle instead of R [21, 37]. An alternative is the steady state of R (R_S), which approximates the average. R_S can be simply calculated by solving $\frac{dM}{dt} = \frac{dC}{dt} = \frac{dR}{dt} = 0$, which is equivalent to $A(R)/A_T = R$ [21]. By solving this equation, for each choice for the value of A_T and K_d, we can get a single corresponding value of R_S. Thus, by adjusting the value of A_T and K_d, we can draw a heat map whose X-axis and Y-axis represent the value of R_S/A_T and $K_d/$

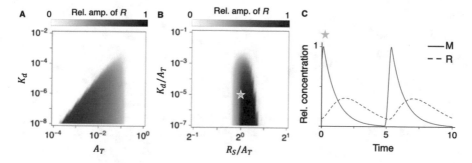

Fig. 4 The conditions required for the Kim-Forger model to generate sustained oscillations. (**a**) The heat map of relative amplitude, where the X-axis and Y-axis represent the concentration of total activator (A_T) and the dissociation constant (K_d), respectively. The color represents the relative amplitude of the repressor (R) oscillation. (**b**) The heat map of the relative amplitude, where the X-axis and Y-axis represent the relative concentration of total nuclear repressor at the steady state (R_S) and the dissociation constant (K_d) compared to the concentration of total activator (A_T), respectively. The oscillation occurs when the molar ratio between the repressor at the steady state (R_S) and the total activator (A_T) is around 1:1 and the binding between repressor and activator becomes tight enough (i.e., low K_d). R_S can be obtained by solving $\frac{dM}{dt} = \frac{dC}{dt} = \frac{dR}{dt} = 0$ of Eq. (4), which is equivalent to $A(R)/A_R = R$. (**c**) Simulated trajectories of M and R with A_T=0.00316 and $K_d = 10^{-5} A_T$ (star mark in the heat map (B)). Here, the trajectories are normalized to the maximum M

A_T, respectively (Fig. 4b). With this heat map, we can easily identify two conditions required for the model to generate sustained oscillations.

- The oscillation occurs only when the dissociation constant (K_d) is low enough compared to the total activator concentration (A_T), indicating that tight binding between the repressor and activator is required for oscillation.
- The oscillation occurs only when the molar ratio between R_S and A_T is around 1:1. This indicates that on average the molar ratio between repressor and activator in the nucleus should be around 1:1 throughout the cycle to maintain sustained oscillation (see [21] for the analysis using the actual average). Specifically, if the molar ratio is much less than 1:1, there is excess activator relative to the repressor, and thus the gene is fully expressed nearly all time. On the other hand, if the molar ratio is much larger than 1:1, the majority of the activator is bound to the repressor, and thus the transcription of the gene is nearly permanently repressed.

Finally, we would like to emphasize another advantage of choosing R/A_T and K_d/A_T (Fig. 4b) over A_T and K_d (Fig. 4a). That is, as A_T and K_d are scaled parameters (Eq. 3), they are a relative total activator concentration and dissociation constant compared to $\frac{\alpha_M \alpha_C \alpha_R}{\beta^3}$. Due to this complex scale, we cannot interpret them as total activator concentration and dissociation constant. On the other hand, the complex

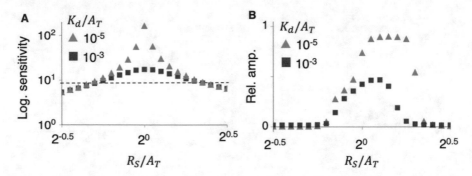

Fig. 5 The logarithmic sensitivity of the transcription repression function should be large enough to generate sustained oscillations. (**a**) The logarithmic sensitivity increases as the dissociation constant (K_d) decreases or the molar ratio between repressor and activator at the steady state (R_S/A_T) becomes closer to 1:1. (**b**) Only when the logarithmic sensitivity is greater than 8 (dashed line; A) the oscillation occurs

scale is cancelled in R/A_T and K_d/A_T, and thus, the above natural biological interpretation can be made.

As can be seen in the simulated trajectories (Fig. 4c), the amplitude of the oscillatory mRNA concentration (M) is much larger than the amplitude of the oscillatory repressor concentration (R). This is because a small change in R leads to a large change in transcriptional activity ($f(R) = \frac{A(R)}{A_T}$), which determines the amplitude of M. This amplification prevents the oscillation from dampening to the steady state. Specifically, to generate sustained oscillation, a small change in relative repressor concentration needs to trigger a large change in relative transcription activity, which can be measured by logarithmic sensitivity ($|dlogf(R)/dlogR| = |(df(R)/dR)(R/f(R))|$). The logarithmic sensitivity should be greater than 8 at the steady state (i.e., above the dashed line in Fig. 5a) for the model to escape the steady state and generate sustained oscillation (Fig. 5b) [17, 21, 37–40]. The logarithmic sensitivity (i.e., the amplification level of the transcription) increases as the dissociation constant (K_d) decreases and the molar ratio at the steady state (R_S/A_T) becomes closer to 1:1 (Fig. 5a) [17, 21]. This explains why these two conditions are required for the model to simulate sustained oscillations (Fig. 5b). Another way to analyse the model is based on bifurcation analysis (see [41] for details).

While the Kim-Forger model (Eq. 4) consists of three steps, we can add an intermediate molecule (I) between C and R, which can be considered a phosphorylated repressor before entering the nucleus:

$$\frac{dM}{dt} = \frac{1 - R/A_T - K_d/A_T + \sqrt{\left(1 - R/A_T - K_d/A_T\right)^2 + 4K_d/A_T}}{2} - M,$$

$$\frac{dC}{dt} = M - C,$$

$$\frac{dI}{dt} = C - I,$$

$$\frac{dR}{dt} = I - R.$$

$$(5)$$

For this model with four steps to generate sustained oscillation, the logarithmic sensitivity (Fig. 5a) needs to be larger than 4 rather than 8. Thus, the model can generate sustained oscillations for a wider range of parameter regions compared to the three-step model (Exercise 1). In fact, for the model with n steps, the logarithmic sensitivity needs to be greater than $\sec(\pi/n)^n$, which is known as the secant condition [37, 39, 40]. When $n=3$ and 4, $\sec(\pi/n)^n$ is 8 and 4, respectively.

- The $\sec(\pi/n)^n$ is a decreasing function of n, indicating that as more intermediate steps are included and thus more time delay exists in the feedback loop, the feedback loop is more likely to generate sustained oscillations. Furthermore, when $n=2$, $\sec(\pi/n)^n$ is infinity, and thus, the model cannot simulate sustained oscillations, which highlights the necessity of time delay in generating oscillations.

Discussion

The *simple* Kim-Forger model predicts that the transcriptional negative feedback loop based on protein sequestration can generate sustained oscillations under three conditions: (1) a 1:1 molar ratio between repressor and activator, (2) their tight binding, and (3) time delay in the feedback loop. These conditions are also required for the *detailed* Kim-Forger model to generate sustained oscillations [21]. In particular, when some of the core clock molecules have mutated, and thus the molar ratio between the repressors (PER1/2-CRY1/2) and the activator (BMAL1-CLOCK) is far from 1:1 in the detailed model, the simulated mutant phenotypes become arrhythmic.

Consistent with the model prediction, the 1:1 molar ratio between the repressor and activator is critical for the mammalian circadian clock to generate sustained rhythms. Specifically, the molar ratio between the repressors (PER1/2) and the activator (BMAL1) is maintained at around 1:1 in the liver tissue of mammals [3]. Importantly, as the molar ratio becomes closer to 1:1, the amplitude and sustainability of circadian rhythms are enhanced [14, 27]. Furthermore, the tight binding between repressor and activator has been recently identified in the

mammalian circadian clock [42]. When the binding affinity becomes weaker, the
Drosophila circadian clock cannot generate sustained rhythms [43]. Finally, the
transcriptional negative feedback loop of the mammalian circadian clock contains a
long-time delay. The time delay is mainly generated by the multi-step phosphorylation of PER by CK1δ/ε, which regulates entry to the nucleus, follow-up transcriptional suppression and release of PER [3, 4, 24].

While the mechanisms for the self-sustained rhythms have been mainly discussed
in this chapter, the circadian clock has many interesting dynamic properties. For
instance, the period of self-sustained rhythms is maintained over a range of physiological temperatures although the temperature alters the biochemical reaction rates
[44]. Recently, molecular mechanisms underlying the temperature compensation
were identified with a combination of model simulations and follow-up experiments
[24, 30, 45]. Furthermore, circadian rhythms are entrained by external signals such
as light [46, 47]. This is how we can overcome the jet lag caused by misalignment
between internal circadian time and external time. The mechanism underlying the
entrainment of circadian rhythms by the external light-dark cycle has been widely
investigated with mathematical models [48–53] (Exercise 3). Furthermore, recently,
systems pharmacology models have been used to investigate pharmacological
entrainment of circadian rhythms, which can be used to treat circadian rhythm
sleep disorders [26, 28].

Exercises

1. Using the non-dimensionalized Kim-Forger model (Eq. 4), simulate trajectories
   of mRNA (M) with $A_T = 0.067$ and varying $K_d = 10^{-3}$, $10^{-3.5}$ and 10^{-4}. With
   the same parameters, simulate trajectories with the non-dimensionalized
   Kim-Forger model with four steps (Eq. 5). By comparing the simulated trajectories, discuss how the time delay induced by an additional step affects oscillations
   (i.e., generation of oscillation, amplitude, and period).

2. In the Kim-Forger model (Eq. 4), except for transcription, all of the other
   reactions are described as linear reactions. A recent study found that the degradation of nuclear PER (R) occurs in a non-linear manner, which can be described
   with the Michaelis-Menten equation as follows [31]:

$$\frac{dM}{dt} = \frac{1 - R/A_T - K_d/A_T + \sqrt{(1 - R/A_T - K_d/A_T)^2 + 4K_d/A_T}}{2} - M,$$

$$\frac{dC}{dt} = M - C,$$

$$\frac{dR}{dt} = C - \frac{R}{\frac{R}{K_R} + 1},$$

where $K_R = 0.067$. With the same values of A_T and K_d as in exercise 1, simulate
trajectories and then compare them with the ones simulated with the three-step
model in exercise 1. Discuss the role of having additional non-linearity in generating

oscillation. See [31] for experimental data comparing circadian rhythms in the presence and absence of non-linear PER degradation.

3. The light signal transmitted to the master circadian clock in the suprachiasmatic nucleus induces the transcription of *Per* gene, which leads to the entrainment of the circadian rhythms by the external light-dark cycle. This light-induced *Per* transcription can be incorporated in the model by adding L as follows:

$$\frac{dM}{dt} = \frac{1 - R/A_T - K_d/A_T + \sqrt{(1 - R/A_T - K_d/A_T)^2 + 4K_d/A_T}}{2} - M + L,$$

$$\frac{dC}{dt} = M - C,$$

$$\frac{dR}{dt} = C - R,$$

where L is 0.05 during the day and 0 during the night.

(1) Calculate the intrinsic circadian period (T_{int}) of a simulated trajectory with $A_T = 0.067$ and $K_d = 10^{-5}$ in the absence of light (i.e., $L=0$). Note that as the time is unitless (Eq. 3), T_{int} does not need to be ~24.

(2) Compared to the intrinsic circadian period (T_{int}), which is calculated in (1), the period of the external day-night cycle (T_{ext}) is 10% longer (i.e., $T_{ext} = 1.1 T_{int}$). The length of the night and day are the same as $T_{ext}/2$. When the external day-night cycle is given, calculate the period of a simulated trajectory and compare it with T_{ext}. Is the circadian rhythm entrained to the external day-night cycle?

(3) A genetic mutation (e.g. familial advanced sleep phase syndrome) can shorten the intrinsic circadian period, and thus, the difference between T_{ext} and T_{int} can become larger. If T_{ext} is 50% longer than the T_{int} calculated in (1), can the circadian rhythm be entrained to the external day-night cycle?

Matlab Code
Exercise 1

```
tspan=0:0.1:500; %Time domain
IC1=[0 0 0]; IC2=[0 0 0 0]; %Initial conditions
AT=0.067; Kd=10^-3;
[time1,sol1]= ode45(@(t,y)
KF3step(t,y,AT,Kd),tspan,IC1);
Kd=10^-3.5;
[time2,sol2]= ode45(@(t,y)
KF3step(t,y,AT,Kd),tspan,IC1);
Kd=10^-4;
[time3,sol3]= ode45(@(t,y)
```

```
KF3step(t,y,AT,Kd),tspan,IC1);

plot(time1(end-200:end),sol1(end-200:end,3),time2(end-
200:end),sol2(end-200:end,3),time3(end-
200:end),sol3(end-200:end,3),'linewidth',2)
title('Three Steps');xlabel('Time');ylabel('R');leg-
end({'Kd=10e-3','Kd=10e-3.5','Kd=10e-4'},'Loca-
tion','southwest');axis([480 500 0
0.15]);set(gca,'font-
size',14,'xtick',480:10:500,'ytick',0:0.05:0.15);

function dy = KF3step(t,y,AT,Kd)
    M = y(1);   C = y(2);      R = y(3);
    dM = (AT-R-Kd+sqrt((AT-R-Kd)^2+4*AT*Kd))/(2*AT)-M;
    dC = M-C;
    dR = C-R;
    dy=[dM dC dR]';
end

%Use KF4step for the simulations of the four-step
model.
function dy = KF4step(t,y,AT,Kd)
    M = y(1);    C = y(2);       I = y(3);      R = y(4);
    dM = (AT-R-Kd+sqrt((AT-R-Kd)^2+4*AT*Kd))/(2*AT)-M;
    dC = M-C;
    dI = C-I;
    dR = I-R;
    dy=[dM dC dI dR]';
end
```

Exercise 2

```
AT=0.067;Kd=10^-3; KR=0.067;

function dy = KFnonlinear(t,y,AT,Kd,KR)
    M = y(1); C = y(2); R = y(3);
    dM = (AT-R-Kd+sqrt((AT-R-Kd)^2+4*AT*Kd))/(2*AT)-M;
    dC = M-C;
    dR = C-R/(R/KR+1);
    dy=[dM dC dR]';
end
```

Exercise 3

```
tspan=0:0.001:200;
IC=[0 0 0];
```

```
AT=0.067;Kd=10^-5;
[time1,sol1]= ode45(@(t,y)
KFnolight(t,y,AT,Kd),tspan,IC);
Tint=calculateP(sol1(:,3));
fprintf('Period with no light is %d\n',Tint)
L=0.05;Text=1.1*Tint;
[time2,sol2]= ode45(@(t,y)
KFlight(t,y,AT,Kd,L,Text),tspan,IC);
fprintf('Period with light is %d\n',calcu-
lateP(sol2(:,3)))

plot(time1(end-20000:end),sol1(end-
20000:end,3),time2(end-20000:end),sol2(end-
20000:end,3),'linewidth',2)
title('Circadian rhythms under day-night');xla-
bel('Time');ylabel('R')
legend({'Without light','With light'},'Loca-
tion','southwest');axis([180 200 0 0.1]);set(gca,'font-
size',14,'xtick',180:10:200,'ytick',0:0.05:0.1)

function dy = KFnolight(t,y,AT,Kd) %Model without light
input
    M = y(1); C = y(2); R = y(3);
    dM = (AT-R-Kd+sqrt((AT-R-Kd)^2+4*AT*Kd))/(2*AT)-M;
    dC = M-C;
    dR = C-R;
    dy=[dM dC dR]';
end

function dy = KFlight(t,y,AT,Kd,L,Text) %Model with
light input
    M = y(1); C = y(2);  R = y(3);
    dM = (AT-R-Kd+sqrt((AT-R-Kd)^2+4*AT*Kd))/(2*AT)-
M+L*(mod(t,Text)<Text/2);
    dC = M-C;
    dR = C-R;
    dy=[dM dC dR]';
end

function period = calculateP(profile)
    [~, peakT] = findpeaks(profile);
    period=0.001*(peakT(end)-peakT(end-1));
end
```

References

1. Dibner C, Schibler U, Albrecht U (2010) The mammalian circadian timing system: organization and coordination of central and peripheral clocks. Annu Rev Physiol 72:517–549
2. Sulli G, Manoogian EN, Taub PR, Panda S (2018) Training the circadian clock, clocking the drugs, and drugging the clock to prevent, manage, and treat chronic diseases. Trends Pharmacol Sci 39:812–827
3. Lee C, Etchegaray J-P, Cagampang FR, Loudon AS, Reppert SM (2001) Posttranslational mechanisms regulate the mammalian circadian clock. Cell 107:855–867
4. Gallego M, Virshup DM (2007) Post-translational modifications regulate the ticking of the circadian clock. Nat Rev Mol Cell Biol 8:139–148
5. Novák B, Tyson JJ (2008) Design principles of biochemical oscillators. Nat Rev Mol Cell Biol 9:981
6. Forger DB (2017) Biological clocks, rhythms, and oscillations: the theory of biological timekeeping. MIT Press
7. Goldbeter A, Gérard C, Gonze D, Leloup J-C, Dupont G (2012) Systems biology of cellular rhythms. FEBS Lett 586:2955–2965
8. Tibbitt MW, Anseth KS (2012) Dynamic microenvironments: the fourth dimension. Sci Transl Med 4:160ps124–160ps124
9. Zhang R, Lahens NF, Ballance HI, Hughes ME, Hogenesch JB (2014) A circadian gene expression atlas in mammals: implications for biology and medicine. Proc Natl Acad Sci 111:16219–16224
10. Ruben MD, Smith DF, FitzGerald GA, Hogenesch JB (2019) Dosing time matters. Science 365:547–549
11. Panda S (2019) The arrival of circadian medicine. Nat Rev Endocrinol 15:67–69
12. Ballesta A, Innominato PF, Dallmann R, Rand DA, Lévi FA (2017) Systems chronotherapeutics. Pharmacol Rev 69:161–199
13. Montaigne D et al (2018) Daytime variation of perioperative myocardial injury in cardiac surgery and its prevention by rev-Erbα antagonism: a single-Centre propensity-matched cohort study and a randomised study. Lancet 391:59–69
14. Lee Y, Chen R, Lee, H.-m. & Lee, C. (2011) Stoichiometric relationship among clock proteins determines robustness of circadian rhythms. J Biol Chem 286:7033–7042
15. Ye R, Selby CP, Ozturk N, Annayev Y, Sancar A (2011) Biochemical analysis of the canonical model for the mammalian circadian clock. J Biol Chem 286:25891–25902
16. Aryal RP et al (2017) Macromolecular assemblies of the mammalian circadian clock. Mol Cell 67:770–782.e776
17. Kim JK (2016) Protein sequestration versus Hill-type repression in circadian clock models. IET Syst Biol 10:125–135
18. Podkolodnaya OA, Tverdokhleb NN, Podkolodnyy NL (2017) Computational modeling of the cell-autonomous mammalian circadian oscillator. BMC Syst Biol 11:27–42
19. Millius A, Ueda HR (2017) Systems biology derived discoveries of intrinsic clocks. Front Neurol 8:25
20. Asgari-Targhi A, Klerman EB (2019) Mathematical modeling of circadian rhythms. Wiley Interdiscip Rev Syst Biol Med 11:e1439
21. Kim JK, Forger DB (2012) A mechanism for robust circadian timekeeping via stoichiometric balance. Mol Syst Biol 8
22. Goriki A et al (2014) A novel protein, CHRONO, functions as a core component of the mammalian circadian clock. PLoS Biol 12
23. Liberman AR et al (2017) Circadian clock model supports molecular link between PER3 and human anxiety. Sci Rep 7:1–10

24. Zhou M, Kim JK, Eng GWL, Forger DB, Virshup DM (2015) A Period2 phosphoswitch regulates and temperature compensates circadian period. Mol Cell 60:77–88
25. Kurosawa, G., Fujioka, A., Koinuma, S., Mochizuki, A. & Shigeyoshi, Y. (2017) Temperature–amplitude coupling for stable biological rhythms at different temperatures. PLoS Comput Biol 13
26. Kim JK et al (2013) Modeling and validating chronic pharmacological manipulation of circadian rhythms. CPT Pharmacometrics Syst Pharmacol 2:1–11
27. D'Alessandro M et al (2015) A tunable artificial circadian clock in clock-defective mice. Nat Commun 6:8587
28. Kim DW et al (2019) Systems approach reveals photosensitivity and PER2 level as determinants of clock-modulator efficacy. Mol Syst Biol 15
29. Ju D et al (2020) Chemical perturbations reveal that RUVBL2 regulates the circadian phase in mammals. Sci Transl Med 12
30. Narasimamurthy R et al (2018) CK1δ/ε protein kinase primes the PER2 circadian phosphoswitch. Proc Natl Acad Sci 115:5986–5991
31. D'Alessandro M et al (2017) Stability of wake-sleep cycles requires robust degradation of the PERIOD protein. Curr Biol 27:3454–3467.e3458
32. Kim JK, Kilpatrick ZP, Bennett MR, Josić K (2014) Molecular mechanisms that regulate the coupled period of the mammalian circadian clock. Biophys J 106:2071–2081
33. Gotoh T et al (2016) Model-driven experimental approach reveals the complex regulatory distribution of p53 by the circadian factor period 2. Proc Natl Acad Sci 113:13516–13521
34. Zou X et al (2020) A systems biology approach identifies hidden regulatory connections between the circadian and cell-cycle checkpoints. Front Physiol 11
35. Kim JK, Josić K, Bennett MR (2014) The validity of quasi-steady-state approximations in discrete stochastic simulations. Biophys J 107:783–793
36. Fall C, Marland E, Wagner J, Tyson J (2002) Computational cell biology. Springer, New York, NY
37. Forger DB (2011) Signal processing in cellular clocks. Proc Natl Acad Sci 108:4281–4285
38. Griffith J (1968) Mathematics of cellular control processes I. Negative feedback to one gene. J Theor Biol 20:202–208
39. Tyson JJ, Othmer HG (1978) The dynamics of feedback control circuits in biochemical pathways. Prog Theor Biol 5:62
40. Thron C (1991) The secant condition for instability in biochemical feedback control—I. The role of cooperativity and saturability. Bull Math Biol 53:383–401
41. Tyson JJ, Novak B (2020) A dynamical paradigm for molecular cell biology. Trends Cell Biol
42. Fribourgh JL et al (2020) Dynamics at the serine loop underlie differential affinity of cryptochromes for CLOCK: BMAL1 to control circadian timing. Elife 9:e55275
43. Lee E et al (2016) Pacemaker-neuron–dependent disturbance of the molecular clockwork by a Drosophila CLOCK mutant homologous to the mouse Clock mutation. Proc Natl Acad Sci 113: E4904–E4913
44. Hastings JW, Sweeney BM (1957) On the mechanism of temperature independence in a biological clock. Proc Natl Acad Sci U S A 43:804
45. Shinohara Y et al (2017) Temperature-sensitive substrate and product binding underlie temperature-compensated phosphorylation in the clock. Mol Cell 67:783–798.c720
46. Winfree AT (2001) The geometry of biological time, Vol. 12. Springer Science & Business Media, Cham
47. Golombek DA, Rosenstein RE (2010) Physiology of circadian entrainment. Physiol Rev 90:1063–1102
48. Abraham U et al (2010) Coupling governs entrainment range of circadian clocks. Mol Syst Biol 6
49. Serkh K, Forger DB (2014) Optimal schedules of light exposure for rapidly correcting circadian misalignment. PLoS Comput Biol 10

50. Diekman CO, Bose A (2018) Reentrainment of the circadian pacemaker during jet lag: East-west asymmetry and the effects of north-south travel. J Theor Biol 437:261–285
51. Bellman J, Kim JK, Lim S, Hong CI (2018) Modeling reveals a key mechanism for light-dependent phase shifts of Neurospora circadian rhythms. Biophys J 115:1093–1102
52. Gonze D, Ruoff P (2020) The Goodwin oscillator and its legacy. Acta Biotheor 1–18
53. Tokuda IT, Schmal C, Ananthasubramaniam B, Herzel H (2020) Conceptual models of entrainment, jet lag, and seasonality. Front Physiol 11:334

Spruce Budworm and the Forest

Lauren M. Childs

Introduction

Choristoneura funiferana, commonly known as the spruce budworm, is a defoliator of boreal forests in North America [1]. These pests cause major impacts to spruce (*Picea* spp.) and balsam fir (*Abies balsamea*), which are important tree species in the logging industry [2]. While most years the spruce budworm density is low and little harm comes to these trees, periodically throughout history large outbreaks of the spruce budworms occurred [2]. These outbreaks caused major losses to the trees and to economic industries tied to the trees. As a result, the spruce budworm is one of the most well-studied—empirically and theoretically—ecological systems.

In our exploration of the spruce budworm, we will begin with a simple model describing the budworm dynamics alone. Our analysis and simulation of even the simple model will demonstrate the complicated dynamics it can produce, as a function of a parameter that depends on the health and size of the forest. We will then expand the system to include the forest as a variable itself and include feedbacks with the spruce budworm. We will extend our analyses and simulation for this more complete system. Many more additions are possible, but we leave more sophisticated exploration of the system, of which there are many references, to the interested reader.

L. M. Childs (✉)
Department of Mathematics, Virginia Polytechnic Institute and State University, Blacksburg, VA, USA
e-mail: lchilds@vt.edu

© Springer Nature Switzerland AG 2021
P. Kraikivski (ed.), *Case Studies in Systems Biology*,
https://doi.org/10.1007/978-3-030-67742-8_7

Ecological Mechanisms

Observed budworm populations show extreme fluctuations with similar external conditions leading to very different levels of budworms [1]. High densities can see a single branch covered with hundreds of larvae [2]. Thus, the spruce budworm, *Choristoneura funiferana,* can cause significant ecological and economic damage during outbreak seasons through defoliation of spruce and balsam fir, which are its main source of food.

As is evident from empirical data, the budworm has a complicated life cycle, which we review briefly here. Female moths lay egg masses, typically in batches of about 20, on the needles of spruce and fir. After about 10 days, the eggs hatch, and larvae emerge. These larva progress through six stages over the course of six to nine months. It is during these extended stages where they can experience significant predation by parasitic wasps and birds. Following the larval stages, the budworms become pupa and eventually enclose as moths, and the cycle begins anew. Predation in the larval and, to a lesser extent, the pupal stages, is the primary determining factor of changes to local budworm populations. The larvae use the needles of the trees, which they occupy, as food, leading to substantial loss of leaf surface area. For a more detailed description of the life history of the spruce budworm, we refer the reader to Morris [1].

We do not focus on the details of the budworm life cycle as our models will take a simplistic view: when food, i.e. balsam fir and spruce, are available, the budworm population will grow. This is balanced by loss of the budworm due to predation. This simple version does not include the intricacies, of which there are many, in the life cycle and predation. Many details have been explored in other works (See *Selected additional reading resources*).

While the budworm population can change significantly in only a few years, the forest itself changes on a much longer time scale: on the order of decades to centuries in the absence of the budworm. Replacement of foliage takes about a decade [3]. The predators of budworms, particularly birds, also exhibit a much slower time scale for changes in their population. Thus, when we consider inclusion of the forest in our system of equations, we scale such that the changes to the forest occur much slower than changes to the budworm population.

Extensive empirical studies have tracked local budworm populations across years and decades [1]. The Green River Project, spearheaded by Dr. Morris, led to an extensive data set on the spruce budworm including life history data at various locations in New Brunswick, Canada across several decades, beginning in 1945. A tome of data related to the budworms was published from this study in [1]. These data include information on the known cyclic behavior of the population. Over long time periods large fluctuations in the budworm population were observed. See Fig. 1 for a compilation of several data sources over the course of more than two centuries. The peaks in budworm density, or outbreaks, cause significant damage and much time, money and effort has been expended to better predict the timing and location of outbreaks.

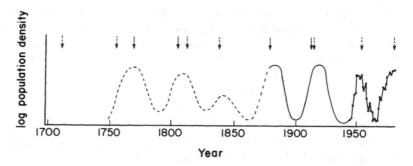

Fig. 1 Spruce budworm population cycles from circa 1750 to 1980 on a logarithmic scale. Data is from radial growth-ring analysis of surviving trees (dashed line), historical records since 1878 (solid line), and restored sampling data (solid line with data points). Solid and dotted arrows are budworm peaks for New Brunswick and Quebec provinces, respectively. Figure reproduced from Royama [4] with permission

Mathematical Model

Spruce Budworm Model

We begin by focusing on the budworm population alone, denoted by B, as in Ludwig et al. [3]. As described above, we assume that the change in the budworm density, our equation for $\frac{dB}{dt}$, is a balance between growth and predation. We represent growth as a logistic function. When the budworm density is small, growth is nearly linearly related to the density of budworms present, but as the population increases, the rate of growth diminishes. This decrease may be due to, for example, competition for resources. We denote the intrinsic growth rate, r_B, and the maximum population density, i.e. carrying capacity, by K_B. In the case of budworms, this would be related to leaves in the forest to feed on. As growth creates more budworms, this term positively contributes to our equation for the change in budworm density as

$$\frac{dB}{dt} = r_B B \left(1 - \frac{B}{K_B} \right).$$

Budworms, however, are also subject to predation by birds, among other species. Predation is particularly important when the budworm population is large, such that budworms are a common and easy food source for the birds. In order to avoid complicating the model, we do not model the predatory birds directly. Instead, we include predation through a loss term in our equation for the change in the budworm density. We choose the form of this loss term such that at high prey density the predator effect saturates, as there is an upper limit to the rate that birds can consume budworms even when a large number are present. This upper limit is represented by β. At very low budworm density, there is also a decrease in the effectiveness of predation. This is because budworms are not the only food source available to birds.

When budworms are scarce, birds focus on other more readily available prey. We build this into the model through an S-shaped response curve, i.e. Type-III Holling curve [5]. Here, the parameter α determines the budworm density when saturation of predators begins to take place. This contributes to our change in budworm density by

$$\frac{dB}{dt} = -\beta \frac{B^2}{\alpha^2 + B^2} .$$

We combine these two terms to form the following ordinary differential equation (ODE) to represent the change in budworm density by

$$\frac{dB}{dt} = r_B B\left(1 - \frac{B}{K_B}\right) - \beta \frac{B^2}{\alpha^2 + B^2} . \tag{1}$$

When focusing solely on the budworm, it is reasonable to consider each of the four parameters as constant. In reality, however, some of these parameters are truly functions of slowly varying quantities such as the size of the forest.

Spruce Budworm and Forest Model

We extend the budworm model to include a representation of the size of the forest via the average surface area of leaves per tree, which we denote by S. We assume that both the carrying capacity of the budworm population, K_B, and the density of budworms when saturation of predators begins, α, are functions of the forest size, S. Thus, we let $K_B = f(S)$ and $\alpha = g(S)$ and reformulate the budworm equation as

$$\frac{dB}{dt} = r_B B\left(1 - \frac{B}{f(S)}\right) - \beta \frac{B^2}{g(S)^2 + B^2} .$$

We also include an equation for how the average leaf surface area is changing in size. We assume, similar to the budworms, the forest grows via a logistic function. The growth rate, r_S, for the forest, however, is much slower than the growth of the budworm as it takes years to decades for trees to grow. For the logistic function, we include a carrying capacity for the forest, K_S, to form the positive term in the change in leaf surface area as

$$\frac{dS}{dt} = r_S S\left(1 - \frac{S}{K_S}\right).$$

Similar to the budworm equation, we include a loss term. We assume that the most important loss to consider here is due to the budworms themselves. Thus, our term involves destruction of the leaves in the forest by budworms occurring at rate e_s by

$$\frac{dS}{dt} = -e_S B.$$

Thus, our full two ODE system for the budworm density, B, and average leaf surface area per tree, S, is written as

$$\frac{dB}{dt} = r_B B \left(1 - \frac{B}{f(S)} \right) - \beta \frac{B^2}{g(S)^2 + B^2} \quad,$$

$$\frac{dS}{dt} = r_S S (1 - \frac{S}{K_S}) - e_s B. \tag{2}$$

This system is similar to the structure presented in May [6], but the form in that paper focuses on a scaled version of the forest equation.

Results

Spruce Budworm Model

From System (1), the budworm population will reach an equilibrium density, a state where the budworm density is no longer changing. This occurs when the two terms on the right hand side exactly balance. Mathematically, we can find this equilibrium by setting $\frac{dB}{dt} = 0$ which yields

$$r_B B \left(1 - \frac{B}{K_B} \right) - \beta \frac{B^2}{\alpha^2 + B^2} = 0.$$

To find the equilibrium density, we must solve for B. From observation, we see that if $B = 0$, then this equation is satisfied. We refer to this equilibrium biologically as the extinction equilibrium and mathematically as the trivial equilibrium. Beyond this, however, solving for B is rather tricky. Factoring out the B (to account for the $B = 0$ solution) and trying to solve, yields a nasty cubic equation. Although computer algebra programs can provide solutions analytically, we will consider them graphically.

System (1) was formulated from biological principles and includes a single ODE equation with four parameters: r_B, K_B, β and α. Before we proceed further, let us perform the mathematical process of non-dimensionalization, where we scale the equation, to reduce the parameter space. First, we scale the budworm density, such that $x = \frac{B}{\alpha}$. Thus, System (1) at equilibrium in terms of x becomes

$$r_B\left(1 - \frac{\alpha x}{K_B}\right) - \beta\frac{\alpha x}{\alpha^2(1+x^2)} = 0.$$

We also scale time by $\frac{\alpha}{\beta}$ which is equivalent to multiplying each term in the above equation by this quantity, resulting in

$$\frac{\alpha r_B}{\beta}\left(1 - \frac{\alpha}{K_B}x\right) - \frac{x}{1+x^2} = 0.$$

While all four parameters remain, they now only appear in two groups. Thus, we rename our parameter groups, such that $R = \frac{\alpha\, r_B}{\beta}$ and $Q = \frac{K_B}{\alpha}$, which results in, following minor rearrangement, the two-parameter equation

$$R\left(1 - \frac{x}{Q}\right) = \frac{x}{1+x^2}. \tag{3}$$

The left hand side is a line which intersects the x-axis at Q and the y-axis at R. The right hand side is a humped shaped curve that rises linearly near zero, peaks at a value of 0.5 at $x = 1$, and approaches 0 as x gets large (see Fig. 2). The intersection of these two curves are exactly the solution to Eq. (3), which is equivalent to equilibrium solutions to System (1). As the right hand side contains no parameters, the number and value of the solutions entirely depends on the position of R and Q. We graphically depict this in Fig. 2. The solid lines show three intersections. Changes to R (and Q), as shown by the dotted (and dashed) lines, depict only a single intersection. An important question is: what are R and Q in a realistic biological setting? For this, we must return to our original parameterization to find appropriate values.

Some of our parameters may be directly measurable, such as the growth rate, r_B, and carrying capacity, K_B. Others, however, given our simplifications are more

Fig. 2 Graphical solutions to non-dimensionalized System (1). Solid black line depicts the left hand side of Eq. (3) with $S = 1100$. Dashed lines vary Q through variation of $K_B = f(S)$, with $K_B = 2S$, $K_B = 8S$. Dotted lines vary R by changing S with $S = 550$ and $S = 1800$. The solid curve depicts the right hand side of Eq. (3). Parameters: $r_B = 2$, $\beta = 2000$, $K_B = 4S$, and $\alpha = S/2$

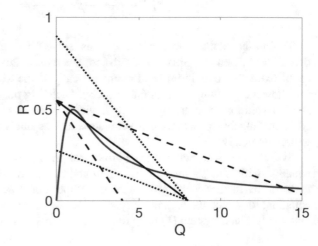

phenomenological and, thus, harder to determine empirically, such as the measure of predation, β. We choose fixed values for some parameters such that $r_B = 2$ and $\beta = 2000$. Upon further reflection, however, the carrying capacity, K_B, and the beginning of saturation for predation, α, depend on the surface area of the leaves of the forest, S. Thus, we choose $K_B = f(S) = 4S$ and $\alpha = g(S) = 0.5S$. As a result, our values of R and Q are actually functions of the underlying trees and are described by

$$R = \frac{\alpha r_B}{\beta} = \frac{(0.5\,S)\,2}{2000} = \frac{S}{2000},$$

$$Q = \frac{K_B}{\alpha} = \frac{4\,S}{0.5\,S} = 8.$$

Notice that for Q, we return to a fixed value, despite the dependence on the leaf surface area, because S appears in both the numerator and denominator and, thus, cancels out. Figure 2 utilizes the parameters given here, except as noted in the caption.

Observing Fig. 2, it is readily apparent that as we vary S, the number of solutions can also vary. For low values of S, the only possible solution is a low value of x. Recall that x is simply a scaled version of the budworm density, B. Biologically, when the forest leaf surface area is small, only a small population of budworms are sustainable. As S is increased, there is a region where three solutions occur; one for each of low, intermediate and high budworm density. In this case, the system exhibits bistability, with two possible stable states. In other words, the final budworm density depends on the initial value, and similar environments can lead to drastically different budworm populations. Although an intermediate budworm density solution exists, it is unstable and, practically, never reached. As the surface area per tree S grows even larger, we return to a single solution, but now with a large budworm density. Biologically, a forest with a large leaf surface area can support a lot of budworms.

Spruce Budworm and Forest Model

Up until now, we have considered the leaf surface area in the forest as a fixed quantity and shown that an equilibrium budworm population depends on that value. Recall, that we also want to consider when the forest is able to grow over time, increasing the average surface area of leaves per tree. Simultaneously, these leaves can be eaten by the budworms, as in System (2). With the expansion of the System (1), we will continue with the same parameters ($r_B = 2$, $\beta = 2000$, $K_B = f(S) = 4S$, $\alpha = g(S) = S/2$) used for the budworm population alone. However, with the addition of the equation for the average leaf surface area to form System (2), three other parameters – the growth of the leaves, r_S, the carrying capacity of the forest, K_S, and the loss of leaf surface area due to budworms, e_S – must be considered.

Importantly, in System (2), the equations are coupled. As the surface area grows, the carrying capacity of the budworm population also grows via $f(S)$. Furthermore, the beginning of when saturation of predation occurs, $g(S)$, grows with S. The leaf surface area, however, decreases more rapidly as the budworm population grows. Thus, the surface area of leaves, S, naturally changes, altering the budworm population.

We expect the forest and the budworm to grow at very different rates, and set $r_S = r_B/100$. Given appropriate choices of other parameters, this creates cyclic dynamics over the course of decades. When the average leaf surface area per tree is small, regardless of the initial budworm population density, the budworm population will crash towards the low equilibrium value. With this small budworm density, the budworms are unable to significantly harm the leaves and the forest is left to grow slowly over time. This pushes the budworm population equilibrium higher, akin, in Fig. 2, to increasing R but remaining with the same fixed Q. Although two additional equilibria (one unstable and one stable) appear, the budworm population stays at the low equilibrium until that equilibrium disappears. Then, the population rapidly grows towards the larger equilibrium. At this point, the budworm population is so large, that now it begins to negatively impact the forest, and the leaf surface area declines. In turn, this slowly reduces the carrying capacity of the budworm. As the leaf surface area decreases, the budworm population remains at the high equilibrium value even as two additional equilibria (one unstable and one stable) reappear. It is not until the larger budworm equilibrium disappears, because the leaf surface area shrank so considerably, that the budworm population crashes down to the low equilibrium. Now, the forest can recover, and the cycle can repeat.

Figure 3 shows an example of such population cycles. Here, the destruction of the leaf surface area, e_S, is set equal to the growth rate of the forest, r_S. The carrying capacity of the forest, K_S, is set at 40,000 leaves per tree. In this case, we observe population cycles roughly every 60 years with peaks in the budworm density of about 3000 occurring just after peaks in the average leaf surface area. As the budworm population grows, it rapidly destroys the leaves. The loss in forest is quickly followed by a crash in the budworm population. As the forest grows much slower, it takes many years of steady forest growth before the budworm population can begin to increase. Figure 3 (right panel) depicts the average leaf surface area versus the budworm population. Notice that once the budworm density collapses onto the cycle (Fig. 3, left panel) for any given average leaf surface area, two budworm densities are possible.

The size and frequency of the cycles depends on the choice of parameters. Given data for a specific region, it is possible to find parameters that mimic the timing of these cycles [3]. Furthermore, it is important to note that such cyclic behavior does not occur for all parameter choices. A single stable equilibrium with a large budworm population and a large leaf surface area is possible. This occurs, for example, when predation is very strong and controls the budworm population. See Hollis [7] for an excellent computational example with a tunable parameter for predation. While the equations in Hollis [7] are identical in structure to the ones found here, they are scaled differently and, thus, none of the parameter values are comparable.

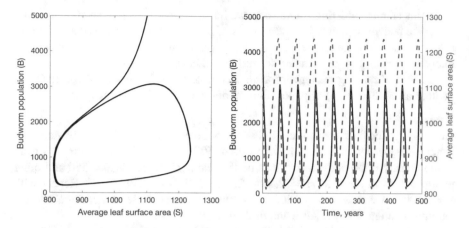

Fig. 3 Dynamics of System (2) with the budworm and forest. Left: Population cycling of the forest, represented by the average leaf surface area per tree (S), and the budworm population (B). Right: Population dynamics over 500 years of the budworm population (solid black lines) and the average leaf surface area per tree (red dashed lines). Parameters: $r_B = 2$, $\beta = 2000$, $K_B = 4S$, $\alpha = S/2$, $r_S = e_S = 2/100$, and $K_S = 40{,}000$. Initial conditions: $B(0) = 5000$, $S(0) = 1100$

Discussion

The spruce budworm system described here provides an ecological example with rich dynamics. The system is scalable to many levels of complexity, of which, we have only explored two: the budworm population alone and the budworm population coupled to the average leaf surface area per tree in the forest. Considering the budworm population alone, we can find solutions that exhibit either of the two disparate population densities exhibited in a given year: very small or very large. In order to observe both densities, we need to alter parameters implicitly dependent on the size of the forest. With the leaf surface area as the tunable parameter, we do not observe cycles, but we do observe bistability. For a fuller picture of the budworm system, we added the leaf surface area as a dynamic variable as well. In this case, the feedbacks between the growth of the leaf surface area in the forest, which provides food for the budworm, and the loss of leaf surface area due to the budworm, led to cycling of population sizes on the order of decades.

Empirically, the budworm population exhibited large outbreaks in regular intervals throughout history (Fig. 1). These extreme outbreaks are followed by many years with very low population density. The simple models presented here provide qualitative dynamics replicating what is seen in the data. With finer tuning of parameters, better quantitative agreement is possible [3].

One of the most commonly cited early models describing the spruce budworm population, published by [3], considered three species: the spruce budworm, the leaf surface area and the energy reserve of trees. With this three dimensional ODE system, parameterized by empirical observations of the New Brunswick budworm, they were able to replicate the forty-year cycle, as observed in nature.

In this simple modeling framework, we have ignored multiple factors that could affect the budworm population or the leaf surface area or both, such as the variability in the predator population or the energy reserve of trees. Both of these, as well as many other additions, or different formulations, have been explored elsewhere. See the additional reading resources for an incomplete list (there are far too many to include all of them here) of papers that model the spruce budworm system.

Exercises

Exercise 1. Consider the spruce budworm population by itself, as in System (1). Determine the equilibrium population after 500 years of the budworm using an initial density of 50 budworms. Assume the leaf surface area is $S = 550$ and other parameters as in the text: $r_B = 2$, $\beta = 2000$, $K_B = 4S$, and $\alpha = S/2$. Repeat with $S = 1100$ and then with $S = 1800$. You may notice that these choices of S and the parameters align with the solid and dashed lines in Fig. 2.

Exercise 2. Repeat Exercise 1, except consider an initial budworm population density of 10,000 budworms. How do the results differ from those in Exercise 1 for each S value?

Exercise 3. Consider the combined spruce budworm and leaf surface area model, as in System (2). Assume the following parameters: $r_B = 2$, $\beta = 2000$, $K_B = 4S$, $\alpha = S/2$, $K_S = 40,000$, and $r_S = e_S = 2/100$. Describe the dynamics over 500 years when using a density of 50 budworms and 1100 as the leaf surface area for the initial condition.

Exercise 4. Repeat Exercise 3 with 10,000 budworms and surface area 1100 as an initial condition. How do the results differ from Exercise 3? Contrast these changes to what was observed in Exercise 2.

Exercise 5. Consider the combined spruce budworm and leaf surface area model, as in System (2). Assume the following parameters: $r_B = 2$, $\beta = 2000$, $K_B = 4S$, $\alpha = S/2$, $K_S = 40,000$, and $r_S = e_S = 2/100$. How do variations in the growth rates of the budworm population, r_B, and the surface area of leaves, r_S, alter the dynamics? For example, double and halve r_B, and observe how the dynamics change. Set $r_B = 2$ again and repeat with r_S, i.e. double and halve the value. Which variations have a large impact on the dynamics? Why?

Method

We provide four Matlab functions (Matlab R2019b). To use these functions, you will need to save the four files with their function titles as the file names with a ".m" in the same directory. The four included functions are:

```
SimulateBudworm
Budworm
SimulateBudwormForest
BudwormForest
```

To simulate System (1), use the function SimulateBudworm which calls the function Budworm. To run SimulateBudworm, you will need to include the first value for the initial budworm density and the second value as the leaf surface area. For example, for an initial budworm density of 30 and a fixed leaf surface area of 1000, run:

```
SimulateBudworm(30,1000)
```

If no values are included, the function will still run and assume an initial budworm density of 50 and a fixed leaf surface area of 1100.

To simulate System (2), use the function SimulateBudwormForest which calls the function BudwormForest. To run SimulateBudwormForest, you will need to include the first value for the initial budworm density and the second value as the initial leaf surface area. For example, for an initial budworm density of 30 and an initial leaf surface area of 1000, run:

```
SimulateBudwormForest(30,1000)
```

If no values are included, the file will still run and assume an initial budworm density of 50 and a fixed leaf surface area of 1100.

```
function SimulateBudworm(bud,leaf)

if nargin==0
   bud = 50;
   leaf = 1100;
elseif nargin==1
   leaf = 1100;
end

tmax = 500;
tspan = [0 tmax];
IC = bud; % Initial conditions

[t,B] = ode45(@Budworm,tspan,IC,[],leaf);

figure(1)
clf
set(gcf, 'Position', [1, 1, 500, 400])
plot(t,B,'linewidth',2)
ylabel('Budworm population')
xlabel('Time')
set(gca,'fontsize',14)
end
```

```
function dB = Budworm(t,B,S)

rb = 2; % growth rate of budworm population
beta = 2000; % predation rate
alpha = 1/2*S; % beginning of predation saturation
Kb = 4*S; % carrying capacity of budworm population

dB = rb*B*(1-B/Kb)-beta*B^2/(alpha^2+B^2);

end

function SimulateBudwormForest(bud,leaf)

if nargin==0
  bud = 50;
  leaf = 1100;
elseif nargin==1
  leaf = 1100;
end

tmax = 500;
tspan = [0 tmax];
IC = [bud leaf]; % Initial conditions

[t,y] = ode45(@BudwormForest,tspan,IC);

B = y(:,1);
S = y(:,2);

figure(1)
clf
set(gcf, 'Position', [1, 1, 900, 400])
subplot(1,2,1)
plot(S,B,'k','linewidth',2)
xlabel('Average leaf surface area')
ylabel('Budworm population')
set(gca,'fontsize',14,'ytick',0:1000:5000)
subplot(1,2,2)
[ax, ay1, ay2] = plotyy(t,B,t,S);
ylabel(ax(1),'Budworm population','color','k')
ylabel(ax(2),'Average leaf surface area')
set(ay1(1),'linewidth',2,'color','k')
set(ay2(1),'linewidth',2,'linestyle','--')
set(ax(1),'ycolor','k')
set(ax(2),'fontsize',14)
xlabel('Time')
set(gca,'fontsize',14)

end

function dy = BudwormForest(t,y)
```

```
B = y(1);
S = y(2);


rb = 2; % growth rate of budworm population
rs = 1/100*rb; % growth rate of forest (leaf surface area)
es = rs; % loss of leaf surface area due to budworm population
beta = 2000; % predation rate
Ks = 40000; % carrying capacity of forest (leaf surface area)


alpha = 1/2*S; % beginning of predation saturation
Kb = 4*S; % carrying capacity of budworm population


dB = rb*B*(1-B/Kb)-beta*B^2/(alpha^2+B^2);
dS = rs*S*(1-S/Ks)-es*B;


dy = [dB; dS];
end
```

References

1. Morris RF (1963) The dynamics of epidemic spruce budworm populations. Memoirs Entomol Soc Canada 95(S31):1–12
2. Royama T (2012) Analytical population dynamics. Vol. 10. Springer Science & Business Media, Chapman and Hall, London.
3. Ludwig D, Jones DD, Holling CS (1978) Qualitative analysis of insect outbreak systems: the spruce budworm and forest. J Anim Ecol 47:315–332
4. Royama T (1984) Population dynamics of the spruce budworm *Choristoneura fumiferana*. Ecol Monogr 54(4):429–462
5. Holling CS (1973) Resilience and stability of ecological systems. Annu Rev Ecol Syst 4(1):1–23
6. May RM (1977) Thresholds and breakpoints in ecosystems with a multiplicity of stable states. Nature 269(5628):471–477
7. Hollis S (2011) Bifurcation in a Model of Spruce budworm populations. Wolfram Demonstrations Project. http://demonstrations.wolfram.com/ BifurcationInAModelOfSpruceBudwormPopulations/ Accessed 3 May 2020

Selected Additional Reading Resources

Bergeron Y, Leduc A (1998) Relationships between change in fire frequency and mortality due to spruce budworm outbreak in the southeastern Canadian boreal forest. J Veg Sci 9(4):492–500
Candau J-N, Fleming RA (2005) Landscape-scale spatial distribution of spruce budworm defoliation in relation to bioclimatic conditions. Can J For Res 35(9):2218–2232
Fleming RA, Shoemaker CA (1992) Evaluating models for spruce budworm-Forest management: comparing output with regional field data. Ecol Appl 2(4):460–477
Ludwig D, Aronson DG, Weinberger HF (1979) Spatial patterning of the spruce budworm. J Math Biol 8(3):217–258
MacLean DA (1996) The role of a stand dynamics model in the spruce budworm decision support system. Can J For Res 26(10):1731–1741

Rasmussen A, Wyller J, Vik JO (2011) Relaxation oscillations in spruce–budworm interactions. Nonlinear Anal Real World Appl 12(1):304–319

Régnière J, You M (1991) A simulation model of spruce budworm (Lepidoptera: Tortricidae) feeding on balsam fir and white spruce. Ecol Model 54(3-4):277–297

Régnière J, Lysyk TJ (1995) Population dynamics of the spruce budworm, *Choristoneura fumiferana*. In: Armstrong JA, Ives WGH (eds) Forest insect pests in Canada. Canadian Forest Service, Ottawa, pp 95–105

Royama T et al (2005) Analysis of spruce budworm outbreak cycles in New Brunswick, Canada, since 1952. Ecology 86(5):1212–1224

Stedinger JR (1984) A spruce budworm-forest model and its implications for suppression programs. For Sci 30(3):597–615

Swetnam TW, Lynch AM (1993) Multicentury, regional-scale patterns of western spruce budworm outbreaks. Ecol Monogr 63(4):399–424

Williams DW, Liebhold AM (2000) Spatial synchrony of spruce budworm outbreaks in eastern North America. Ecology 81(10):2753–2766

Modeling cAMP Oscillations in Budding Yeast

Amogh Jalihal

Introduction

Glucose is the preferred source of energy for budding yeast, which is rapidly utilized by fermentation. It is vital for cells to sense glucose abundance in the environment in order to modulate their metabolism, which is in turn vital to maintain an optimal growth rate. The cAMP/PKA signal transduction pathway performs this task: the pathway 'senses' environmental glucose and carries out a variety of biological roles, including activation of growth responses, and inhibition of stress responses in response to glucose starvation. This particular pathway is interesting because it demonstrates a commonly observed feature of signaling pathways, that of negative feedback. In fact, as discussed in the next section, there are *two* negative feedback loops in the cAMP/PKA pathway. These characteristics make it difficult to reason about what the dynamics of this pathway will be in response to a particular environmental input. Gonzales et al. 2011 carried out an exploration of this pathway and its short-term responses to relief from glucose starvation (i.e. glucose addition to starved cells) by creating a mathematical model of the molecular interactions [1]. The motivation for the construction of this model were twofold:

1. explain a set of experimental observations in wildtype cells as well as strains containing mutations in regulators in the cAMP/PKA pathway, and
2. to further explore the space of possible dynamics exhibited by the pathway [2].

Here, we will briefly describe the structure of the cAMP/PKA pathway, present a mechanistic model describing its response to glucose readdition, and conclude with an analysis of cAMP oscillations.

A. Jalihal (✉)
Virginia Polytechnic Institute and State University, Blacksburg, VA, USA
e-mail: jamogh@vt.edu

© Springer Nature Switzerland AG 2021
P. Kraikivski (ed.), *Case Studies in Systems Biology*,
https://doi.org/10.1007/978-3-030-67742-8_8

Molecular Mechanisms and Physiology

The cAMP/PKA Pathway

The cAMP/PKA pathway is illustrated in Fig. 1. Extracellular glucose is sensed by the Gpr1/Gpa2 G-proteins. Upon binding to glucose, the Gpa2 subunit is guanidylated and activated. Independently, by a mechanism that is currently unknown, Ras2 is guanidylated by Cdc25. Both Gpa2-GTP and Ras2-GTP activate the enzyme Adenylate Cyclase Cyr1 which converts ATP to cyclic AMP (cAMP). cAMP serves an important role of activating Protein Kinase A (PKA), which is regulator of cellular growth. PKA in its inactive state is a heterotetramer bound to a regulatory unit Bcy1. cAMP binds to, and frees Bcy1 from the PKA catalytic subunit, thereby activating PKA. Finally, PKA implements negative feedback loops at two levels:

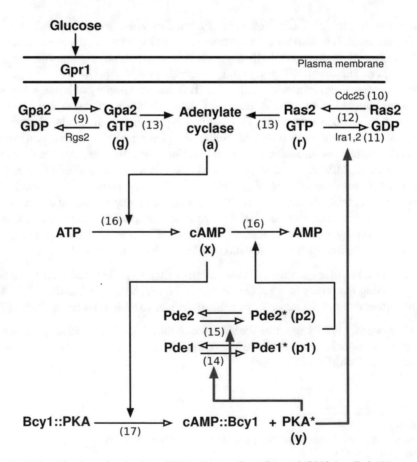

Fig. 1 Full mechanism of activation of PKA. Source: Gonzales et al. 2013 (see Ref. [2])

1. PKA activates phosphodiesterases Pde1/2 which degrade cAMP to AMP
2. PKA, via Ira1/2 inactivates Ras2

Experimental Observations of the cAMP/PKA Pathway Response

While there is strong biochemical evidence for the individual interactions composing the cAMP/PKA pathway, it is an important question as to whether this mechanism is sufficient in explaining the diverse experimental observations related to this pathway. Specifically

1. Ma et al. 1999 studied the activation of the PKA pathway in yeast cells starved for glucose. Date from this study shows a rapid spike in cAMP levels in the first 60–90 s after glucose induction [3].
2. Garmendia-Torres et al. 2007 were interested in the behavior of a general stress response transcription factor Msn2 [4]. This transcription factor, which is a target of PKA, demonstrates "oscillatory nucleocytoplasmic shuttling", or a rapid translocation between the nucleus and cytoplasm over intermediate levels of cellular stresses. Garmendia-Torres propose a model where Msn2 oscillations are potentially caused by long term sustained cAMP oscillations [4].

Fig. 2 depicts these two sets of observations.

In order to understand the role of the negative feedback loops on the behaviors of cAMP, Gonzales et al. created a mathematical model [1, 2]. A major objective of this model is that it should be able to explain both the observations depicted in Fig. 2. This model was investigated not only for the behavior of wildtype cells, but also for a

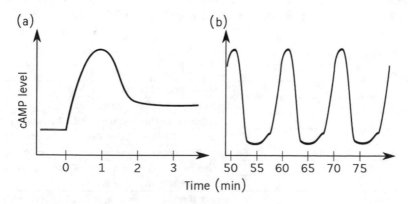

Fig. 2 Cartoons representing cAMP short term and long-term dynamics under different conditions. Note the times-scales on the x-axis. (**a**) depicts data from Ref. [3], when glucose is added at minute zero to glucose starved cells. (**b**) depicts long term oscillations cAMP for yeast cells experiencing "intermediate" stresses, predicted by Garmendia-Torres et al. 2007 in Ref. [4]

range of gene deletion mutants which displayed altered cAMP dynamics. This model is described in the next section.

Model of cAMP/PKA Pathway

Gonzales et al. start with a full description of the cAMP/PKA pathway, which contains nine ODEs [2]. Since this system is too large, the authors make a series of biochemically realistic assumptions to reduce the system to four ODEs described below.

$$[\text{Ras2} - \text{GTP}] \quad \frac{dr}{dt} = \frac{A(1-r)}{\Gamma_1 + 1 - r} - \frac{Bzr}{\Gamma_1 + r} \tag{1}$$

$$[\text{Ira1}] \quad \frac{dz}{dt} = N(x^2 - z) \tag{2}$$

$$[\text{Pde}] \quad \frac{dp}{dt} = M(x^2 - p) \tag{3}$$

$$[\text{cAMP}] \quad \frac{dx}{dt} = C + r - D_0 x - \frac{Dpx}{\Gamma + x} \tag{4}$$

where A, Γ_1, B, N, M, C, G, D_0, D, Γ are model parameters that are described in Table 1.

PKA: Two molecules of cAMP bind to one Bcy1 subunit, thus activating one catalytic subunit of PKA. Thus in the model, PKA activity is approximated by $[\text{cAMP}]^2$ instead of explicitly tracking the PKA activity.

Ras2: Both the activation and inactivation of Ras2 are denoted by Michealis-Menten terms. Ras2 activity r is denoted as a fraction between 0 and 1. Ras2 activation is dependent on the amount of inactive, or GDP-bound Ras2. This is

Table 1 Model parameters

Parameter	Interpretation
A	Stress-mediated signal that activates Ras2
	The larger the stress, the smaller the A
Γ_1	Michealis-Menten constant for Ras2 activation
B	Strength of Ira1 mediated Ras2 inactivation
N	Effective time scale parameter of Ira1 activation
M	Effective time scale parameter of Pde activation
C	Basal production rate of cAMP
G	Rate of cAMP production induced by Ras2 activity
D_0	Basal rate of cAMP degradation
D	Strength of Pde mediated cAMP degradation
Γ	Michealis-Menten constant for cAMP degradation

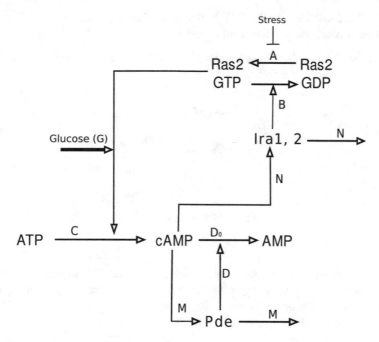

Fig. 3 Illustration of four-variable model. Source: Gonzales et al. 2013 (see Ref. [2])

denoted by the term $(1 - r)$. Moreover, Ras2 inactivation is depicted as a Michaelis-Menten law governed by the activity of Ira1 (variable z) (Eq. (1)).

Ira1 and Pde: Both Ira1 (variable z) and Pde (variable p) are assumed to be directly activated by PKA, at time scales N and M respectively (Eqs. (2) and (3)).

cAMP: Instead of modeling Adenylate Cyclase activity, cAMP (variable x) is assumed to be produced at a rate proportional to the activity of Ras2-GTP. Moreover, cAMP degradation is assumed to happen via a Pde dependent and Pde-independent modes. The Pde-independent basal degradation is represented by the parameter D_0. The Pde-dependent degradation is represented by a Michaelis-Menten expression shown in Eq. (4).

The interactions represented in the four-variable model are shown in Fig. 3

Results

Examining the Reduced Two-Variable System

We can further simplify the four-variable system by considering the extreme Ras2 states:

1. If Ras2 is completely in the GTP bound state, then $r_{ss} = 1$.
2. If Ras2 is completely GDP-bound, then $r_{ss} = 0$.

Since the Ira1 variable doesn't appear in the Pde and cAMP equations, the model can be represented in the following form:

$$\frac{dp}{dt} = M\left(x^2 - p\right) \tag{5}$$

$$\frac{dx}{dt} = C_0 - D_0 x - \frac{Dpx}{\Gamma + x} \tag{6}$$

where

$$C_0 = 1 + C \text{ if } r_{ss} = 1 \tag{7}$$

$$C_0 = C \text{ if } r_{ss} = 0 \tag{8}$$

We are interested in the parameter regimes where cAMP displays oscillations. We can investigate the steady state behavior of this system by examining the eigenvalues of the Jacobian matrix at steady state

The Jacobian of the system is given by

$$\mathcal{J} = \begin{bmatrix} \frac{\partial f_x}{\partial x} & \frac{\partial f_x}{\partial p} \\ \frac{\partial f_p}{\partial x} & \frac{\partial f_p}{\partial p} \end{bmatrix} = \begin{bmatrix} -D_0 - \left(\frac{Dp}{\Gamma + x} - \frac{Dpx}{(\Gamma + x)^2}\right) & -\frac{Dx}{\Gamma + x} \\ 2Mx & -M \end{bmatrix} \tag{9}$$

We evaluate the eigenvalues of $\mathcal{J}$ at the steady states of Eqs. (5) and (6). For Eq. (5)

$$M\left(x^2 - p\right) = 0 \tag{10}$$

$$\Rightarrow p_{ss} = x_{ss}^2 \tag{11}$$

Substituting in Eq. (5)

$$C_0 - D_0 x_{ss} - \frac{Dx_{ss}^3}{\Gamma + x_s s} = 0 \tag{12}$$

While it is not easy to get analytical solutions for x_{ss}, we analyze the general behavior exhibited by $\mathcal{J}$ by substituting in Eq. (11) in Eq. (13).

$$\mathcal{J} = \begin{bmatrix} -D_0 - \dfrac{\Gamma D x_{ss}^2}{(\Gamma + x_{ss})^2} & -\dfrac{D x_{ss}}{\Gamma + x_{ss}} \\ 2M x_{ss} & -M \end{bmatrix} \tag{13}$$

Examining the trace and determinant of $\mathcal{J}$

$$\mathrm{Tr}(\mathcal{J}) = \qquad\qquad -M - D_0 - \dfrac{\Gamma D x_{ss}^2}{(\Gamma + x_{ss})^2} < 0$$

$$\det(\mathcal{J}) = \quad (-M)\left(-D_0 - \dfrac{\Gamma D x_{ss}^2}{(\Gamma + x_{ss})^2}\right) - (2M x_{ss})\left(-\dfrac{D x_{ss}}{\Gamma + x_{ss}}\right)$$
$$> 0 \quad \forall x_{ss} > 0$$

The trace of the Jacobian is negative and the determinant is positive for all positive (realistic) solutions of x_{ss}. This is sufficient information to conclude that the eigenvalues of the Jacobian will have negative real part. This means that the steady state of the system will be asymptotically stable.

From this analysis we can conclude that in the limit of the four variable system where Ras2 is either fully active or fully inactive, the reduced two variable system *is not sufficient* to explain sustained cAMP oscillations.

Exploring Oscillatory Regimes of cAMP

Having shown that the reduced two variable model is insufficient to explain the oscillations observed in cAMP, we return to the four-variable model. From the previous section, it is clear that the special cases of active and inactive Ras2 will lead to asymptotic solutions. Thus, we next focus on intermediate activities of r.

As mentioned in the introduction, Garmendia-Torres concluded that for intermediate levels of stress, cAMP will demonstrate sustained oscillations [4]. In the four-variable model, stress is introduced into the model via the parameter A. High stress will result in low A, whereas low stress will result in high A. In order to investigate this, Gonzales et al. performed bifurcation analysis, using the parameter A as the bifurcation parameter [1]. Fig. 4(Right) shows the result of this analysis.

Observe that a Hopf bifurcation occurs around $A = 10^{-2}$. In this parameter region, cAMP exhibits limit cycle oscillations.

We can now use appropriate parameter values that characterize the low, intermediate, and high stress regimes. Sample values are provided in Table 2. Using these values and simulating the four-variable model, we can demonstrate that we indeed observe oscillations at intermediate stress levels, as shown in Fig. 4(left), as reported by Garmendia-Torres et al. in Ref. [4]. Note that for the low stress regime (green line), the model demonstrates the rapid peak in the first few minutes post glucose

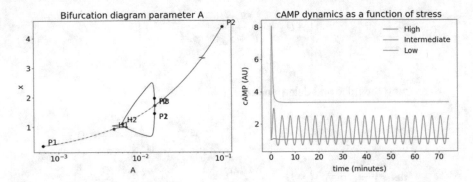

Fig. 4 Dependence of cAMP (variable *x*) dynamics on the value of stress *A*. Left: Bifurcation diagram showing the Hopf bifurcation points, indicating the region of oscillations. Note the logarithmic scale on the x -axis. Right: Time courses of cAMP for different values of stress (A). Intermediate values of stress lead to sustained oscillations

Table 2 Values of stress parameter *A* used to produce Fig.4 (right)

A value	Stress
0.005	High
0.014	Intermediate
1.4	Low

Table 3 Kinetic parameter values used in time course simulations

Parameter name	Value
A	1.45
Γ_1	0.0004
N	0.032
C	0.044
D_0	0.013
M	0.01
D	1.0
Γ	33.6
B	0.0051
G	1.0

addition, as observed by Ma et al. in Ref. [3]. Thus, the model proposed by Gonzales et al. captures both the short term as well long term cAMP responses as described previously in the literature (Table 3).

Discussion

Once the membrane-bound glucose receptors sense environmental glucose, the cAMP/PKA pathway is activated rapidly on time-scales of a few minutes. What is the significance of this early acting signaling pathway? Mutational studies of this pathway have revealed wide-ranging effects on yeast survival and proliferation in response to carbon starvation. Interestingly, these early signaling events seen to be very important not just for fast growth, but also for growth during starvation. For examples, Kayikci and Magwene 2018 [5] studied the cAMP/PKA pathway in related yeast species, and showed that mutations in the molecular players in this pathway affected the "filamentous growth" phenotype which yeasts exhibit when starved for nutrients. This phenotype is crucial for yeasts to penetrate the nutrient substrate and has important consequences for survival during starvation. More broadly, defects in the cAMP/PKA pathway have been shown to decrease the virulence of pathogenic fungal species [6].

In light of these far reaching consequences, it is important for us to first understand the roles played by the various molecular components of the signaling pathway. The model proposed by Gonzales et al. provides us with a tool to make quantitative predictions on the behavior of this pathway. Furthermore, this model allows us to simulate not just wildtype or mutant lines in laboratory strains, but related *Saccharomyces* species by choosing appropriate parameter values in order to discover the conserved functions of this pathway across evolution.

Exercises

Exercise 1: The Model of cAMP/PKA pathway section discusses a four-variable model of the cAMP/PKA pathway. However, Fig. 2 shows that the full pathway has at least nine molecular components. First, write out an ODE model describing the full nine-variable system. Next, following the description of model simplification in the Model of cAMP/PKA pathway section, reduce the nine-variable model to the four-variable model. In carrying out this simplification, describe how the parameters in the four-variable model relate to the variables and parameters in the nine-variable model.

Exercise 2: There are two negative feedback loops present in the cAMP/PKA system. What could be the significance of these feedbacks in the context of the underlying biology? First, without using the model, reason about the role of each of the feedback loops, namely, (1) the activation of the phosphodiesterases by PKA, and (2) the inhibition of Ras2 by PKA. Next, modify the four-variable system (Eqs. (1)–(4)) to reflect a strain with disrupted negative feedback from PKA via (1) Ira1, and (2) Pde.

Exercise 3: Results section (Examining the reduced two-variable system) examines the two extreme cases of fully GTP-bound or fully GDP-bound Ras2. Using the code for two-variable model provided below, simulate the model and examine the cAMP time course for the two extreme cases of Ras2. Hint: Examine Eqs. (7) and (8) for the definition of the parameter C_0. Identify the value of C_0 in the model definition and modify it to simulate the two conditions. Finally, use the script in Methods Section (Part C) to import, simulate, and plot the time courses)

Methods

The analysis presented about is implemented in python, and depends on the PyDSTool library. This library provides programmatic access to the bifurcation analysis tool AUTO, which is also used by XPPAUT (see the Computational Software chapter). The following sections present the code used to

1. Define the four-variable model
2. The bifurcation analysis using the four-variable model
3. The time course simulations under various stress levels

Note that the latter two scripts save plots to a folder ./img. This folder will have to be created manually, otherwise the script will throw an error and will not save the plot.

In order to run the scripts, all the dependencies must be properly installed. Run the following at a terminal prompt. (If you are using jupyter notebooks, open a new Python terminal.)

```
pip install --user pydstool matplotlib numpy
```

Part A: Model Definition—Two Variable

Definition of two-variable model. Save this code in gonzales2011_2v.py

```python
import PyDSTool as dst
import matplotlib.pyplot as plt

DSargs = dst.args(name='test',
varspecs={
'p' : 'M*(x^2 - p)',
'x' : 'C0 - D0*x - (D*p*x)/(gamma1 + x)',
},
pars={
'M' : 0.01,
'D' : 1.0,
'gamma1' : 33.6,
'C0' : 0.044, # C = 0.044
'D0' : 0.013,
},
ics={
'x' : 0.5,
'p' : 1.0,
},
fnspecs={'shs':(['sig','summation'],'1/(1+e^(-sig*summation))')},
tdata=[0,10])

DS = dst.Vode_ODEsystem(DSargs)
p = DS.compute('test').sample()
```

Part B: Model Definition—Four Variable

Definition of four-variable model. Save this code in gonzales2011_4v.py

```
import PyDSTool as dst

DSargs = dst.args(name='test',
 varspecs={
 'r' : 'A*(1-r)/(gamma1 + 1 - r) - (B*z*r)/(gamma1 + r)',
 'z' : 'N*(x^2 - z)',
 'p' : 'M*(x^2 - p)',
 'x' : 'C + G*r - D0*x - (D*p*x)/(gamma0 + x)',
 },
 pars={
 'A' : 1.45,
 'gamma1' : 0.0004,
 'N' : 0.032,
 'C' : 0.044,
 'D0' : 0.013,
 'M' : 0.01,
 'D' : 1.0,
 'gamma0' : 33.6,
 'B' : 0.0051,
 'G' : 1.0,
 },
 ics={
 'r' : 1.0,
 'z' : 0.1,
 'p' : 0.1,
 'x' : 1.0,
 },
 tdata=[0,100])

DS = dst.Vode_ODEsystem(DSargs)
```

Part C: Stress Parameter A Induces Oscillations in Intermediate Regime

The following code produces Fig. 4(right).

```
import matplotlib
font = {'size' : 14}

matplotlib.rc('font', **font)
matplotlib.rcParams.update({'font.size': 22})
import matplotlib.pyplot as plt
from gonzales2011_4v import DS
```

```
# Set initial conditions to low glucose, get steady state
DS.set(pars={'G':0.0})

preshift = DS.compute('test').sample() # This is a python dictionary
preshiftics = {k:preshift[k][-1] for k in DS.variables}

# For varying stresses
Avals = [0.005, 0.014, 1.4]
labels = ['High', 'Intermediate', 'Low']

plt.figure(figsize=(8,6))
timescale = 0.037
stepsize = 0.1
start = int(50.0/(0.037*stepsize))
for A,lab in zip(Avals, labels):
 print(lab)
 # Glucose upshift
 DS.set(ics=preshiftics, pars={'G':1.0,'A':A},tdata=[0,2000])
 p = DS.compute('test').sample(dt=stepsize)
 plt.plot([timescale*t for t in p['t'][start:]], p['x'][start:],
 label=lab)
plt.legend(frameon=False)
plt.xlabel('time (minutes)')
plt.ylabel('cAMP (AU)')
plt.title('cAMP dynamics as a function of stress')
plt.tight_layout()
plt.savefig('./img/stress-regimes.png')
```

Part D: Bifurcation Analysis

The following code is used for the bifurcation analysis. Note that the Hopf bifurcation points might not be found each time. Please rerun the code until there are no errors and the plot is saved successfully.

```
import PyDSTool as dst
import numpy as np
import matplotlib.pyplot as plt
import matplotlib
font = {'family' : 'normal',
 'size' : 18}
matplotlib.rc('font', **font)
plt.figure(figsize=(8, 6))
# Import model
from gonzales2011_4v import DSargs

g4v = dst.Vode_ODEsystem(DSargs)
g4v.set(tdata=[0,300], pars={'A':0.0001})
```

```
PC = dst.ContClass(g4v)
PCargs = dst.args(name='EQ1', type='EP-C')
freepar = 'A'
PCargs.freepars= [freepar]

# Bifurcation analysis settings
PCargs.MaxNumPoints = 300
PCargs.MaxStepSize = 1e-1
PCargs.MinStepSize = 1e-5
PCargs.StepSize = 1e-3

PCargs.LocBifPoints = 'all'
PCargs.SaveEigen = True

PC.newCurve(PCargs)
PC['EQ1'].forward()
PC['EQ1'].info()

PCargs = dst.args(name='LC1', type='LC-C')
PCargs.freepars = ['A']

# Once the Hopf point H2 is found,
#continue and identify other special points
PCargs.name = 'LC1'
PCargs.type = 'LC-C'
PCargs.initpoint = 'EQ1:H2'
PCargs.MinStepSize = 0.000001
PCargs.MaxStepSize = .01
PCargs.StepSize = 0.00001
PCargs.MaxNumPoints = 420
PCargs.NumSPOut = 10000;
PCargs.LocBifPoints = []
PCargs.verbosity = 2
PCargs.SolutionMeasures = 'avg'
PCargs.SaveEigen = True
PC.newCurve(PCargs)

PC['LC1'].backward()
PC['LC1'].forward()

PC.display((freepar,'x'),stability=True, figure=4)
PC['LC1'].display((freepar,'x_min'),stability=True, figure=4)

plt.title('Bifurcation diagram parameter A')
plt.xscale('log')
plt.tight_layout()
plt.savefig('./img/bifurcation-points.png', dpi=200)
```

Additional Material

Extensive derivations and analysis of parameter regimes of this model are presented in the PhD Thesis of Kevin Gonzales [1].

References

1. Gonzales KE (2011) PhD thesis: modeling oscillations in the cAMP-PKA network within budding yeast. https://dukespace.lib.duke.edu/dspace/bitstream/handle/10161/5648/Gonzales_ duke_0066D_11006.pdf?sequence=1
2. Gonzales K, Kayikçi Ö, Schaeffer DG, Magwene PM (2013) Modeling mutant phenotypes and oscillatory dynamics in the saccharomyces cerevisiae Camp-Pka pathway. BMC Syst Biol 7 (1):40
3. Ma P, Wera S, Van Dijck P, Thevelein JM (1999) The Pde1-encoded low-affinity phosphodies- terase in the yeast *Saccharomyces cerevisiae* has a specific function in controlling agonist- induced camp signaling. Mol Biol Cell 10(1):91–104
4. Garmendia-Torres C, Goldbeter A, Jacquet M (2007) Nucleocytoplasmic oscillations of the yeast transcription factor Msn2: evidence for periodic Pka activation. Curr Biol 17(12):1044–1049
5. Kayikci, Magwene PM (2018) Divergent roles for camp-Pka signaling in the regulation of filamentous growth in *Saccharomyces cerevisiae* and *Saccharomyces bayanus*. G3 Genes Genomes Genet 8(11):3529–3538
6. Fuller KK, Rhodes JC (2012) Protein kinase A and fungal virulence. Virulence 3(2):109–121

Synchronization of Oscillatory Gene Networks

Pavel Kraikivski

Introduction

Periodic gene expression in living cells could be subject to synchronization during processes such as cell-to-cell communication, quorum sensing, and entrainment of circadian rhythms. Synchronization occurs when two oscillatory gene networks coordinate the expression of their genes, and subsequently perform in-phase gene transcription. Coordination in gene expression of two oscillatory gene networks can occur when these networks are coupled by a cell signaling pathway. Cell signaling is common in all multicellular systems to perform cell-to-cell signaling and communication. For example, bacteria use a cell communication process known as quorum sensing to coordinate gene expression according to the density of their local population. Bacteria implement quorum sensing by secreting a signal molecule (autoinducer) into the environment. The autoinducer molecules can be taken up by bacteria, which activates gene expression, and also includes production of the autoinducer itself. Bacterial quorum sensing is used to engineer cell-to-cell communication systems that are composed of populations of "sender cells" and "receiver cells," performing cell-to-cell communication that can be experimentally controlled. If sender and receiver cells exhibit oscillatory gene expression dynamics, then cell-to-cell communication may lead to synchronization of their oscillatory gene networks. By studying this synchronization, we can better understand how cells coordinate the decision-making process.

Cells can also receive signals from the environment, such as those induced by light or changes in temperature. These signals can influence expression of certain genes. For example, the expression of genes that regulate circadian rhythms

P. Kraikivski (✉)
Academy of Integrated Science, Division of Systems Biology, Virginia Polytechnic Institute and State University, Blacksburg, VA, USA
e-mail: pavelkr@vt.edu

© Springer Nature Switzerland AG 2021
P. Kraikivski (ed.), *Case Studies in Systems Biology*,
https://doi.org/10.1007/978-3-030-67742-8_9

is affected by the light-dark cycle and temperature changes [1]. Circadian rhythms typically occur in accordance with nature's light-dark cycle; however, if the light-dark cycle is artificially changed, the circadian rhythm can reset to synchronize with the new cycle. The synchronization of circadian rhythms with environmental oscillations is known as entrainment. This term is used because the circadian rhythms persist even in the absence of environmental oscillating signals; however, organisms have a mechanism to adjust the period of the circadian rhythms to match the period of the environmental oscillation. Thus, entrainment is an important mechanism that allows organisms to maintain an adaptive phase relationship with the environment. Circadian rhythms are controlled by rhythmic expression of "clock genes." The molecular mechanism of circadian rhythms is usually modeled using oscillatory gene networks [2–5]. To understand the entrainment of rhythms, we can analyze synchronization of an oscillatory gene network, perturbing it by a periodic signal that influences gene transcription, and obtaining the information on how the period of gene expression is entrained.

There are several models of gene regulatory networks that exhibit oscillatory dynamic behavior. The most studied is the Goodwin oscillatory gene network [6], a synthetic relaxation oscillator, and a synthetic delay oscillator known as the repressilator. These oscillatory gene networks are used to study entrainment and synchronization of oscillators. In this chapter, we will explore the process of synchronization of oscillatory gene networks.

Molecular Mechanism

A gene regulatory network may produce oscillatory dynamic behavior if the structure of the network has the following three properties: (i) the network is based on a negative feedback loop regulatory circuit, (ii) action of the negative feedback loop is delayed, and (iii) there is sufficient non-linearity in the negative feedback loop. A classic example of an oscillatory gene network that encompasses all these properties was proposed in 1965 by Brian Goodwin, and is known as the Goodwin oscillator [6]. The Goodwin model describes dynamics of three variables: X, Y, and Z, where X describes the concentration of mRNA transcribed from a gene that is negatively regulated by Z, X is translated into protein Y, and Z is a protein that is activated by Y (see Fig. 1, left). The intermediate Y is introduced to impose a delay in the negative feedback loop: X having an effect on Z and then Z having a negative effect on X. Thus, the model describes a system with the delayed negative feedback loop. J. S. Griffith analyzed the Goodwin model and showed that the model exhibits sustained oscillations only if the negative regulation of Z on X is described by a function with a high degree of nonlinearity [7]. The Goodwin oscillator is often used to model circadian rhythms [2]. To study entrainment of rhythms or synchronization, we can extend the Goodwin model by adding a periodic variable u that affects X, where u represents an external periodic signal that has an influence on gene transcription (see Fig. 1, left). This new model shown in Fig. 1, left, will be named Model 1.

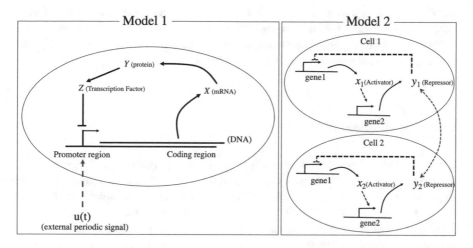

Fig. 1 Schematic illustration of oscillatory gene regulatory networks. Left: Model 1 represents a Goodwin oscillatory network coupled to an external periodic signal. The red dashed arrow indicates regulation of the Goodwin oscillator by an external periodic signal. Right: Model 2 represents two coupled relaxation oscillators, each consisting of two genes, where gene1 is expressed into an activator of gene2 and gene2 is expressed into an inhibitor (repressor) of gene 1. The red dashed line indicates coupling between oscillators

Synchronization can also occur in a system of coupled oscillators. As an example, oscillators can represent cells that are able to communicate their states to one another, allowing this system to exhibit synchronized population-wide oscillations. To analyze synchronization of two oscillating cells that communicate their states to each other, we will consider two coupled relaxation oscillators and refer to this model as Model 2 (see Fig. 1, right).

Relaxation oscillators produce more robust periodic behavior than delay oscillators implemented in the Goodwin model. Also, without the need to impose a delay, two variables are enough to describe the relaxation oscillator. To construct a relaxation oscillator, we need two alternating processes running on different time scales. One process should have a short impulsive period (burst), which promotes the second process that has a long relaxation period. The corresponding oscillatory gene network that behaves as a relaxation oscillator can be constructed using two genes, where the product of the first gene x acts as an activator for the second gene y, that in turn produces a repressor blocking x. The expression of x grows sharply until it is blocked by the repressor. Then the concentration of the repressor slowly decays to a level that allows the next burst of expression of x to begin. This dynamic behavior of x and y genes will be demonstrated in Mathematical Models section.

Two cells containing such oscillatory networks can be coupled by a signal that modulates gene transcription in these two cells. For example, the signal can be a diffusion of the repressor molecules between two cells. Each cell will then have its own repressor molecules, plus the repressor molecules from another cell. This interaction scheme between two cells that exchange their repressor molecules is shown in Fig. 1, right (red dashed line). Cells will synchronize the expression of their genes when the gene expression depends on the repressors from both cells. Synchronization of the gene expression in two cells can also be achieved if cells exchange activator molecules instead of repressors.

Mathematical Models

We will study synchronization using two different mathematical frameworks: (i) entrainment of oscillations by an external periodic signal to mimic circadian rhythm entrainment, and (ii) synchronization of two identical coupled oscillators to mimic synchronization of oscillatory gene networks in two communicating cells. Firstly, we will analyze synchronization of the Goodwin oscillator driven by an external sinusoidal oscillator, as shown in the scheme in Fig. 1, left. The mathematical representation of the Goodwin model can be described by the following system of differential equations:

$$\frac{dX}{dt} = \alpha \frac{1}{1 + Z^n} - X$$

$$\frac{dY}{dt} = X - Y$$

$$\frac{dZ}{dt} = Y - Z \tag{1}$$

where α is the maximum rate of transcription, and the power n introduces nonlinearity into the negative-feedback loop. All variables and time in System (1) are rescaled to be dimensionless. The numerical solution of the Goodwin model (1) is shown in Fig. 2a.

As was shown by Griffith, the Goodwin model produces sustained oscillations if n is > 8. This condition can be derived by performing a linear stability analysis and finding eigenvalues that are used to determine the stability of the Goodwin system close to a steady state. To find a steady state solution, all derivatives in System (1) must be set to zero. Thus, the set of equations determining the steady state is

$$\alpha \frac{1}{1 + Z_{ss}{}^n} - X_{ss} = 0, \quad X_{ss} - Y_{ss} = 0, \quad Y_{ss} - Z_{ss} = 0.$$

The steady state value of Z (denoted as Z_{ss}) is a solution of $Z_{ss}{}^{n+1} + Z_{ss} - \alpha = 0$ equation. Also, for other components, we have $X_{ss} = Y_{ss} = Z_{ss}$ at the steady state. It can be shown that the steady state value Z_{ss} satisfies the following condition, $Z_{ss}/\alpha < 1$ (see Exercise 1). The steady state solution is stable only if all eigenvalues have negative real parts. If even one eigenvalue has a positive real part, then the steady state solution is unstable. It can be shown that the real parts of some eigenvalues are

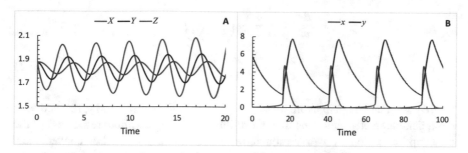

Fig. 2 Dynamics exhibited by Goodwin (A), and Relaxation (B), oscillators. (a) Simulation of the Goodwin oscillator (1) is performed for parameters $n = 8.5$ (Hopf bifurcation occurs at n $= 8.1$), $\alpha = 300$, and initial values $X = Y = Z = 1.8768$. The period of oscillations on the plot is 3.6, which agrees with the analytical prediction $T_0 = \frac{2\pi}{\sqrt{3}} \approx 3.6$. (b) Simulation of the relaxation oscillator (2). The temporal evolution of variables for the relaxation oscillator (2) are obtained using the following parameter values: $\gamma = 50$, $\beta = 20$, $\mu = 0.1$, $\varepsilon = 0.02$. Initial values: $x = 0.052$ and $y = 5.7$

positive if $n > 8/(1 - Z_{ss}/\alpha)$, (see Exercise 2). Therefore, the steady state is unstable when $n \gg 8$, and in that condition, the solution of the system relaxes to a limit-cycle (oscillations). The Hopf bifurcation (the point at which a steady state solution switches to a periodic solution) occurs when $\left(1 - \frac{Z_{ss}}{\alpha}\right) = \frac{8}{n}$. The imaginary part of eigenvalues (see Exercise 2) defines the frequency of oscillations: $\omega_0 = \frac{\sqrt{3}}{2}\left(n\left(1 - \frac{Z_{ss}}{\alpha}\right)\right)^{1/3}$. Therefore, at Hopf bifurcation, the frequency $\omega_0 = \sqrt{3}$ and the period of oscillations $T_0 = \frac{2\pi}{\omega_0} = \frac{2\pi}{\sqrt{3}} \approx 3.6$ are independent of the parameters α and n. The numerical simulation results, shown in Fig. 2a, confirm that the period of oscillations is in fact ~3.6 (as mentioned, time in this model (1) is dimensionless).

To study synchronization of the Goodwin oscillator, we can add an external periodic signal u(t) that affects gene transcription in the Goodwin model. Then, the first equation in system (1) takes the following form:

$$\frac{dX}{dt} = \alpha \frac{1}{1 + Z^n} + u(t) - X \qquad (M1)$$

We will refer to this model as Mathematical Model 1 (M1). u(t) can be chosen as u(t) $= A \cos^2(\omega t)$, where cos() is the cosine function, A represents amplitude, and ω describes the frequency of oscillations. The cosine function in u(t) is squared to produce only positive values, thus, having a positive effect on the transcription of X. This simple model can be used to investigate how an external periodic signal affects an oscillator. We will use this model to simulate a process of entrainment to better understand the mechanism of circadian rhythm entrainment.

We will also analyze synchronization in the system of two coupled relaxation oscillators. An example of the gene regulatory mechanism is shown as Model 2 in Fig. 1, right. The mathematical representation of the relaxation oscillator is described by the following system of equations:

$$\frac{dx}{dt} = \gamma \frac{\varepsilon + x^4}{1 + x^4} \left(\frac{1}{1 + y^2} \right) - x$$

$$\frac{dy}{dt} = \mu \left(\beta \frac{\varepsilon + x^4}{1 + x^4} - y \right) \tag{2}$$

where γ and β are the maximum transcription rates of the genes x and y, respectively; ε is a lick parameter indicating that transcription can occur without regulation of the transcription region by x or y (non-zero rate for x and $y = 0$). For convenience, equations contain only dimensionless time, variables, and parameters. The temporal evolution of x and y variables are shown in Fig. 2b. In each oscillation cycle, the variable x exhibits a sharp increase, followed by a slow decay, which is a common property of the relaxation oscillator.

Two relaxation oscillators can be coupled as:

$$\frac{dx_1}{dt} = \gamma_1 \frac{\varepsilon + x_1^4}{1 + x_1^4} \left(\frac{1}{1 + y_1^2} \right) - x_1$$

$$\frac{dy_1}{dt} = \mu \left(\beta_1 \frac{\varepsilon + x_1^4}{1 + x_1^4} - y_1 \right) + C(y_2 - y_1)$$

$$\frac{dx_2}{dt} = \gamma_2 \frac{\varepsilon + x_2^4}{1 + x_2^4} \left(\frac{1}{1 + y_2^2} \right) - x_2$$

$$\frac{dy_2}{dt} = \mu \left(\beta_2 \frac{\varepsilon + x_2^4}{1 + x_2^4} - y_2 \right) + C(y_1 - y_2) \tag{M2}$$

where constant C represents the coupling strength. We will refer to this model as Mathematical Model 2 (M2). The period of oscillations of the relaxation oscillators can be controlled by β constants. The coupling C is the parameter that controls the synchronization of oscillators. If C is close to zero, then oscillators are weakly coupled, and synchronization should not occur. If the coupling is strong and thus C is large enough, then oscillators can synchronize.

Synchronization of Oscillators

We will use the XPPAUT tool (see Computational Software chapter in this book) to simulate the M1 and M2 mathematical models and to analyze synchronization in gene oscillatory networks. The corresponding XPP code for the M1 model is given at the end of the chapter. By running this code in XPP software, we can reproduce Fig. 2a, which shows simulation results of the Goodwin model without an external signal. We can also analyze how synchronization of the Goodwin model oscillation depends on frequency and amplitude of the external periodic signal $u(t)$. It is important to note that, for the external signal in model M1, which is described by

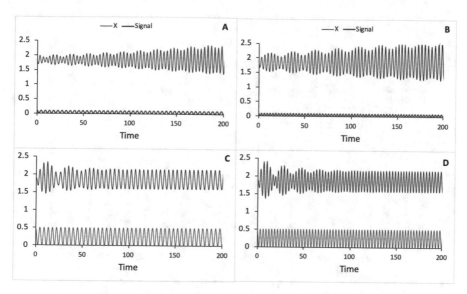

Fig. 3 Simulation of the model M1 applying the external signal u(t) with different frequencies and amplitudes. (**a, b**) Oscillations of the X variable are not synchronized by a signal u(t) that has the amplitude $A = 0.1$ and the period $T = 4.5$ (corresponding frequency $\omega = 0.7$) in (**a**), and the same amplitude but the period $T = 3.15$ (corresponding frequency $\omega = 1$) in (**b**). The period of the X oscillation in (**a**) is measured to be 3.7, which is close to the period $T_0 = \frac{2\pi}{\sqrt{3}} \approx 3.6$ of the Goodwin oscillator. The period of the X oscillation in (**b**) remains ~3.65. Thus, the external signal with an amplitude $A = 0.1$ is unable to entrain the Goodwin oscillator. By contrast, (**c, d**) show oscillations of X, which are entrained by the signal u(t) that has the amplitude $A = 0.5$ and the period $T = 4.5$ in (**c**), and the same amplitude but the period $T = 3.15$ in (**d**). In (**c**), the period of the X oscillation is measured to reach 4.45, which is close to the period of the signal. In (**d**), the period of the X oscillation is measured to be 3.1, which is again, close to the period of the signal. Thus, the external signal with the amplitude $A = 0.5$ is strong enough to entrain the Goodwin oscillator. In all these simulation results, the periods for X oscillations were measured at time ≈ 1000

the cosine squared: $A \cos^2(\omega t) = A(1 + \cos (2\omega t))/2$, the frequency of oscillations is equal to 2ω. When the frequency of external periodic signal is zero, the model produces oscillations with the period $T_0 = \frac{2\pi}{\sqrt{3}}$ of the Goodwin oscillator. Also, if the external signal is not strong enough (amplitude A is small), then oscillations of the Goodwin oscillator are not entrained and the period stays near $T_0 = \frac{2\pi}{\sqrt{3}} \approx 3.6$ value (see Fig. 3a, b). However, if the frequency of the external signal 2ω is not zero, and amplitude of the signal is high, then the Goodwin oscillator is entrained by the external signal (see Fig. 3c, d). During entrainment, the period of the oscillator T_0 approaches the period of the external signal $T = 2\pi/2\omega$. The time of entrainment depends on the difference between period T_0 and period of the external signal T. If the difference between periods is significant, then the entrainment may not be observed. Thus, we can also say that, when the external signal is strong enough, and its frequency is close to the intrinsic frequency of the Goodwin oscillator, ω_0, then the entrainment of the oscillator can occur.

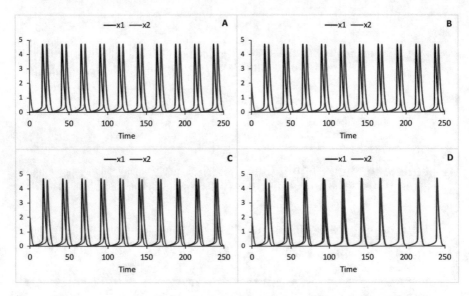

Fig. 4 Synchronization of two relaxation oscillators (model M2) that start initially out of phase with respect to each other, depending on the coupling strength. Only x_1 (black) and x_2 (red line) variables of model M2 are shown. (**a**) Simulation results with the coupling constant $C = 0$; (**b**) C = 0.001 is not sufficient for oscillators to synchronize; (**c**) phases of oscillators get closer when $C = 0.003$; (**d**) $C = 0.01$ is sufficient coupling strength for oscillators to be synchronized after time >150. Simulation results are obtained using the following parameter values: $\gamma_1 = 50$, $\gamma_2 = 50$, $\beta_1 = \beta_2 = 20$, $\mu = 0.1$, $\varepsilon = 0.02$, and initial values: $x_1 = 0.052$, $y_1 = 5.7$ and $x_2 = 1.98$, $y_2 = 6.58$

Next, we simulate and discuss the M2 model of two coupled oscillators that oscillate out of phase, and that also have different oscillation periods. Synchronization of these oscillators depends on the value of the parameter C that describes the strength of the coupling between oscillators. Consider a scenario where oscillators have the same frequency, but start out of phase, as shown in Fig. 4a. If we gradually increase the value of the coupling constant C, then we find that the oscillators become synchronized when the coupling constant is sufficiently large (see Fig. 4b–d). For example, two coupled relaxation oscillators are synchronized at $C = 0.01$ after a short period of time, (see Fig. 4d). In another scenario, two oscillators can also have different oscillating periods.

It is also reasonable to assume that the synchronization of two oscillators should occur faster if their oscillation periods are equal, or close to each other (e.g., Fig. 4). However, if two coupled oscillators also have different frequencies, then the larger coupling constant would be required to synchronize them. We can test this assumption by simulating the model M2 with different values of β_1 and β_2 constants, thus generating different oscillation periods for two relaxation oscillators (see Fig. 5a). Indeed, the simulation results confirm our initial hypothesis that a stronger coupling constant is needed to synchronize oscillators that have different oscillation frequencies and also start out of phase. The bigger the difference in the periods of oscillators, the bigger the value of the coupling constant C is required to synchronize the

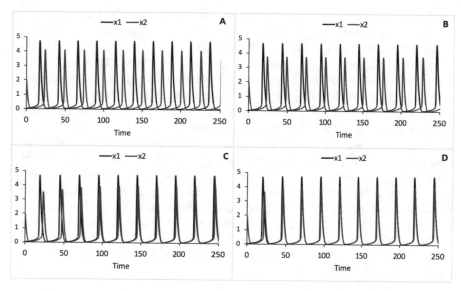

Fig. 5 Synchronization of two relaxation oscillators that initially start out of phase with respect to each other and also have different frequencies depends on the coupling strength. The frequency of the second oscillator is changed using parameter $\beta_2 = 25$. (**a**) Simulation results obtained using the coupling constant $C = 0$; (**b**) $C = 0.01$ is not sufficient for oscillators to synchronize; (**c**) the oscillators are almost synchronized when $C = 0.02$; (**d**) $C = 0.04$ is sufficient for oscillators to be synchronized after time >100. The simulation results are obtained using the following parameter values: $\gamma_1 = 50$, $\gamma_2 = 50$, $\beta_1 = 20$, $\mu = 0.1$, $\varepsilon = 0.02$, and initial values: $x_1 = 0.052$, $y_1 = 5.7$ and $x_2 = 1.98$, $y_2 = 6.58$

oscillators. The simulation results for different values of the coupling constant C is shown in Fig. 5b–d. Synchronization of oscillators is not observed if the coupling parameter C is small (see Fig. 5b). By contrast, two oscillators become synchronized (their initially different periods become identical) when the coupling parameter C is large enough (Fig. 5c, d). The larger the coupling constant C, the faster two oscillators get synchronized (see the difference in Fig. 5c, d).

Discussion

We studied the synchronization of two different models: (i) an oscillatory gene network that is influenced by an external periodic process, and (ii) two coupled oscillatory gene networks. We used the first model to study network entrainment. We observed how cells can entrain their intrinsic gene oscillatory dynamics in synchrony with an external periodic signal. We used the second model to understand how living cells can mutually coordinate their gene expression processes. A cell's ability to coordinate its molecular processes with other cells and with environmental cyclic signals, is an important function that allows cells to perform crucial physiological processes, such as quorum sensing and entrainment of circadian rhythms.

Synchronization occurs when oscillating processes of interacting systems affect each other's phases, such that they spontaneously lock to a certain frequency or phase. In the case of two gene oscillatory networks, we observed that synchronization depends on the coupling parameter describing the interaction between these networks. When the coupling between gene oscillatory networks is strong enough, they can synchronize their gene expression processes.

Also, we learned that the entrainment of an oscillator depends on the difference between the intrinsic frequency of the oscillator and the frequency of an external signal that affects the oscillator. If the difference between the frequencies is small, then the entrainment of the oscillator by the external signal can be observed. This modeling result agrees with the observation made by studying the entrainment of circadian rhythms in fruit flies. In living organisms, for example, circadian rhythms can be entrained by an artificially changed light-dark cycle. More information about circadian biology, entrainment, and mathematical models of circadian rhythms can be found in Refs. [1–5].

Exercises

Exercise 1. The steady state value of Z in the Goodwin Model is defined by the following equation: $Z^{n+1} + Z - \alpha = 0$, which can be written as $\frac{1}{1+Z^n} = \frac{Z}{\alpha}$. Thus, the solution of this equation corresponds to an intersection point of the left-hand-side function of the equation, $\frac{1}{1+Z^n}$, and the right-hand-side function of the equation, $\frac{Z}{\alpha}$. Sketch these functions as a function of variable Z to show that the intersection point (solution) satisfies the following condition: $Z/\alpha < 1$.

Exercise 2.

(a) Show that the Jacobian of system (1) evaluated at the steady state has the following form:

$$\begin{pmatrix} -1 & 0 & -\alpha \dfrac{nZ_{ss}^{n-1}}{\left(1 + Z_{ss}^n\right)^2} \\ 1 & -1 & 0 \\ 0 & 1 & -1 \end{pmatrix}$$

(b) Use the Jacobian and the following equations for steady state $\frac{1}{1+Z_{ss}^n} = \frac{Z_{ss}}{\alpha}$ and $Z_{ss}^{n+1} = \alpha - Z_{ss}$ from Exercise 1 to show that the equation for eigenvalues is:

$$(\lambda + 1)^3 = -\frac{n}{\alpha} Z_{ss}^{n+1} \text{ or } (\lambda + 1)^3 = -n\left(1 - \frac{Z_{ss}}{\alpha}\right)$$

(c) Confirm that the equation for eigenvalues has the following three roots:

$$\lambda_1 = -1 - \left(n\left(1 - \frac{Z_{ss}}{\alpha}\right)\right)^{\frac{1}{3}}$$

$$\lambda_2 = -1 + \left(\frac{1}{2} + i\frac{\sqrt{3}}{2}\right)\left(n\left(1 - \frac{Z_{ss}}{\alpha}\right)\right)^{\frac{1}{3}}$$

$$\lambda_3 = -1 + \left(\frac{1}{2} - i\frac{\sqrt{3}}{2}\right)\left(n\left(1 - \frac{Z_{ss}}{\alpha}\right)\right)^{\frac{1}{3}}$$

where i is the imaginary unit that is defined by the following property $i^2 = -1$.

(d) Show that the real parts of two eigenvalues λ_2 and λ_3 are positive if $n > 8/(1 - Z_{ss}/\alpha)$.

Exercise 3. Consider model M2 and assume that activator molecules, instead of repressor molecules, are exchanged between cells, and add $C(x_2 - x_1)$ and $C(x_1 - x_2)$ terms to the right-hand sides of $\frac{dx_1}{dt} = \ldots$ and $\frac{dx_2}{dt} = \ldots$ equations of coupled relaxation oscillators. Write the corresponding code for XPPAUT. Run your code in XPP and show that these terms can also induce synchronization of oscillators.

Exercise 4. Run the coupled relaxation oscillators code (provided in Method section) in XXPAUT to synchronize oscillators with different g1 and g2 parameters (for example, g1=50 and g2=60). Find the coupling constant C values for which the two oscillators are synchronized. Set values for beta1 and beta2 differently and retest the synchronization.

Method

The corresponding XPPAUT description of the model M1 has to be written in a text file with file extension, .ode, and has the following form:

```
# Goodwin Oscillator with a periodic signal u(t)

u(t)=A*(cos(w*t))^2
dX/dt=alpha/(1+Z^n)-X+u(t)
dY/dt=X-Y
dZ/dt=Y-Z

aux Sig=u(t)

par n=8.5, alpha=300, w=0.7, A=0.0
init X=1.8768, Y=1.8768, Z=1.8768
@ total=1000

done
```

To run this code in XXPAUT, use the following steps:

(1) On the top of the main XPP Window, click ICs and also Param buttons. Initial Data and Parameters windows will appear. It is always useful to keep Initial Data and Parameters windows open.

(2) Open Viewaxes and in 2D set Ymin=1.5, Ymax=2.1. In the Initial Data window click the GO button; the plot of the X variable as a function of time should appear in the Graphics area, similar to the plot shown in Fig. 2a. By using Graphic stuff, (A)dd curve, Y and Z variables can be added to the graphics area, use a different number in Color space to plot curves with different colors.

(3) To simulate the entrainment of the oscillator, as in Fig. 3, first click Erase to clean the graphics area, then set amplitude parameter A to 0.1 in the Parameters window and click OK and GO in this window. In Viewaxes->2D set Ymin=0, Xmax=200, Ymax=2.5. In Graphic stuff, (A)dd curve in Y-axis, add Sig variable to plot the signal. To simulate A=0.5, click Erase then set A to 0.5 in the Parameters window, click OK and Go. Set a different value for the w parameter in the Parameters window to simulate the effect of the different frequency value.

The corresponding XPPAUT description of the model M2 has the following form:

```
# Two coupled Relaxation Oscillators

dx1/dt=g1*((e+x1^4)/(1+x1^4))*(1/(1+y1^2))-x1
dy1/dt=mu*(beta1*(e+x1^4)/(1+x1^4)-y1)+C*(y2-y1)

dx2/dt=g2*((e+x2^4)/(1+x2^4))*(1/(1+y2^2))-x2
dy2/dt=mu*(beta2*(e+x2^4)/(1+x2^4)-y2)+C*(y1-y2)

par g1=50, beta1=20, g2=50, beta2=20, mu=0.1, e=0.02, C=0.0
init x1=0.052, y1=5.7, x2=1.98, y2=6.58

@ total=1000

done
```

To run this code in XXPAUT, use the following steps:

(1) In the XPP Window, click ICs and also Param buttons. The Initial Data and Parameters windows will appear.

(2) Open Viewaxes and in 2D set Ymin=0, Xmax=100, Ymax=8. In the Initial Data window click the GO button; the plot of the x1 variable as a function of time should appear in the Graphics area, similar to the plot shown in Fig. 2b. By using Graphic stuff, (A)dd curve, the y1 variable can be added to the graphics area, put 1 in the Color space to plot a red curve.

(3) To simulate synchronization, open `Viewaxes` and in `2D` set `Ymin=0`, `Xmax=200`, `Ymax=5`. In the `Initial Data` window click the `GO` button, the plot of the `x1` variable as a function of time will appear in the Graphics area. Then, open `Graphic stuff`, `(A)dd curve`, put `x2` in `Y-axis` and `1` in the `Color` space to plot the `x2` variable curve in red color. `x1` and `x2` curves will appear out of phase. Set `C =0.001`, then `0.003`, and `0.01` to observe the synchronization effect as in Fig. 4.

(4) Set `beta2=25` to analyze synchronization of oscillators with different frequencies, to reproduce Fig. 5.

Additional Reading

The biological principles of cell communication are frequently described in biology textbooks, for example in Refs. [8, 9]. For studying the modeling of circadian rhythms in bacteria see Refs. [10, 11], and in eukaryotes see Refs. [12–16].

References

1. Takahashi JS et al (1989) The avian pineal, a vertebrate model system of the circadian oscillator: cellular regulation of circadian rhythms by light, second messengers, and macromolecular synthesis. Recent Prog Horm Res 45:279–348. discussion 348-52
2. Ruoff P et al (1999) The Goodwin oscillator: on the importance of degradation reactions in the circadian clock. J Biol Rhythm 14(6):469–479
3. Gerard C, Goldbeter A (2012) Entrainment of the mammalian cell cycle by the circadian clock: modeling two coupled cellular rhythms. PLoS Comput Biol 8(5):e1002516
4. Leloup JC, Goldbeter A (1998) A model for circadian rhythms in *Drosophila* incorporating the formation of a complex between the PER and TIM proteins. J Biol Rhythm 13(1):70–87
5. Tyson JJ et al (1999) A simple model of circadian rhythms based on dimerization and proteolysis of PER and TIM. Biophys J 77(5):2411–2417
6. Goodwin BC (1965) Oscillatory behavior in enzymatic control processes. Adv Enzym Regul 3:425–438
7. Griffith JS (1968) Mathematics of cellular control processes. I. Negative feedback to one gene. J Theor Biol 20(2):202–208
8. Raven, P.H., et al. (2014) Biology, Vol. 1. Tenth Edition, AP Edition. Dubuque, McGraw-Hill, Iowa. (various pagings).
9. Russell, P.J., P.E. Herz, and B. McMillan (2017) Biology: the dynamic science, Vol. 1. Fourth edition. Cengage Learning, Boston. (various pagings)
10. Mehra A et al (2006) Circadian rhythmicity by autocatalysis. PLoS Comput Biol 2(7):e96
11. Kurosawa G, Aihara K, Iwasa Y (2006) A model for the circadian rhythm of cyanobacteria that maintains oscillation without gene expression. Biophys J 91(6):2015–2023
12. Leloup JC, Goldbeter A (2003) Toward a detailed computational model for the mammalian circadian clock. Proc Natl Acad Sci U S A 100(12):7051–7056

13. Forger DB, Peskin CS (2003) A detailed predictive model of the mammalian circadian clock. Proc Natl Acad Sci U S A 100(25):14806–14811
14. Kim JK, Forger DB (2012) A mechanism for robust circadian timekeeping via stoichiometric balance. Mol Syst Biol 8:630
15. Locke JC et al (2006) Experimental validation of a predicted feedback loop in the multi-oscillator clock of Arabidopsis thaliana. Mol Syst Biol 2:59
16. Locke JC et al (2008) Global parameter search reveals design principles of the mammalian circadian clock. BMC Syst Biol 2:22

The *lac* Operon

Pavel Kraikivski

Introduction

The *lac* operon regulatory mechanism allows bacteria to sense external nutrient conditions and, accordingly, regulate gene transcription that assists in the digestion of available nutrients. The *lac* system produces a switch-like dynamic behavior, where the switching between 'on' and 'off' states of *lac* gene transcription depends on the specific thresholds of nutrient concentrations. As *lac* operon regulation has been extensively studied and is one of the most well-understood gene regulatory systems, it is often used as an example to demonstrate the functional principles of genetic switches. By studying this case, we learn the role of nonlinearity and positive feedback loops in gene regulatory networks, and how gene circuits can generate an all-or-nothing response.

Many gene and protein regulatory networks in living cells exhibit a switch-like dynamic behavior that resembles the one observed for *lac* operon regulation. Molecular switches with all-or-nothing responses are common in the regulation of cell cycle checkpoints, morphogenesis, and other vital processes that occur in living cells and organisms. Therefore, the *lac* operon case study teaches us the common principles governing many diverse regulations realized in living cells. The model of the *lac* operon also assists with understanding the relationship between circuit design and its dynamic behavior, which is helpful and necessary for formulating engineering principles that can be used to design biological circuits for applications in synthetic biology and biotechnology.

P. Kraikivski (✉)
Academy of Integrated Science, Division of Systems Biology, Virginia Polytechnic Institute and State University, Blacksburg, VA, USA
e-mail: pavelkr@vt.edu

© Springer Nature Switzerland AG 2021
P. Kraikivski (ed.), *Case Studies in Systems Biology*,
https://doi.org/10.1007/978-3-030-67742-8_10

Physiology

The *lac* operon includes a set of genes whose products allow *E. coli* bacteria to use lactose (milk sugar) as an energy source. *E. coli* commonly inhabit mammalian guts, where glucose is a more abundant energy source than lactose. In this natural condition, the expression of genes from the *lac* operon is repressed to maintain the resources that otherwise would be consumed in the process of lactose metabolism. Thus, bacteria activate the *lac* operon when lactose is the only available sugar source. The *lac* operon genes then produce enzymes that break down lactose into glucose and galactose molecules. Therefore, *E. coli* has a genetic mechanism that allows it to produce a preferable energy source (glucose) from lactose.

The genes in the *lac* operon share the same transcriptional regulation. The activation of *lac* operon transcription results in production of a single mRNA transcript that includes all enzyme sequences encoded in the *lac* operon. Then, the mRNA transcript is translated into proteins that are, in turn, involved in the regulation of *lac* operon gene transcription. Protein translation in prokaryotic gene expression is a faster and less regulated process than gene transcription because ribosomes can directly access mRNA transcripts due to the lack of a nuclear membrane in prokaryotes. Thus, gene transcription is the rate-limiting (slowest) step in prokaryotic gene expression and is also highly regulated by various transcriptional activators and repressors.

In 1961, Jacob and Monod studied *E. coli* mutants, in which expression of *lac* genes was abnormal, so as to identify components of *lac* operon regulation [1, 2]. The *lac* operon contains three coding genes that are expressed into enzymes involved in lactose metabolism. A single mutation in any of these genes leads to a deficiency of the corresponding enzyme production. However, mutations affecting the DNA sequence near the promoter of *lac* genes disrupt expression of all three enzymes. Based on the analysis of mutations, Jacob and Monod proposed the *lac* operon model, which explains observed physiological behavior of *E. coli* mutants in glucose-depleted and lactose-rich nutrient conditions. They discovered that the operator region and *lac* repressor play the ultimate role in *lac* operon regulation. The *lac* repressor implements a negative control on the initiation of transcription. In the absence of lactose, the *lac* repressor binds specifically to the operator sequence, which prevents transcription of protein-coding genes of the *lac* operon. Thus, the operator works as a switch and the repressor keeps the switch in the 'off' state.

The *lac* operon regulation was the first example of a gene regulatory network that regulates the expression of a group of adjacent genes. However, later it turned out that the operon model is also applicable to explain the regulation of many genes in prokaryotes and their viruses. In 1965, Jacob and Monod received the Nobel Prize for Physiology or Medicine for discoveries concerning regulatory activities in bacteria.

Molecular Mechanism

The basic structure of the *lac* operon and the molecular mechanism of its activation are shown in Fig. 1. Repression of the *lac* operon is controlled by the *lac* repressor, which is constructively active when other sugars are more abundant than lactose. The *lac* repressor can tightly bind the operator region of the *lac* operon. Bound repressor works as an obstacle for the motion of RNA polymerase and, thus, prevents the transcription of coding genes called *lacZ*, *lacY*, and *lacA*. If the repressor is removed, RNA polymerase can transcribe these three coding genes into mRNA, which is then translated into corresponding proteins. The three coding *lac* genes function as follows:

1. *lacZ* encodes the β-galactosidase enzyme, which catalyzes lactose into glucose and galactose but can also catalyze the conversion of lactose to allolactose. Allolactose inhibits the *lac* repressor and, thus, enhances the activation of the *lac* operon. Therefore, the product of the *lacZ* gene has a positive effect on *lac* operon activity and its own production.

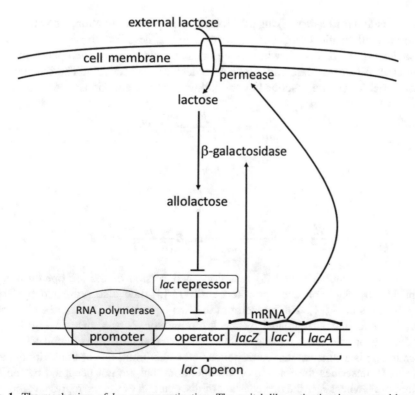

Fig. 1 The mechanism of *lac* operon activation. The switch-like activation is governed by two positive feedback loops: (i) the production of permease that further increases the lactose uptake, and (ii) the production of β-galactosidase that can be converted into the inducer (allolactose) which then continues to support the transcription of *lac* genes

2. *lac*Y encodes β-galactoside permease, which is a membrane transport protein that catalyzes the accumulation and transportation of lactose across the cell membrane.
3. *lacA* encodes β-galactoside transacetylase, which may play an important cellular role in acetylating, and thus detoxifying non-metabolizable pyranosides. However, the role of this enzyme in *lac* operon regulation is still unclear, and we will exclude it from the *lac* operon mathematical model.

Therefore, lactose positively controls the expression of enzymes that are necessary to keep the *lac* operon in an active state. In addition, activation of the *lac* operon allows *E. coli* to metabolize lactose into glucose and galactose, and, thus, use lactose as an energy source (not shown in Fig 1). However, the main goal of this chapter is to present a model of regulation of the *lac* operon to demonstrate the working principles of genetic switches.

Mathematical Model

To understand the basic working principles of *lac* operon regulation, we will analyze a simple mathematical model of the *lac* operon that describes the dynamics of four variables: M, P, B, and L, which represent concentrations of *lac* operon mRNA, permease, β-galactosidase and intracellular lactose, respectively. The dynamics of these variables can be described by the following set of equations:

$$\frac{dM}{dt} = \alpha_1 \frac{K}{(K+R)} - \delta_1 M$$

$$\frac{dP}{dt} = \alpha_2 M - \delta_2 P \tag{1}$$

$$B(t) = \frac{1}{4}P(t)$$

$$\frac{dL}{dt} = \alpha_3 P \frac{L_e}{(K_1 + L_e)} - 2\alpha_4 B \frac{L}{(K_2 + L)} - \delta_3 L$$

This mathematical model (1) is a simplified version of the *lac* operon model, published by Santillan and coworkers in 2007 [3], and also presented by Brian Ingalls in his book [4]. The first term in the first equation of (1) describes the transcription rate that is regulated by the *lac* repressor R, as shown in Fig. 1. The production of mRNA is repressed when the *lac* repressor, R, is bound to the *lac* operator. α_1 is a constant that corresponds to the maximum rate of transcription, and $K/(K + R)$ describes the fraction of *lac* operators that are not occupied by the *lac* repressor R, where K is the rate constant of dissociation of the *lac* repressor from the *lac* operator at equilibrium (dissociation constant). When the concentration of the *lac* repressor, R, is high (much greater than K), the ratio $K/(K + R)$ is very small and, thus, there is no, or very little, production of mRNA from *lac* genes. When the concentration of R is much less than K, then $K/(K + R)$ is close to 1 and transcription proceeds at the maximum rate, α_1. The second term in the first equation of (1) describes mRNA degradation.

The activity of *lac* repressor R depends on the allolactose concentration. Thus, the repressor activity can be explicitly formulated as a function of lactose, so that the system of four equations (1) contains only four variables. The *lac* repressor is inactivated by allolactose, which is a derivative of lactose, as shown in Fig. 1. To simplify the mathematical description of *lac* repressor inhibition, we can assume that the inhibition directly depends on lactose concentration. The *lac* repressor is a complex of four subunits, and allolactose inactivates the repressor by binding to any of its subunits. Thus, the concentration of the active repressor tetramer can be described as $R_0 \left(\frac{K_0}{K_0+A}\right)^4$, where the ratio $\frac{K_0}{K_0+A}$ is the fraction of repressor subunits that are not inhibited by allolactose A, power 4 is due to the four subunits in the repressor, R_0 is the total concentration of repressor protein, and K_0 is a dissociation constant. The fraction of active repressor $\left(\frac{K_0}{K_0+A}\right)^4$ is small when allolactose concentration, A, is high. Furthermore, this fraction approaches 1 at a low concentration of allolactose. Assuming that allolactose dilution is negligible, it can be shown that, in a quasi-steady state, the concentration of allolactose equals the concentration of lactose: $A(t) = L(t)$. Therefore, the transcription rate of the *lac* operon can be written as the function of lactose concentration, i.e., $\dfrac{\alpha_1 K}{K+R_0\left(\frac{K_0}{K_0+L}\right)^4}$.

The second equation in (1) describes the dynamics of permease P, where the first term describes the translation of permease P from mRNA, and the negative term describes permease degradation/dilution. α_2 and δ_2 are the rate constants of permease translation and degradation (and/or dilution), respectively.

In the third equation in (1), we assume that β-galactosidase enzyme B is produced at a comparable rate to permease P from the same *lac* operon mRNA transcript. ¼ factor takes into consideration that the β-galactosidase enzyme is a tetrameric complex, whereas permease is a monomer.

The last equation in (1) describes the change in the intracellular lactose level L, due to the transport of external lactose L_e inside the cell by permease P (the first term in the last equation of (1)), lactose catalysis by β-galactosidase enzyme B (the second term), and lactose dilution (the third term) described by the dilution rate constant δ_3. Where α_3 is the maximum uptake rate of external lactose L_e (per permease), α_4 is the maximum rate (per β-galactosidase enzyme) at which lactose is metabolized into glucose and galactose, or converted to allolactose. The lactose uptake mediated by permease (the first term in the last equation of (1)) and the action of β-galactosidase enzyme on the lactose (the second term) are modeled using Michaelis-Menten kinetics. K_1, K_2 are Michaelis constants corresponding to concentrations of substrates (L_e, L) at which rates of lactose transportation and lactose metabolism are equal to half of the corresponding maximal rates.

We can simplify the system of equations (1) by introducing three dimensionless variables: $m(t) = (\delta_1/\alpha_1)M(t)$, $p = (\delta_1\delta_2/(\alpha_2\alpha_1))P(t)$, $l(t) = L(t)/K_0$, and dimensionless time: $\tau = \delta_3 t$. Then the *lac* operon model takes the following form:

$$\frac{dm}{d\tau} = \mu\left(\frac{1}{1 + \rho\left(\frac{1}{1+l}\right)^4} - m\right)$$

$$\frac{dp}{d\tau} = \eta(m - p) \tag{2}$$

$$\frac{dl}{d\tau} = \alpha\left(\frac{L_e}{(K_1 + L_e)} - \sigma\frac{l}{(\kappa + l)}\right)p - l,$$

where six new dimensionless parameters are defined through eleven old parameters as: $\mu = \delta_1/\delta_3$, $\rho = R_0/K$, $\eta = \delta_2\backslash\delta_3$, $\alpha = \alpha_1\alpha_2\alpha_3/(\delta_1\delta_2\delta_3 K_0)$, $\sigma = \alpha_4/(2\alpha_3)$, and $\kappa = K_2/K_0$. Only two parameters, K_1 and L_e, in the model (2) remain unchanged and have units of concentration as in the model (1). The model (2) is simpler and significantly easier to analyze than the *lac* operon model described by system (1).

Results

The *lac* operon model (2) includes 8 different parameters that can be varied to investigate different dynamic behaviors of the *lac* system. However, in order to compare model results with experimental data, we choose to analyze the *lac* operon response to the external level of lactose L_e, which can be controlled easily in a wet lab experiment with *E. coli*. The experimental data show that the *lac* operon transcription switch is 'off' when the external level of lactose is low, and turns to 'on' when the lactose level exceeds a certain threshold. Thus, the first test for model (2) will be to reproduce this switch-like behavior.

System (2) can be solved using the XPPAUT computational tool (see Computational Software chapter in this book) for a range of L_e concentrations. First, we can explore how different values of the external level of lactose L_e affect the dynamic behavior of model components. Simulation results in Fig. 2a, b show how different external lactose levels change the temporal behavior of *lac* operon mRNA, permease, and intracellular lactose concentrations. When the external lactose level $L_e = 40\mu M$, the intracellular lactose concentration reaches a steady state level that is not sufficient to activate the *lac* operon promoter and induce the production of mRNA and permease. By contrast, at $L_e = 80\mu M$, the intracellular lactose concentration increases to a level that is sufficient to activate the *lac* operon promoter and induce rapid production of mRNA and permease. This result indicates that there is a threshold value for the external lactose level L_e, which can activate the *lac* operon

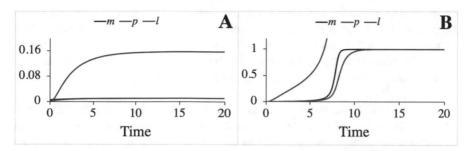

Fig. 2 Temporal evolution of model (2) variables depending on the external level of lactose L_e. (**a**) The external level of lactose $L_e = 40 \, \mu$M, which is not sufficient to activate the *lac* operon, and thus the level of mRNA remains near zero. (**b**) The external level of lactose $L_e = 80 \, \mu$M, which is sufficient to activate the *lac* operon and initiate the production of mRNA (black curve) and permease (blue curve) when the internal level of lactose (red curve) is above a certain level. Other model parameters in these simulatoins are: $K_1 = 700\mu$M, $\alpha = 400$, $\eta = 1.5$, $\kappa = 0.25$, $\mu = 0.5$, and $\rho = 200$, $\sigma = 10$

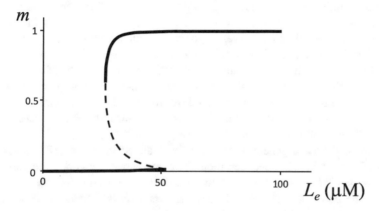

Fig. 3 One-parameter bifurcation diagram. When the external lactose concentration L_e is between 26.4μM and 51.2μM, there are three coexisting steady states: two stable steady states (solid lines) and one unstable steady state (dashed line)

promoter. This threshold concentration value is a number between 40 μM and 80 μM.

The response of the *lac* operon to the external level of lactose L_e can be characterized by the level of mRNA that is transcribed from the operon as a function of L_e. The resulting one-parameter bifurcation diagram representing the signal-response curve, is shown in Fig. 3. The model predicts that there is a physiologically relevant range for the external lactose concentration in which the *lac* operon can have three coexisting steady states. If the external level of lactose L_e is below ~51 μM, the level of mRNA transcripts from the *lac* operon remains near zero, whereas if L_e exceeds ~51 μM, the *lac* operon transcription will begin. The expression of permease and β-galactosidase enzymes enhances lactose uptake and keeps the *lac* operon in the active state. A further increase in external lactose concentration has no effect on the transcription level of *lac* genes, as the production of mRNA transcripts has already reached the maximum level. Therefore, the *lac* operon model (2) reproduces an all-or-nothing gene transcription response to varied nutrient conditions.

This model demonstrates how the *lac* gene regulatory network allows *E. coli* to realize a sensory ability. *E. coli* can sense external nutrient conditions and then generate a specific transcriptional response that allows *E. coli* to change its metabolism.

Discussion

We have learned that the *lac* gene regulatory network can produce an all-or-nothing gene transcription response tuned to a particular concentration of the nutrient, lactose. Molecular components of the *lac* gene regulatory network interact such that the molecular circuit involves a positive feedback loop when the *lac* gene products activate transcription from the *lac* operon. The *lac* operon model demonstrates how the circuit with a positive feedback loop leads to a switch-like dynamic behavior of the *lac* operon system, resulting in all-or-nothing gene transcription responses. Similar switches are realized in many other molecular and genetic mechanisms regulating different checkpoints and all-or-nothing responses of living cells to internal and external cues.

We learned how a genetic switch can be realized in living cells and we also identified the relationship between molecular interaction network structure and the type of dynamic behavior that this network exhibits. Understanding the relationship between the network structure and network dynamics is crucial for understanding how to design and build synthetic biological systems. Thus, the *lac* operon is one of those models that can teach us the engineering and design principles that can be applied in synthetic biology and biotechnology.

The mathematical model (2) is one of the simplest models of the *lac* operon. We have used several assumptions, including (a) that β-galactosidase has similar kinetics to permease, and (b) the concentration of allolactose is equivalent to the concentration of lactose. There are published *lac* operon models that include more components and details, yet with a similar goal to demonstrate the switch-like dynamic behavior of *lac* operon regulation. Model (2) can be further simplified by assuming the quasi-steady state for permease translation. This assumption will lead to a model consisting of only two equations for mRNA and lactose variables. The simplified model can also demonstrate bistability in gene translation regulation, depending on the level of the external lactose, see Exercise 1.

It should be noted that the model, based on ordinary differential equations, has some explanatory limitations. The model predicts that once a stable state is attained, then the system remains at the steady state. The state can be switched only by changing the signal (external lactose concentration). However, some experimental results demonstrate spontaneous transition between steady states, and the coexistence of both states [5]. This indicates that stochastic processes in the *lac* operon may play an important role. Therefore, a stochastic version of the *lac* operon model must be used to explain spontaneous transition between steady states.

To continue learning about *lac* operon regulation, review the literature section that provides links and references to corresponding resources.

Exercises

Exercise 1. Assume that the translation of mRNA into permease is faster than the transcription regulation, lactose uptake, and metabolism, and simplify model (2) by applying $dp/d\tau \approx 0$. The model of the *lac* operon is then described by two equations with two variables: m and l. The resulting two differential equations then can be analyzed using a phase plot (m vs. l). Simulate the simplified model of the *lac* operon and reproduce nullclines and trajectories, shown in Fig. 4.

Exercise 2. Explore how other parameters affect the dynamic behavior of mathematical model (2). Identify parameters that induce bistable model behavior.

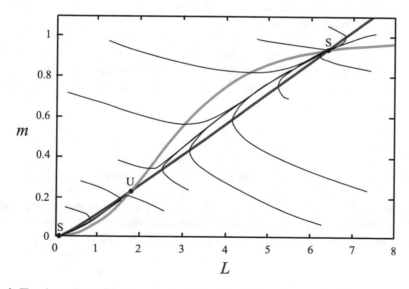

Fig. 4 The phase plane of the *lac* operon model with nullclines (green and red curves), showing three steady states labeled as S (stable) or U (unstable). Thin black lines are trajectories starting at different initial conditions and approaching either one stable steady state corresponding to active transcription and mRNA production (m is close to 1), or another stable state corresponding to repressed transcription ($m \approx 0$). These results were obtained for the external level of lactose $L_e=30\mu M$

Method

The *lac* operon model (2) was simulated using XPPAUT computational software. The XPP code below can be used to obtain all results shown in Figs. 2 and 3. This code has to be written in a text file, with file extension .ode as:

```
# lacoperon model

dm/dt=mu*(1/(1+rho*(1/(1+l))^4)-m)
dp/dt=eta*(m-p)
dl/dt=(alph*Le/(K1+Le)-sig*l/(kappa+l))*p-l

par Le=80, mu=25, eta=1.5, rho=200, alph=400, K1=700, sig=10,
kappa=0.25

done
```

To plot the temporal evolution of the variable for two different values of the Le parameter, first open Viewaxis->2D and set Ymin:0, Ymax:0.2, click OK. Next, click Param button and set Le equals to 40 in the Parameters window, then click OK button. Open the Initial Data window by clicking the ICs button, and then click GO. The black curve for the mRNA variable should appear in the XPP graphics area, as in Fig. 2a. To add another curve, open Graphic stuff->(A)dd curve and set Y-axis:L and Color:1 then click the OK button. The red curve corresponding to the intracellular lactose variable should be added in the XPP graphics area. To add the permease variable, set Y-axis:p and Color:9 in the Graphic stuff->(A)dd curve window, then click the OK button. The blue curve should be added to the graphics area. Similarly, to generate Fig. 2b results, set Le = 80 in the Parameters window, click OK and then the GO button; all three curves in the graphics area will be updated. Set Ymax:2 in the Viewaxis->2D window to expand the y axis view.

To obtain the one-parameter bifurcation diagram, as in Fig. 3, set Le equals to 10 in the Parameters window and click the OK and GO buttons. Click Initialconds->(L)ast to make sure that the steady state solution is reached. Open File->Auto, set Par1:Le in the Parameters window. In the Axes->hI-lo window, set Xmin:0, Ymin:0, Xmax:100, Ymax:1.2. Set Par Min:10 and Par Max:100 in the Numerics window. Click the Run and then Steady state buttons, and the one-parameter bifurcation diagram (identical to that in Fig. 3) should appear in the graphics area of AUTO window.

Additional Reading

The description of biological principles of *lac* operon regulation and other similar gene regulation examples can be found in all Biology textbooks usually in a regulation of gene expression chapter, for example in Refs. [6, 7].

References

1. Jacob F et al (1960) The operon: a group of genes with the expression coordinated by an operator. C R Hebd Seances Acad Sci 250:1727–1729
2. Jacob F, Ullman A, Monod J (1964) The promotor, a genetic element necessary to the expression of an operon. C R Hebd Seances Acad Sci 258:3125–3128
3. Santillan M, Mackey MC, Zeron ES (2007) Origin of bistability in the lac Operon. Biophys J 92 (11):3830–3842
4. Ingalls, B.P.(2013) Mathematical modeling in systems biology an introduction. MIT Press, Cambridge, p 1 online resource (xiv, 408 pages).
5. Ozbudak EM et al (2004) Multistability in the lactose utilization network of *Escherichia coli*. Nature 427(6976):737–740
6. Raven, P.H., et al. (2014) Biology, Vol. 1. Tenth Edition, AP Edition. McGraw-Hill, Dubuque, Iowa (various pagings)
7. Russell, P.J., P.E. Herz, and B. McMillan (2017) Biology: the dynamic science, Vol. 1. Fourth edition. Cengage Learning, Boston, MA, (various pagings)

Fate Decisions of CD4$^+$ T Cells

Andrew Willems and Tian Hong

Introduction

In vertebrates, the immune response against pathogens is mounted by the innate immune system and the adaptive immune system [1–3]. The latter form of immune response involves immunological memory which allows the system to recognize subsequent invasions of the pathogens following their initial response and activate enhanced responses. This memory endows long-term protection from pathogens and serves as the basis for vaccination. T cells, or T lymphocytes, play key roles in the adaptive immune system. They are named based on their developmental origin: their T cell lineage is determined in the thymus gland [4]. The antigen receptors on the surface of T cells, known as T cell receptors (TCR), are highly specific to the type of pathogen that they are combating. The diversity of these receptors is achieved by somatic hypermutation [5, 6] and genetic recombination, which generate a large number of antigen receptors. Specific recognition of antigens by TCR initiates the activation of the T cell and subsequent responses, including direct or indirect elimination of the infected cells. In addition to the receptors that recognize antigens from pathogens, T cells also have co-receptors that facilitate the antigen-specific activation. Some of these co-receptors serve as surface markers that distinguish two major types of T cells. T cells that express co-receptor CD8 on their surfaces (CD8$^+$ T cells) are cytotoxic T cells which are responsible for killing infected cells directly. CD4$^+$ T cells are helper T cells that eliminates pathogens indirectly by activating

A. Willems
Genome Science and Technology Program, The University of Tennessee, Knoxville, TN, USA

T. Hong (✉)
Department of Biochemistry & Cellular and Molecular Biology, The University of Tennessee, Knoxville, TN, USA
e-mail: hongtian@utk.edu

© Springer Nature Switzerland AG 2021
P. Kraikivski (ed.), *Case Studies in Systems Biology*,
https://doi.org/10.1007/978-3-030-67742-8_11

other parts of the immune system, including $CD8^+$ T cells and innate immune cells such as macrophages and natural killer cells [7–9]. Within the class of $CD4^+$ T cells, there exists several types of T helper cells, e.g. T helper 1 cells (Th1), T helper 2 cells (Th2), and T helper 17 (Th17) cells [10]. The differentiation (i.e. lineage specification) of these T helper cell types marks the final maturation of these T cells upon which they become functional. The differentiation of $CD4^+$ T cells is triggered by TCR activation, but the specification towards specific lineages is also determined by other signals such as cytokines [11], which are small immunomodulating molecules released from immune cells. For example, the differentiation of Th2 cells requires activations of TCR (antigen-specific receptor), CD3 (co-receptor), CD28 (secondary stimulatory receptor) and interleukin-4 (IL-4, a cytokine) receptor, whereas IL-12 and interferon-γ (IFN-γ) bias the differentiation toward Th1 cells in the absence of IL-4 [12].

The goal of this chapter is to use the differentiation of $CD4^+$ T cells as a paradigm to illustrate how gene regulatory networks (GRNs) control the dynamics of cell differentiation. We will address the following questions in terms of the differentiation using simple mathematical models: (1) how robust fate decisions of T cells are made with the help of GRN; (2) how a population of undifferentiated T cells differentiate to a population of cells with diverse molecular signatures and functions; and (3) how strengths of the TCR and cytokine signals influence the differentiated population of T cells. Insights into these questions will not only help to elucidate the understanding of T cell biology, but also shed light on the fundamental mechanisms underlying the complex immune responses.

Physiology and Molecular Mechanisms

$CD4^+$ T cells coordinate the immune systems by communicating with a wide range of immune cells. These T cells perform this function primarily by releasing cytokines. Each type of $CD4^+$ T cells produce and release a unique set of cytokines, known as signature cytokines, which specifically activate or inhibit the activities of certain groups of immune cells. There are at least four types of $CD4^+$ T cells. Th1 cells are primarily responsible for clearing intracellular infection. Th2 cells lead to response against extracellular parasites. Th17 cells are used to eliminate extracellular pathogens and fungi. There is a distinct group of $CD4^+$ T cells called regulatory T cells (Treg cells) that serve as immunosuppressors, which modulate the immune system and prevent autoimmune diseases. More recently, other types of $CD4^+$ T cells, such as follicular T cells (Tfh cells), Th9 cells and Th22 cells, were also discovered [13, 14]. In addition to their unique functions and cytokine profiles, different types of $CD4^+$ T cells also exhibit distinct gene expression patterns in general, indicating that large gene regulatory programs are required to achieve cell-type specification. It has been shown that the differentiation of each type of $CD4^+$ T cell is governed by the upregulation of at least one transcription factor (TF) that is considered the master regulator of the cell type. Well-known master regulators

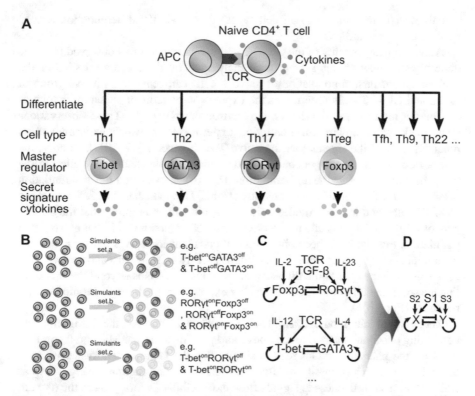

Fig. 1 Overview of CD4⁺ T cell differentiation and network diagram for modeling. (**a**) Illustration of cell fate choice of a naïve CD4⁺ T cell upon the simulation of TCR by an antigen-presenting cell (APC) and the influence of surrounding cytokines. Four examples of differentiated cell types and their master regulators are shown. (**b**) Examples of scenarios in which a population of naïve CD4⁺ T cells are differentiated into a heterogeneous population. Red and green cells represent single-positive cells where one master regulator is upregulated. Yellow cells represent double-positive cells where two master regulators are upregulated. (**c**) Examples of regulatory networks controlling dual-master-regulator differentiation systems. (**d**) A generic network summarizing gene networks for CD4⁺ T cell differentiation. X and Y represent two master regulators. S1 represents the common activator (e.g. TCR). S2 and S3 are polarizing signals representing two sets of cytokines

include T-bet (Th1), Gata3 (Th2), RORγt (Th17) and Foxp3 (Treg) [12]. Upon the stimulation of TCR and other receptors on an undifferentiated CD4⁺ T cell, known as the naïve T cell, a specific master regulator is upregulated, and this activation leads to changes of the transcriptional profile in the T cell, including the upregulation of the genes that encode the signature cytokines of a T cell type (Fig. 1a). The differentiation is irreversible with respect to the withdrawal of the stimulants, suggesting the robustness of the cell type specification [15, 16]. The microenvironment of the naïve T cell, which includes the types and amounts of cytokines available during the differentiation, is critical for determining the generated cell type. These cytokines are known as the polarizing signals (as opposed to TCR signals which are common to all cell types) that guide the differentiation toward specific cell types [13].

Although irreversibility of the CD4$^+$ T cell differentiation was observed [16], the deterministic view of this system was challenged by three characteristics of the differentiation. First, a population of naïve T cells can generate a diverse group of functional CD4$^+$ T cells in terms of the expression patterns of signature cytokines and master regulators in the same microenvironment (Fig. 1b). These observations were made with in vitro differentiation experiments in which experimentalists precisely control the signals that the naïve T cells receive [17–20]. Secondly, the composition of this heterogeneous cell population can be changed by altering the polarizing signals before the differentiation (Fig. 1b), or even after the differentiation is completed if the signals are sufficiently strong. This indicates the plasticity of the CD4$^+$ T cells, and it also provides evidence against the possibility that the heterogeneous differentiation is due to the predetermined diversity within the naïve T cell population. Finally, it has been observed that cells that co-express master regulators of multiple cell types (e.g. RORγt and Foxp3) stably exist in vitro and in vivo [21–23]. This further suggests the diversity of the CD4$^+$ T cells is beyond the model of lineage specification toward cells that stably express a single lineage-defining master regulator (Fig. 1b).

The goal of our models for CD4$^+$ T cells is to capture the intriguing and perplexing observations mentioned above and provide a mechanistic and theoretical basis for the phenomena. Our idea is simply that the gene regulatory networks in these T cells, which primarily involve interactions among the master regulators as well as their responsiveness to signals from extracellular factors, govern the dynamics of the differentiation, including heterogeneity of the derived cell population, the tunability of the population composition based on the signal types and strengths, and the existence of stably co-expressed master regulators.

How do different master regulators of different CD4$^+$ T cell types interact with each other? It was found that two master regulators typically have a cross-repressive relationship: they inhibit the expression of each other. For example, it was shown that T-bet and Gata3 mutually inhibit each other and form a double-negative (i.e. positive) feedback loop. In addition to this feedback loop, the master regulators often involve other forms of positive feedback, including self-activation through transcriptional upregulation or autocrine regulation (master regulator activated by cytokine released from the same cells). These gene regulations form a cross-repression-autoactivation network motif (Fig. 1c, d) that is common in CD4$^+$ T cells as well as other systems involving binary cell fate decisions. In addition to this network motif involving master regulators, CD4$^+$ T cells have a recurring signaling structure connecting extracellular simulants to master regulators: the TCR stimulation, which serves as the main activation signal for differentiation, activates all master regulators, whereas each cytokine in the microenvironment of the cells polarizes the differentiation toward specific lineages by activating a unique master regulator. This network structure is illustrated in Fig. 1d in the dual-master-regulator paradigm. A common activator TCR signal upregulates two master regulators, and two representative polarizing signals (cytokines) activate the two master regulators respectively. In the next section, we build simple mathematical models based on this recurring network structure in CD4⁺ T cells and examine how this regulatory network controls the dynamics of differentiation in general.

Mathematical Models and Results

As shown in the influence diagram (Fig. 1d), we start our analysis with the generic differentiation network motif which serves as a unifying framework for understanding CD4⁺ T cell differentiation. Once we have obtained a basic understanding on this motif, we can connect the framework to specific models and simulations that can be compared with experiments.

Master Regulator Cross-Repression with TCR Signal

We first consider a core subnetwork within the proposed cell differentiation motif: a mutual inhibition, double-negative feedback loop between master regulators X and Y, both of which are activated by TCR signal. This simplified system can be described by the following ODEs:

$$\frac{dX}{dt} = \gamma_X \left(F \left(\omega_X^o + \omega_{XY} Y + s_1 + s_2 \right) - X \right)$$

$$\frac{dY}{dt} = \gamma_Y \left(F \left(\omega_Y^o + \omega_{YX} Y + s_1 + s_3 \right) - Y \right)$$

$$F(W) = \frac{1}{\left(1 + e^{-\sigma W} \right)} \tag{1}$$

where, X and Y represent the dimensionless activities of factors X and Y respectively. The nonlinear function $F(W)$ is a sigmoidal function that describes the activation of a factor under the influence of other factors, and the ODE indicates that this function also describes the steady state activity of the factor. Parameter γ is the timescale of the change of a factor. σ describes the nonlinearity of the function F. W denotes the overall influence (weight) of other factors on a factor, which has taken an offset parameter ω^o into account. ω^o also determines the basal activity of the factor in the absence of any other factors. Importantly, ω_{XY} is a weight parameter describing the strength of the inhibition of X by Y, and $\omega_{XY}Y$ is the effective weight with given activity of Y. ω_{YX} represents the strength of the other inhibitory regulation in the feedback loop. s_1 is a parameter representing the strength of the TCR signal. In this particular model, the overall weight W for X is the summation of ω_X^o, $\omega_{XY}Y$ and s_1, reflecting the assumption that gene X is only influenced by Y and signal S_1. We provide a set of generic parameter values to this model, and they are

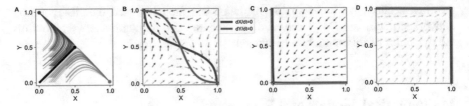

Fig. 2 Simulation and analysis of the cross-repression model. (**a**). Model was simulated with TCR signal $s_1 = 2$. Black: simulation trajectory under a perfectly symmetrical (X = 0.01, Y = 0.01) initial condition. Red and green: simulation trajectories under randomly chosen initial conditions in the $X^{low}Y^{low}$ region representing naïve T cells. Dots show the steady states of the simulations. (**b**) Phase plane of the system. Red and green arrows represent vectors in two basins of attractions. Yellow arrows show vectors at the separatrix. Blue and indian red curves show nullclines of X and Y respectively. (**c**) Phase plane without TCR signal ($s_1 = 0$). (**d**) Phase plane with strong TCR signal ($s_1 = 5$)

symmetrical with respect to X and Y: $\gamma_X = \gamma_Y = 1, \omega_X^o = \omega_Y^o = -1, \omega_{YX} = \omega_{XY} = -2, \sigma_X = \sigma_Y = -2$. We set the initial conditions to be a state where X and Y are equally low ($X = Y = 0.01$), and we set the strength of the signal S1 to 2 ($s_1 = 2$), mimicking the stimulation of the TCR upon antigen binding. Note that this form of equations is similar to those with Hill functions, and it has been used to describe complex gene regulatory networks for processes such as epithelial-mesenchymal transition [24].

Solving the ODEs numerically with these parameter values and initial conditions gives rise to trajectories of X and Y as functions of time, but it is helpful to visualize the trajectories in the space of X and Y (Fig. 2a). Under the conditions described above, the trajectory of the simulation starts from an $X^{off}Y^{off}$ state and stays in the region before S1 stimulation. When the S1 signal is raised to 1, the cell state starts to change with increasing X and Y, and it is stabilized at a state with intermediate levels of X and Y at the end of the simulation (Fig. 2a). We simply define that a cell is differentiated when either X or Y is above 0.5. In this simulation, we have obtained a differentiated cell that co-express X and Y. Does this cellular phenotype match any experimental observation? Before we make a conclusion, we examine the robustness of the steady state phenotype with respect to slight changes of the initial conditions. To perform this analysis in XPPAUT, we use the 'mice' option under 'Initialconds'. By repeatedly clicking in the bottom-left region in the X-Y space, we simulate a small population of cells, each of which express low levels of X and Y with a slight cell-to-cell variability in terms of the initial states, reflecting a biologically plausible scenario. With this procedure, we observe that the simulated cells are stabilized at either the $X^{on}Y^{off}$ state or the $X^{off}Y^{on}$ state, representing two canonical types of CD4$^+$ T cell, at the end of the simulations (Fig. 2a). This is an interesting and biologically meaningful observation, because it shows that with the same simulant, a population of naïve CD4$^+$ T cells that are essentially identical can be differentiated into a population with remarkable diversity in terms of the expression of master regulators. This scenario was observed experimentally: activating

TCR without supplying additional cytokine led to a heterogeneous population containing both Th1 and Th2 cells [25].

What is the mechanistic basis of this phenomenon from the viewpoint of dynamical systems? To answer this question, we take the advantage of the model's simplicity, specifically the amenable features of the 2-dimensional system, and use phase plane analysis (see the XPPAUT section of Computational Software chapter of this textbook) to identify key elements in this dynamical system, including vector field, nullclines, steady states and their stabilities. To find the nullclines in XPPAUT, we first erase the existing trajectories by clicking on 'Erase'. Suppose we would like to understand the features of the system in the presence of the TCR signal. We set the S1 signal to 1. We then click on 'Nullcline' and 'New' buttons, and two nonlinear curves are shown in the X-Y plane (Fig. 2b). These two curves represent the positions where dX/dt=0, and dY/dt=0, respectively, and their intersections represent the steady states of the system. In this case, three steady states are governed by the structure of the nullclines. To view the vector field of the system, we first click on the 'Dir.field/flow' and the 'Direct field' buttons, and then we choose a suitable number of elements in the grid (e.g. 20). As a result, a vector field is shown, and each arrow in the field represents the direction of the solution (i.e. the trajectory as the system moves forward with time) at a particular state in the X-Y space (Fig. 2b). The stabilities of the three steady states can be inferred from the vector field: most of the vectors close to the two 'polarized' $X^{on}Y^{off}$ and $X^{off}Y^{on}$ states are pointing toward them, whereas most of the vectors close to the co-expressing state are pointing away from it. Therefore, $X^{on}Y^{off}$ and $X^{off}Y^{on}$ states are stable, representing long term behaviors of cellular phenotype, whereas the $X^{on}Y^{on}$ state is unstable and therefore only represents a transient state of cells. We can confirm the inferred stabilities in a more rigorous manner by using 'Sing pts' and 'Mouse' buttons in XPPAUT to examine the interactions one at a time. The output will show us the stability of each steady state by calculating the eigenvalues of the Jacobian matrix at the state. The two stable steady states are known as 'attractors' of the system, and their basins of attractions are separated by a curve called separatrix which happens to pass the $X^{low}Y^{low}$ region in this model (Fig. 2b, yellow). As such, slight cell-to-cell variability in the naïve cells tips them into distinct basins of attractors, which leads the cell differentiation to two stable 'polarized' phenotypes.

Our initial model successfully explains the basis of heterogeneous differentiation, but how can we explain the observed stable hybrid cell states in CD4$^+$ cells (e.g. cells co-expressing T-bet and Gata3 and those co-expressing RORγt and Foxp3)? To address this question, we vary the TCR signal in a wide range of strengths (e.g. 0–3). Unsurprisingly, lowering the TCR signal leads to the disappearance of the $X^{on}Y^{off}$ and $X^{off}Y^{on}$ states and a stable $X^{off}Y^{off}$ state, representing the naïve cell, is generated (Fig. 2c). In contrast, increasing the TCR signal generates a system with a single stable state with a $X^{on}Y^{on}$ phenotype, which may correspond to the 'double-positive' cells observed experimentally [17, 23] (Fig. 2d). However, this model would predict that the double-positive cells only appear in the presence of strong TCR signals, which do not allow the formation of single-positive cells. This prediction is inconsistent with the observations that the double-positive CD4$^+$ cells often coexist with the single-positive ones under uniform culturing conditions [17]. To capture this phenomenon, we need to consider a model that contains additional elements included in the network motif (Fig. 1d), namely the autoactivation of the master regulators.

Master Regulator Cross-Repression and Autoactivation with TCR Signal and Polarizing Cytokines

To add the autoactivation to our initial model, we modify the weight W in both ODEs:

$$W_X = \omega_X^o + \omega_{XY}Y + \omega_{XX}X + s_1$$

$$W_Y = \omega_Y^o + \omega_{YX}Y + \omega_{YY}Y + s_1 \tag{2}$$

where, $\omega_{XX}X$ describes the influence of factor X by itself. To examine the effect of the autoactivation, we set the weight parameters ω_{XX} and ω_{YY} to 1.5, and we perform the analysis of the nullclines of this new system. By varying the S1 signal strength (s_1), the two nullclines may have 1, 3, or 5 intersections, depending on the choice of s_1. While the scenarios of 1 and 3 intersections (i.e. 1 or 2 stable steady states) are similar to what we observed with our first model, there are two ranges of s_1 that allows for 5 interactions of the nullclines, an observation unique to this model (e.g. Fig. 3a). By examining the stability of the steady states, we find that certain

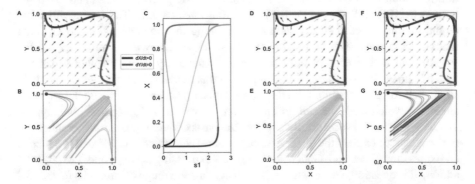

Fig. 3 Simulation and analysis of the cross-repression and autoactivation model. (**a**) Phase plane under strong TCR signal condition ($s_1 = 2.05$). (**b**) Simulation trajectories under strong TCR signal condition. Initial conditions were randomly chosen in the $X^{low}Y^{low}$ region. (**c**) Bifurcation diagram showing the steady state of X with respect to s1. Thin curves represent unstable steady states. Thick curves represent unstable steady states. (**d**) Phase plane under strong TCR signal condition in the presence of polarizing signal S2 ($s_2 = 0.05$). (**e**) Simulation trajectories under strong TCR signal condition in the presence of polarizing signal s2. Initial conditions were randomly chosen in the $X^{low}Y^{low}$ region. (**f**) Phase plane under strong TCR signal condition in the presence of polarizing signal S3 ($s_3 = 0.05$). (**g**) Simulation trajectories under strong TCR signal condition in the presence of polarizing signal s_3. Initial conditions were randomly chosen in the $X^{low}Y^{low}$ region

low levels of s_1 allow the coexistence of $X^{off}Y^{off}$, $X^{on}Y^{off}$ and $X^{off}Y^{on}$ stable states (see Exercises), whereas certain high levels of s_1 allow the coexistence of $X^{on}Y^{on}$, $X^{on}Y^{off}$ and $X^{off}Y^{on}$ stable states (Fig. 3a) (i.e. two types of 'tri-stable' systems). The tri-stable system with low levels of s_1 may be important for the irreversibility of T cell differentiation (see Exercises), and the tri-stable system with high levels of s_1 may provide a mechanistic basis of for the observations that double-positive CD4$^+$ cells coexist with the single-positive ones. The latter statement is supported by the fact that the separatrices (note that we have three basins of attractors in this scenario) are still adjacent to the $X^{low}Y^{low}$ region. Therefore, when we simulate a cell population with slight cell-to-cell variability in terms of initial conditions, we obtain a diverse population of differentiated cells containing three phenotypes, due to the small initial variations which tip the cells to different basins of attractions (Fig. 3b).

The changes of the steady states of the system with respect to S1 signal can be succinctly delineated by a one-parameter bifurcation diagram (see the XPPAUT section in Computational Software chapter of this textbook). In XPPAUT, bifurcation diagrams can be generated in the 'AUTO' window under the 'File' option. Note that to get these bifurcation diagrams we need to use an equilibrium point that has been calculated from your solution of ODEs. In the 'AUTO' window, click on 'Run' 'Steady state' buttons. In Fig. 3c for example, steady state levels of X (i.e. intersections of the nullclines) are plotted with respect to s_1. Thick curves in the diagram represent stable steady states, whereas thin curves represent unstable ones. With the increase of s_1, the system makes four switches in terms of the number of stable steady states: (1) 1-to-3 ($s_1 = 0.07164$); (2) 3-to-2 ($s_1 = 0.4831$); (3) 2-to-3 ($s_1 = 2.017$); and (4) 3-to-1 ($s_1 = 2.428$). This is consistent with our phase-plane analysis, and this diagram captures all qualitative changes of the system without the need of changing s_1 manually.

What is the role of the polarizing signals (cytokines) in shaping the dynamical system and the cell differentiation? This question can be addressed by simplify making another modification to the weight W in the ODEs

$$W_X = \omega_X^o + \omega_{XY} Y + \omega_{XX} X + s_1 + s_2$$
$$W_Y = \omega_Y^o + \omega_{YX} Y + \omega_{YY} Y + s_1 + s_3 \tag{3}$$

where, parameters s_2 and s_3 represent polarizing signals that bias differentiation toward X expression and Y expression respectively. We use phase-plane and simulations again to show the effect of adding a polarizing signal, e.g. s_2, to the system. Raising s_2 from 0 to 0.05 results in a slight shift of vector field and basins of attractions which favor X expression (Fig. 3d). This symmetry breaking effect gives rise to a biased cell population if we randomly choose the initial conditions for the naïve in the $X^{low}Y^{low}$ region (Fig. 3e). Increasing s_3 while holding s_2 constant has the opposite effect (Fig. 3f, g). This analysis shows that the polarizing signals (cytokines) can change the compositions of the heterogeneous cell population by breaking the symmetry of the system.

Our model parameters are assumed to be symmetrical with respect to X and Y, except for the polarizing signals. Perfectly symmetrical gene regulation is an unlikely scenario in biology. In fact, any changes to the internal gene regulatory parameters (e.g. ω_{YX}) will give rise to an asymmetrical system in a manner similar to the effect of polarizing signals. This explains the observation that the double-positive cells often coexist with only one of the two types of single-positive cells.

Discussion

In this case study, we have used mathematical models to show how CD4$^+$ T cells differentiate to functional cell types. Specifically, we show that a simple gene regulatory network containing cross-repression and autoactivation of two lineage defining master regulators can govern differentiation of naive CD4$^+$ T cells into a heterogeneous population with distinct subsets of cells in terms of the master regulator expression. In addition, the model captures the existence of the double-positive cell phenotype which has high expression levels of more than one master regulator [26, 27]. Although different types of CD4$^+$ T cells have specific roles in eliminating different pathogens, it has been proposed that the diversity and the flexibility of the CD4$^+$ T cell populations may be important for the 'agility' of the immune system [28]. Therefore, the robustness formation of the heterogeneous, functional CD4$^+$ T cells, as well as the double-positive phenotype, may serve as the cellular basis of this feature of the immune system. In terms of the extracellular signals, the models show that T cell receptors trigger the heterogeneous differentiation, but the polarizing cytokines can bias the differentiation toward specific types. Therefore, the system has a remarkable capacity of generating either diverse or uniform cell populations depending on the microenvironment of the cells.

We use a generic description of master regulators and signals in our models. This is because the network motif represents several sub-systems of CD4$^+$ T cells with similar structures of gene regulatory network (Fig. 1c). If one needs to obtain a more quantitative model for a specific sub-system, then this generic model can be optimized through, for example, fitting parameters to quantitative measurements. In addition, the cross-repression-autoactivation network motif has been found in several other developmental systems in which binary cell fate decisions are made in a robust manner [29]. Understanding this motif will help to reveal the principles underlying a wide range of biological systems.

In these simple models, only two master regulators are considered. A more realistic description of the CD4$^+$ T cells must include the interactions of more than two master regulators. Previous studies have provided this type of holistic views of CD4$^+$ T cells [30, 31], and these models capture many phenotypes that the simple dual-regulator models cannot describe, including the possible existence of triple-positive phenotypes. In terms of modeling strategy, we have used ODE based models. As an alternative approach, Boolean-network based models have been applied to CD4$^+$ T cell differentiation, with a more comprehensive description of players but less detailed kinetics in the system [32].

More recent studies of CD4⁺ T cells have shown the possibility that these cells are in the continuum of the space of master regulator expression rather than discrete states [33]. The model presented in this chapter cannot describe the expression continuum due to the simplicity of the gene networks. Taking more genes and the stochasticity of their expression into account would provide a more accurate and quantitative model of the CD4⁺ T cells. Nonetheless, our models show that the interactions between master regulators can shape the attractors of the system, and these attractors guide the differentiation of these T cells, which may be diverse at the population level but robust in terms of decisions of individual cells. This may serve as a unifying principle governing differentiation of CD4⁺ T cells and other systems.

Exercises

Exercise 1. Reversibility of cell differentiation.

(1) Generate a one-parameter bifurcation diagram for the first model (the one without autoactivation). This diagram should show the steady state of X with respect to the strength of signal S1 (similar to Fig. 3c). Simply modify the code above by changing the ω_{XX} and ω_{YY} to 0, and increase the range of the parameter for bifurcation analysis to 0-4 (PARMIN=4, AUTOMAX=4) for a better view of the diagram. Run the system to the steady state, and then run bifurcation analysis as described in the Results.

(2) Compare the diagram generated in (1) with Fig. 3c in the low-middle range of s_1. Predict the roles of the autoactivation on the reversibility for the differentiation of $X^{on}Y^{off}$ and $X^{off}Y^{on}$ phenotypes when the signal first increases and then decreases.

(3) Add the following lines of code to the existing code:

```
global 1 {ta-10} {s1=0.1}
ta=t
```

The two lines describe the scenario mentioned in (2): the signal is reduced to 0.1 at Time 10. Run simulations under the conditions $\omega_{XX} = \omega_{YY} = 0$ and $\omega_{XX} = \omega_{YY} = 1.5$ separately, and validate the prediction you make in (2). To run the simulations, use the 'mouse' option and click on an initial condition in the lower-left corner of the X-Y space.

Exercise 2. An alternative description of gene expression variability.

In this chapter we have used different initial conditions to describe cell-to-cell variability. This variability can be described by other methods as well. For example, there exist random fluctuations in gene expression, which can tip the cells into different basins of attractions. Implement this alternative way to describe variability by replacing the two equations in Code 1 with the following lines:

```
dX/dt=1/(1+exp(-sigmaX*(w0X+wXY*Y+s1+wXX*X+s2)))-X+wX*eX
dY/dt=1/(1+exp(-sigmaY*(w0Y+wYX*X+s1+wYY*Y+s3)))-Y+wY*eY
wiener wX, wY
```

Next, add the following parameter values to the bottom of the parameter value list:

par eX=0.02, eY=0.02. The two parameters describe the amplitude of the fluctuations in gene expression. Simulate multiple cells again with the 'Master regulator cross-repression and autoactivation' model. Do you observe heterogeneity in the differentiated population if you start the simulations and the same initial condition? What is the effect of increasing or decreasing the amplitude of the noise on the population composition?

Methods

The XPPAUT code to analyze the models in this case study is shown below.

Code 1

```
dX/dt=1/(1+exp(-sigmaX*(w0X+wXY*Y+s1+wXX*X+s2)))-X
dY/dt=1/(1+exp(-sigmaY*(w0Y+wYX*X+s1+wYY*Y+s3)))-Y

init X=0.001
init Y=0.001

#par s1=2.05
par s1=0
#par s1=2
#par s1=5

par s2=0.00, s3=0.00
par sigmaX=5
par sigmaY=5
par w0X=-1.0
par w0Y=-1.0
par wXY=-2
par wYX=-2

par wXX=1.5
#par wXX=0
# set to 0 for the 1st model

par wYY=1.5
#par wYY=0
# set to 0 for the 1st model

@ dt=0.01, total=100, method=stiff, maxstore=10000000,
bound=100000000
@ XP=X, YP=Y, XLO=-0.1, YLO=-0.1, XHI=1.1, YHI=1.1, NMESH=500
@ AUTOYMIN=-0.1, AUTOYMAX=1.1, PARMIN=0, PARMAX=3, AUTOXMIN=0,
AUTOXMAX=3
@ DSMIN=0.0001, DS=0.002, DSMAX=0.1, NMAX=500, NPR=200

done
```

References

1. Yuan S, Tao X, Huang S, Chen S, Xu A (2014) Comparative immune systems in animals. Annu Rev Anim Biosci 2(1):235–258
2. Medzhitov R, Janeway C Jr (2000) Innate immunity. N Engl J Med 343(5):338–344
3. Bonilla FA, Oettgen HC (2010) Adaptive immunity. J Allergy Clin Immunol 125(2):S33–S40
4. Miller JFAP (2002) The discovery of thymus function and of thymus-derived lymphocytes. Immunol Rev 185(1):7–14
5. Papavasiliou FN, Schatz DG (2002) Somatic hypermutation of immunoglobulin genes: merging mechanisms for genetic diversity. Cell 109(2):S35–S44
6. Tonegawa S (1983) Somatic generation of antibody diversity. Nature 302(5909):575–581
7. Bourgeois C, Tanchot C (2003) Mini-review CD4 T cells are required for CD8 T cell memory generation. Eur J Immunol 33(12):3225–3231
8. Bonecchi R, Sozzani S, Stine JT, Luini W, D'Amico G et al (1998) Divergent effects of interleukin-4 and interferon-γ on macrophage-derived chemokine production: an amplification circuit of polarized T helper 2 responses. Blood, The Journal of the American Society of Hematology 92(8):2668–2671
9. Takahashi H, Amagai M, Tanikawa A, Suzuki S, Ikeda Y et al (2007) T helper type 2-biased natural killer cell phenotype in patients with pemphigus vulgaris. J Invest Dermatol 127 (2):324–330
10. O'Shea JJ, Paul WE (2010) Mechanisms underlying lineage commitment and plasticity of helper CD4+ T cells. Science 327(5969):1098–1102
11. Zhang J-M, An J (2007) Cytokines, inflammation and pain. Int Anesthesiol Clin 45(2):27
12. Zhu J, Paul WE (2010) Peripheral CD4+ T-cell differentiation regulated by networks of cytokines and transcription factors. Immunol Rev 238(1):247–262
13. Zhu J, Yamane H, Paul WE (2009) Differentiation of effector CD4 T cell populations. Annu Rev Immunol 28:445–489
14. Azizi G, Yazdani R, Mirshafiey A (2015) Th22 cells in autoimmunity: a review of current knowledge. Eur Ann Allergy Clin Immunol 47(4):108–117
15. Lee HJ, Takemoto N, Kurata H, Kamogawa Y, Miyatake S et al (2000) GATA-3 induces T helper cell type 2 (Th2) cytokine expression and chromatin remodeling in committed Th1 cells. J Exp Med 192(1):105–116
16. Murphy E, Shibuya K, Hosken N, Openshaw P, Maino V et al (1996) Reversibility of T helper 1 and 2 populations is lost after long-term stimulation. J Exp Med 183(3):901–913
17. Zhou L, Lopes JE, Chong MMW, Ivanov II, Min R et al (2008) TGF-β-induced Foxp3 inhibits TH 17 cell differentiation by antagonizing RORγt function. Nature 453(7192):236–240
18. Ghoreschi K, Laurence A, Yang X-P, Tato CM, McGeachy MJ et al (2010) Generation of pathogenic TH 17 cells in the absence of TGF-β signalling. Nature 467(7318):967–971
19. Lochner M, Peduto L, Cherrier M, Sawa S, Langa F et al (2008) In vivo equilibrium of proinflammatory IL-17+ and regulatory IL-10+ Foxp3+ RORγt+ T cells. J Exp Med 205 (6):1381–1393
20. Antebi YE, Reich-Zeliger S, Hart Y, Mayo A, Eizenberg I et al (2013) Mapping differentiation under mixed culture conditions reveals a tunable continuum of T cell fates. PLoS Biol 11(7): e1001616
21. Hegazy AN, Peine M, Helmstetter C, Panse I, Fröhlich A et al (2010) Interferons direct Th2 cell reprogramming to generate a stable GATA-3+ T-bet+ cell subset with combined Th2 and Th1 cell functions. Immunity 32(1):116–128
22. Voo KS, Wang Y-H, Santori FR, Boggiano C, Wang Y-H et al (2009) Identification of IL-17-producing FOXP3+ regulatory T cells in humans. Proc Natl Acad Sci 106(12):4793–4798
23. Peine M, Rausch S, Helmstetter C, Fröhlich A, Hegazy AN et al (2013) Stable T-bet+ GATA-3+ Th1/Th2 hybrid cells arise in vivo, can develop directly from naive precursors, and limit immunopathologic inflammation. PLoS Biol 11(8):e1001633

24. Watanabe K, Panchy N, Noguchi S, Suzuki H, Hong T (2019) Combinatorial perturbation analysis reveals divergent regulations of mesenchymal genes during epithelial-to-mesenchymal transition. NPJ systems biology and applications 5(1):1–15
25. Yamashita M, Kimura M, Kubo M, Shimizu C, Tada T et al (1999) T cell antigen receptor-mediated activation of the Ras/mitogen-activated protein kinase pathway controls interleukin 4 receptor function and type-2 helper T cell differentiation. Proc Natl Acad Sci 96 (3):1024–1029
26. Hong T, Xing J, Li L, Tyson JJ (2011) A mathematical model for the reciprocal differentiation of T helper 17 cells and induced regulatory T cells. PLoS Comput Biol 7(7):e1002122
27. Hong T, Xing J, Li L, Tyson JJ (2012) A simple theoretical framework for understanding heterogeneous differentiation of CD4+ T cells. BMC Syst Biol 6(1):66
28. Schrom EC, Graham AL (2017) Instructed subsets or agile swarms: how T-helper cells may adaptively counter uncertainty with variability and plasticity. Curr Opin Genet Dev 47:75–82
29. Enver T, Pera M, Peterson C, Andrews PW (2009) Stem cell states, fates, and the rules of attraction. Cell Stem Cell 4(5):387–397
30. Mendoza L, Xenarios I (2006) A method for the generation of standardized qualitative dynamical systems of regulatory networks. Theor Biol Med Model 3(1):13
31. Hong T, Oguz C, Tyson JJ (2015) A mathematical framework for understanding four-dimensional heterogeneous differentiation of CD4+ T cells. Bull Math Biol 77(6):1046–1064
32. Naldi A, Carneiro J, Chaouiya C, Thieffry D (2010) Diversity and plasticity of Th cell types predicted from regulatory network modelling. PLoS Comput Biol 6(9):e1000912
33. Eizenberg-Magar I, Rimer J, Zaretsky I, Lara-Astiaso D, Reich-Zeliger S et al (2017) Diverse continuum of CD4+ T-cell states is determined by hierarchical additive integration of cytokine signals. Proc Natl Acad Sci 114(31):E6447–E6456

Stochastic Gene Expression

Jing Chen

Introduction

Cellular processes are intrinsically noisy. Molecules in a cell constantly jiggle, wiggle and bounce against each other. These erratic movements intensify as temperature rises; they are termed as thermal motion or thermal fluctuation, and the associated energy as thermal energy. The most famous example of the effect of thermal motion is the Brownian motion, where random jostling of small particles (<0.1μm) results from thermal motion of water molecules and their incessant bumping into the particles. Besides random spatial displacements of particles or molecules, thermal motion encompasses all kinds of random movements, such as random conformational fluctuations in molecules and random collision between molecules. Collectively, thermal motion of a molecule causes randomness in the biochemical reactions participated by the molecule. Take the gene expression process as an example. Transcription of a gene does not happen at regular time intervals like on a factory assembly line, but rather at random times. Likewise, decay of the gene product, e.g., mRNAs and proteins, happens at random times. Such randomness ultimately leads to randomness in the dynamics of cellular processes.

Gene expression is a particularly interesting and intensively studied stochastic cellular process, because stochastic effects are especially prominent for gene expression at the single-cell level. Generally speaking, the relative level of randomness in a system scales with the inverse square root of the size of the system (~ number of molecules), which will be shown by one of the results in this chapter. In the macroscopic world of our everyday experience, objects are made of gazillions of molecules and hence the associated stochasticity is negligible. But in the gene

J. Chen (✉)
Department of Biological Sciences, Academy of Integrated Science, Division of Systems Biology, Virginia Tech, Blacksburg, VA, USA
e-mail: chenjing@vt.edu

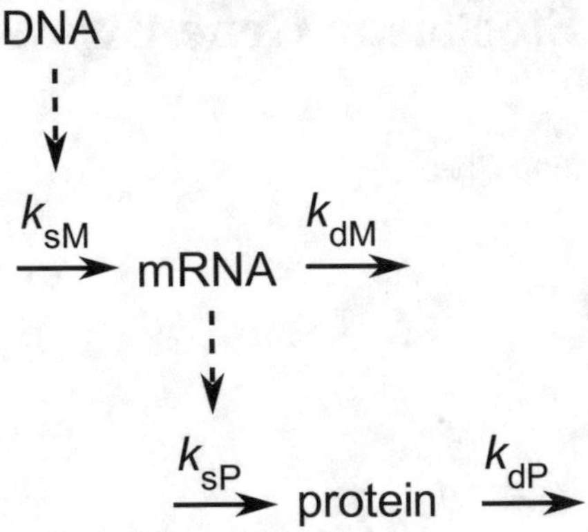

Fig. 1 Gene expression
process posited by the
central dogma. Solid arrows
represent reactions with the
associated rate constants
labeled on top. Dashed
arrows represent regulation;
in both cases here, the
regulation refers to
providing the template of
the reaction

expression process in a single cell, many molecules are present in low numbers.
Particularly, the genes themselves are present in very low numbers (typically 2–4),
and for many genes only a few mRNA molecules are present at any given time.
Therefore, gene expression in a cell is expected to be strongly stochastic. Given the
fundamental role of gene expression in cellular functions, study of stochastic gene
expression lays the cornerstone of understanding the effects of noise on the cell. In
this chapter, we will introduce how to build stochastic models for gene expression,
and how to simulate the stochastic dynamics.[1] Similar approaches can be applied to
molecular regulatory network models to investigate how noise impacts the behavior
and function of the regulatory networks.

Molecular Mechanism

As described by the central dogma of molecular biology, gene expression consists of
two major processes, transcription (DNA $\rightarrow$ mRNA) and translation (mRNA $\rightarrow$
protein). Although transcription and translation each comprise multiple complex
subprocesses, such as initiation, elongation and termination, and each subprocess further involves detailed substeps, for the sake of simplicity, most systems biology
models delineate transcription and translation, as well as degradation of their
products, as single-step processes with a fixed rate constant (Fig. 1). The same
assumption will be adopted in the basic stochastic model for gene expression to be
introduced in this chapter.

---

[1] Prerequisite knowledge of basic probability theory is helpful for reading this chapter. If encoun-
tering unfamiliar concepts in probability theory, the reader can refer to a concise summary of the
basic probability theory on pp. 368-370 of Ingalls [1].

Mathematical Model and Simulation Method

Revisit the Deterministic Model

The wiring diagram in Fig. 1 can be further written as four chemical reactions (1)–(4), where M stands for the mRNA, and P for the protein. One will see later that explicitly writing out the chemical reactions in a system facilitates construction of the stochastic model.

$$\xrightarrow{k_{sM}} \quad M \tag{1}$$

$$M \quad \xrightarrow{k_{dM}} \tag{2}$$

$$M \quad \xrightarrow{k_{sP}} \quad M + P \tag{3}$$

$$P \quad \xrightarrow{k_{dP}} \tag{4}$$

In Reaction (1), the gene is not explicitly shown, because the copy number of the gene is assumed to be constant if DNA replication is ignored, e.g., in a senescent cell or within the G_1 phase. In Reaction (3), M appears on both sides of the arrow, because mRNA serves as the template of translation, and is not produced or consumed in the reaction. Consequently, the reaction rate depends on the abundance of M, but does not appear in the equation for M (Eq. (5)).

The deterministic ODEs for the model read

$$\frac{dM}{dt} = k_{sM} - k_{dM}M, \tag{5}$$

$$\frac{dP}{dt} = k_{sP}M - k_{dP}P, \tag{6}$$

where the italicized *M* and *P* stand for the **molecule numbers** of the mRNA and protein, respectively. Regulatory network models typically use concentrations of the molecules as state variables. In deriving the stochastic model, we need to consider the behavior of individual molecules. For this purpose, it is more convenient to use molecule numbers as the state variables. In this case, the physical units of all reaction rate constants become **time^{-1}**.

Deterministic Model Implies Property of Stochastic Dynamics

Note that Reactions (1)–(4) all assume fixed rate constants. This assumption implies an **exponentially distributed waiting time** for a reaction to occur. To see why, let us consider mRNA degradation alone as an example, described by Eq. (5) with $k_{sM} = 0$.

Suppose the system starts with M_0 copies of the mRNA at time 0. In this case, the solution to Eq. (5) is an exponential function of time:

$$M(t) = M_0 e^{-k_{dM}t}. \tag{7}$$

Equation (7) reflects the overall degradation dynamics of M_0 mRNA molecules. In order for M_0 mRNA molecules to gradually disappear, individual molecules must decay at different (random) times (otherwise $M(t)$ would be a step function). Although the decay time for each mRNA molecule is random, it is reasonable to assume that the decay times for all the mRNA molecules in the system follow the same probability distribution. Now let us use $M(t)$ to deduce this probability distribution. Because $M(t)$ is the number of mRNA molecules left in the system at time t, the probability that an mRNA molecule is _not_ yet degraded by time t, or in other words, the probability that the waiting time for the degradation reaction is longer than t, is

$$P(\text{waiting time for degradation} > t) = \frac{M(t)}{M_0} = e^{-k_{dM}t}. \tag{8}$$

Equation (8) is equivalent to

$$P(\text{waiting time for degradation} \leq t) = 1 - e^{-k_{dM}t}, \tag{9}$$

which is exactly the cumulative distribution function of an exponential distribution with parameter k_{dM}. This shows that the waiting time for degradation of a single mRNA molecule follows an exponential distribution if the deterministic dynamics of mRNA decay is an exponential function of time.

In fact, a mass action rate law of any order with a fixed rate constant implies an exponential distribution for the stochastic waiting time of the reaction. In the following, we will construct a stochastic model for gene expression, based on the assumption of exponentially distributed waiting times for Reactions (1)–(4).

Derive the Chemical Master Equation for Stochastic Dynamics of mRNA

We will first build a stochastic model for the mRNA dynamics alone, to illustrate the conceptual basis for modeling stochastic reaction dynamics. In the next subsection, we will generalize the resulting equations to include protein dynamics, and present a general form of equation applicable for any reaction networks.

The mRNA dynamics is governed by two reactions: synthesis and degradation of the mRNA (Reactions (1) and (2)). Similar to the deterministic model for mRNA (Eq. (5)), the stochastic model also takes the number of the mRNA, M, as its state variable. Unlike the deterministic model, however, the molecule number no longer assumes a definite value at a given time, but rather random values. By definition, the molecule number of the mRNA can assume any non-negative integer value, and each possible mRNA number corresponds to a possible **state of the system**. Our task is to find out the probability distribution of the mRNA number (i.e., the likelihood of all possible states of the system) at any time t. In other words, the relevant quantity to be tracked over time becomes **the probability that m copies of the mRNA exist in the system at time t**, denoted as **P(m, t)**.

To model how P(m, t) evolves over time, let us consider P(m, $t+\Delta t$), for a very small time interval Δt. Δt is sufficiently small such that we can neglect the possibility that more than one reaction event occurs during Δt. In the end we will take the limit $\Delta t \rightarrow 0$, so the exact value of Δt is not important as long as it is sufficiently small to allow at most one reaction. To arrive at m copies of the mRNA at time $t+\Delta t$, one of the following scenarios must happen:

i. $m-1$ copies of the mRNA exist at time t, and one synthesis event occurs within Δt.

ii. $m+1$ copies of the mRNA exist at time t, and one degradation event occurs within Δt.

iii. m copies of the mRNA exist at time t, and neither synthesis nor degradation occurs within Δt.

Hence,

$$
\begin{aligned}
P(m, t + \Delta t) \quad = \quad & P(m - 1, t)P(\text{synthesis occurs in } \Delta t | m - 1, t) \\
+ \quad & P(m + 1, t)P(\text{degradation occurs in } \Delta t | m + 1, t) \\
+ \quad & P(m, t)\{1 - P(\text{synthesis occurs in } \Delta t | m, t) - P(\text{degradation in } \Delta t | m, t)\},
\end{aligned}
$$

$$(10)$$

where the probability terms with a vertical bar in the bracket represent conditional probability (i.e., the probability that the event before the bar occurs given that the event after the bar occurs).

Next, we derive the conditional probabilities in Eq. (10). As posited previously, the waiting times for the reactions follow the exponential distribution. Hence, the probability that one synthesis event occurs during Δt is $\int_0^{\Delta t} k_{sM}e^{-k_{sM}\tau}d\tau$ (the integrand is the probability density function of the exponential distribution). For a very small Δt, $e^{-k_{sM}\tau} \approx 1$, and the probability becomes

$$P(\text{synthesis occurs in } \Delta t | m - 1, t) \approx k_{sM}\Delta t. \tag{11}$$

Similarly, the probability that degradation occurs to a specific mRNA molecule during Δt is approximately $k_{dM}\Delta t$. Because there are $m+1$ copies of the mRNA at time t, the probability that one degradation event occurs, regardless of which mRNA copy, is

$$P(\text{degradation occurs in } \Delta t | m + 1, t) \approx (m + 1)k_{dM}\Delta t. \tag{12}$$

Following the same reasoning,

$$P(\text{synthesis occurs in } \Delta t | m, t) \approx k_{sM}\Delta t, \tag{13}$$

$$P(\text{degradation in } \Delta t | m, t) \approx mk_{dM}\Delta t. \tag{14}$$

Plugging the conditional probabilities (Eqs. (11)–(14)) into Eq. (10) yields

$$\begin{aligned}
P(m, t + \Delta t) &= P(m - 1, t)k_{sM}\Delta t \\
&+ P(m + 1, t)(m + 1)k_{dM}\Delta t \\
&+ P(m, t)\{1 - k_{sM}\Delta t - mk_{dM}\Delta t\},
\end{aligned} \tag{15}$$

which can be rearranged as

$$\begin{aligned}
\frac{P(m, t + \Delta t) - P(m, t)}{\Delta t} &= - P(m, t)k_{sM} + P(m - 1, t)k_{sM} \\
&- P(m, t)k_{dM}m + P(m + 1, t)k_{dM}(m + 1).
\end{aligned} \tag{16}$$

Taking the limit $\Delta t \to 0$, the left-hand side of Eq. (16) becomes the time derivative of $P(m, t)$, resulting in

$$\begin{aligned}
\frac{d}{dt}P(m, t) &= - P(m, t)k_{sM} + P(m - 1, t)k_{sM} \\
&- P(m, t)k_{dM}m + P(m + 1, t)k_{dM}(m + 1).
\end{aligned} \tag{17}$$

Equation (17) is called the **chemical master equation**, or master equation for short, for the simplest stochastic model of mRNA dynamics. Note that Eq. (17) gives the general form of the master equation for a generic state with m copies of the mRNA. The entire model is a system of equations resembling Eq. (17), where $m = 0, 1, 2, \ldots$.

The state with the lowest mRNA number, 0, has only one neighboring state, 1, and its corresponding master equation has fewer terms:

$$\frac{d}{dt}P(0, t) = -P(0, t)k_{sM} + P(1, t)k_{dM}. \tag{18}$$

Note that Eq. (18) is consistent with the general equation, Eq. (17), given that $m = 0$ and $P(-1, t) = 0$ (the state with $m = -1$ is impossible and hence has zero probability).

The terms on the right-hand side of the chemical master equation, Eq. (17), can be understood as "probability fluxes" (Fig. 2) between neighboring states of the system that are accessible through one single reaction. Each flux represents the effect of one

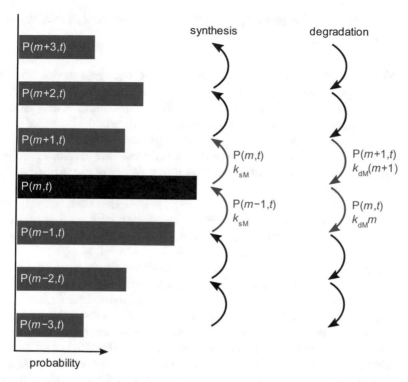

Fig. 2 The chemical master equation can be understood as a probability flux diagram. Horizontal bars illustrate possible states of the system—in this case the molecule number of the mRNA. The length of a bar illustrates the probability of the corresponding state at time t. "Probability fluxes" occur between neighboring states that are accessible via one single reaction. The fluxes connected to a particular state of m mRNAs (black bar) are highlighted by colors. Blue arrows show outfluxes from the state, which appear as negative terms in the master equation (Eq. (17)); red arrows show influxes to the state, which appear as positive terms in the master equation (Eq. (17)). All the flux terms consist of two parts: the probability of the state where the flux originates from (upper rows) and a term reminiscent of the mass action rate law (lower rows)

synthesis or degradation reaction on the probability of the states connected by the flux. Each upward flux represents the effect of a synthesis reaction, which adds one mRNA molecule to the system, hence "moving" probability from the m-th state to the $(m+1)$-th state. Vice versa, each downward flux represents the effect of a degradation reaction, which subtracts one mRNA molecule from the system, hence "moving" probability from the m-th state to the $(m-1)$-th state. For any specific state, its master equation sums up all the influxes to the state as positive terms and outfluxes from the state as negative terms.

Chemical Master Equations for Stochastic Dynamics of mRNA and Protein

Now let us incorporate the protein dynamics into the stochastic model. With this addition, the system has four reactions, synthesis and degradation of the mRNA and protein, respectively (Reactions (1)–(4)). Like mRNA, the number of the protein, p, also assume random, non-negative integer values. The **state** of the expanded system is characterized by a combination of the molecule numbers of the mRNA and protein, (m, p). The relevant quantity to be tracked over time becomes $P(m, p, t)$, **the probability that m copies of the mRNA and p copies of the protein exist in the system at time t.**

The relationship between a state, (m, p), and its neighboring states is illustrated in Fig. 3. Compared to the model for the mRNA alone, each state has four neighbors, with either the mRNA or the protein number ± 1. The neighbors are accessible through a single reaction out of Reactions (1)–(4).

To write the flux terms, let us examine the examples in the model for mRNA alone. In the probability flux diagram in Fig. 2, notice that each flux term consists of two parts: the probability of the state where the flux originates from, and a term reminiscent of the mass action rate law. Because the synthesis reaction is modeled as a zeroth-order reaction, its rate law is just the rate constant and appears identical in the ODE (Eq. (5)) versus in the upward probability fluxes (Fig. 2). For a downward probability flux representing degradation, the abundance of the reactant (mRNA) in the first-order reaction rate law is the mRNA number in the state where the flux originates from (Fig. 2). To understand why, recall that in the derivation of the master equation, it is state of the system at time t (origin state) that determines the

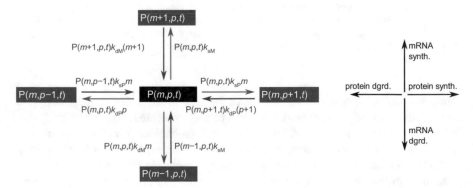

Fig. 3 Probability flux diagram centered around a generic state. Each bin represents a state of the system. Bins in the same row have the same mRNA number and differ in protein number. Bins in the same column have the same protein number and differ in mRNA number. Upward arrows: fluxes caused by mRNA synthesis. Downward arrows: fluxes caused by mRNA degradation. Right arrows: fluxes caused by protein synthesis. Left arrows: fluxes caused by protein degradation. Blue arrows: outfluxes from the state (m, p). Red arrows: influxes to the state (m, p)

probability of reactions during Δt and the resulting probability distribution of states at time $t + \Delta t$.

Applying the same rule, we can arrive at the probability flux terms shown in Fig. 3 (the readers are strongly encouraged to derive the flux terms on their own using the rule). Assembling these fluxes, we obtain the chemical master equation for the simplest stochastic model for mRNA and protein expression of a gene (Eq. (19)). Again, note that Eq. (19) is the equation for a generic state, (m, p), and the entire model consists of infinite number of such equations with $m, p = 0, 1, 2, \ldots$.

$$
\begin{aligned}
\frac{d}{dt}\mathrm{P}(m,p,t) = \quad &- \quad \mathrm{P}(m,p,t)k_{\mathrm{sM}} + \mathrm{P}(m-1,p,t)k_{\mathrm{sM}} \\
&- \quad \mathrm{P}(m,p,t)k_{\mathrm{dM}}m + \mathrm{P}(m+1,p,t)k_{\mathrm{dM}}(m+1) \\
&- \quad \mathrm{P}(m,p,t)k_{\mathrm{sP}}m + \mathrm{P}(m,p-1,t)k_{\mathrm{sP}}m \\
&- \quad \mathrm{P}(m,p,t)k_{\mathrm{dP}}p + \mathrm{P}(m,p+1,t)k_{\mathrm{dP}}(p+1)
\end{aligned}
\tag{19}
$$

A system of master equations is usually hard to solve analytically. But the corresponding steady state equations are algebraic equations and can be more easily solved to obtain the steady state probability distribution of the system. With the steady state probability distribution, one can compute the mean, variance and other statistical quantities of the molecule numbers in the system. These statistical quantities can be compared to the experimental data, e.g., mean and variance of single-cell counts of a fluorescently labeled mRNA. We will learn how to derive the steady state probability distribution in the Results section.

Chemical Master Equations for General Reaction Networks

The procedure of assembling the master equation shown above can be generalized to formulate master equations for any reaction networks:

$$
\frac{d}{dt}\mathrm{P}(\mathbf{N},t) = \sum_k \{-\mathrm{P}(\mathbf{N},t)a_k(\mathbf{N}) + \mathrm{P}(\mathbf{N}-\mathbf{s}_k,t)a_k(\mathbf{N}-\mathbf{s}_k)\}.
\tag{20}
$$

In Eq. (20), the bold-faced $\mathbf{N}$ is the **state vector** containing the copy numbers of all molecular species in the reaction network, i.e., $\mathbf{N} = \{n_1, n_2, \ldots\}$ where n_i is the copy number of the i-th molecular species. k is the index for the reactions. a_k is the **reaction propensity** of the k-th reaction, which is the term that resembles the reaction rate law in ODE models, with the concentration of a molecule replaced by its copy number. $\mathbf{s}_k$ is the **stoichiometry vector** for the k-th reaction, which depicts the impact of the reaction on the number of each molecular species. For example, in the stoichiometry vector for the reaction $O + O \rightarrow O_2$, the entry for O is -2 and that for O_2 is $+1$. Note that the order of molecular species in the stoichiometry vectors must follow the order in the state vector. The master equation for a state $\mathbf{N}$ generally

Table 1 Reaction propensities and stoichiometry vectors
for the stochastic gene
expression model

Reactions	Propensities	Stoichiometry vectors
$\xrightarrow{k_{sM}} M$	$a_1 = k_{sM}$	$s_1 = (+1, 0)$
$M \xrightarrow{k_{dM}}$	$a_2 = k_{dM}m$	$s_2 = (-1, 0)$
$M \xrightarrow{k_{sP}} M + P$	$a_3 = k_{sP}m$	$s_3 = (0, +1)$
$P \xrightarrow{k_{dP}}$	$a_4 = k_{dP}p$	$s_4 = (0, -1)$

includes pairs of negative terms (outfluxes) and positive terms (influxes) associated with the same reaction. The negative term for the k-th reaction appears as a product between the probability of the state N and the reaction propensity based on the state N. The positive term for the k-th reaction appears as a product between the probability of the neighboring state with respect to the k-th reaction, i.e., $N - s_k$, and the reaction propensity based on this neighboring state. In the stochastic model for mRNA and protein expression, for instance, $N = (m, p)$, and the reaction propensities and stoichiometry vectors are given in Table 1.

When deriving the chemical master equation for a reaction network, it is very helpful to first determine the state vector and enumerate all the reactions with the corresponding reaction propensities and stoichiometry vectors, as exemplified by Table 1. Then it is straightforward to plug the terms into the general form in Eq. (20) to obtain the chemical master equations for the system.

Gillespie's Stochastic Simulation Algorithm

The most widely used simulation algorithm for stochastic reaction models was published by Dan Gillespie in 1977 [2]. The Gillespie algorithm generates random time trajectories of the molecule numbers in a reaction system. The algorithm allows exact stochastic simulation of the master equations. Exact simulation means that the distribution of molecule numbers at any time t in the simulated trajectories matches exactly the distribution predicted by the master equations. Therefore, even though the master equations are usually not analytically solvable, the solution can be numerically obtained by simulating a large number of sample trajectories.

To generate stochastic trajectories for a reaction network, the Gillespie algorithm iterates two random sampling steps (Fig. 4):

(1) Sample the waiting time for the next reaction and update the time;
(2) Sample which reaction is next and update the molecule numbers according to the stoichiometry vector of the chosen reaction.

To generate random samples in the two steps above, one needs to draw random numbers using a random number generator function, which is available in most numeric software packages. Most default random number generators generate uniformly distributed random numbers. With appropriate transformation, uniformly

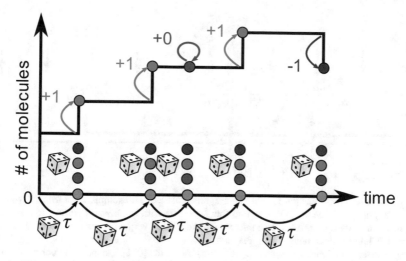

Fig. 4 Gillespie stochastic simulation algorithm. In each iteration, the waiting time for the next reaction (τ) and which reaction it is (red/green/blue dots) are randomly sampled. For each simulated data point (colored dots on black time trajectory), its time value is calculated from the randomly sampled waiting time, τ, and its molecule number(s) is calculated according to which reaction is selected by random sampling. In this illustrative example, the red reaction consumes one molecule, the green reaction produces one molecule, and the blue reaction does not consume or produce the molecule. The simulation simultaneously track the numbers of all molecular species in the system, but only the number of one molecular species is illustrated here

distributed random numbers can be converted to random numbers following any desired probability distribution.

In Step 1, we need to sample the waiting time for the next reaction, which follows an exponential distribution with its parameter equal to the sum of all reaction propensities, $\lambda = \sum_{k} a_k$ (see proof in Appendix B). For the stochastic gene expression model derived above, $\lambda = a_1 + a_2 + a_3 + a_4$, where the reaction propensities a_k are given in Table 1. Note that a_k depends on the current molecule numbers, i.e., $a_k(\mathbf{N})$. A uniformly distributed random number, r, can be converted (see proof in Appendix A) into an exponentially distributed random number, τ, with parameter λ, through

$$\tau = -\frac{1}{\lambda} \ln r. \tag{21}$$

In Step 2, we need to sample which reaction is the next reaction. The index of the next reaction, I, follows the discrete distribution (see proof in Appendix C),

$$P(I = i) = \frac{a_i}{\sum_{k} a_k}. \tag{22}$$

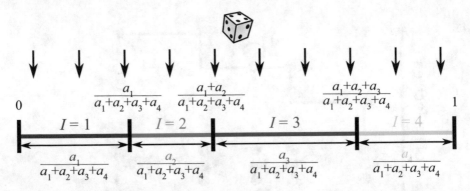

Fig. 5 Sample which reaction is next. This figure illustrates an example for a system with four reactions. The [0, 1] interval is broken into four segments with lengths proportional to each of the four reaction propensities. A uniform random number (black vertical arrows) is drawn, and which segment it falls in determines which reaction occurs next. Fraction terms below the [0, 1] line indicate the length of each segment. Fraction terms above the [0, 1] line indicate coordinate values of the nodes between adjacent segments, which are used in the simulation code (Appendix D)

To draw a random sample from the discrete distribution defined by Eq. (22), one can envision breaking the [0, 1] interval into segments with lengths matching $P(I = i)$ (Fig. 5). For the stochastic gene expression model, the [0, 1] interval is broken into four segments, each with length $a_i / \sum_k a_k$, $i = 1, 2, 3, 4$. Note that by definition, the length of the four segments add up to one. The index of next reaction is chosen according to which segment the uniformly distributed random number falls in (Fig. 5). Intuitively, the probability of choosing the i-th reaction by this method is identical to the length of the i-th segment, i.e., Eq. (22).

Taken together, the steps of Gillespie algorithm are summarized as follows. The MATLAB code for Gillespie simulation of the stochastic gene expression model is given in Appendix D.

1. Initialize the state vector, **N**. Set the initial time at 0 and the initial number of molecules as appropriate.
2. Calculate the reaction propensities, $a_k(\mathbf{N})$.
3. Draw a uniform random number, r_1.
4. Get the time for the next reaction, $\tau = -\dfrac{1}{\sum_k a_k(\mathbf{N})} \ln r_1$.

5. Draw a uniform random number, r_2.
6. Get which reaction is next, $I = i$, if $\dfrac{\sum_{k=1}^{i-1} a_k(\mathbf{N})}{\sum_k a_k(\mathbf{N})} \leq r_2 < \dfrac{\sum_{k=1}^{i} a_k(\mathbf{N})}{\sum_k a_k(\mathbf{N})}$ (c.f. coordinate values

   of nodes in Fig. 5).
7. Update state vector, $\mathbf{N} \rightarrow \mathbf{N} + \mathbf{s}_i$.
8. Update time, $t \rightarrow t + \tau$.
9. Repeat (2)–(8).

Results

Steady State Equations for Stochastic Model

Setting $\frac{dP}{dt}$ on the left-hand side of the master equation in Eq. (20) to be zero, one obtains the steady state equation,

$$0 = \sum_k \{-P_{SS}(\mathbf{N})a_k(\mathbf{N}) + P_{SS}(\mathbf{N} - \mathbf{s}_k)a_k(\mathbf{N} - \mathbf{s}_k)\}. \tag{23}$$

In Eq. (23), $P_{SS}(\mathbf{N}) = P(\mathbf{N}, t = \infty)$ depicts the steady state probability distribution of molecule numbers in the system. Time t is dropped from Eq. (23), because the steady state probability does not depend on time (the probability no longer varies with time). Eq. (23) represents a system of algebraic equations, one equation for a specific state of the system. Unfortunately, this equation system contains one redundant equation and cannot be solved uniquely. To see why, note that the time-dependent master equations add up to zero:

$$
\begin{aligned}
\sum_{\mathbf{N}} \sum_k \{-P(\mathbf{N}, t)a_k(\mathbf{N}) + P(\mathbf{N} - \mathbf{s}_k, t)a_k(\mathbf{N} - \mathbf{s}_k)\} &= \sum_{\mathbf{N}} \frac{d}{dt} P(\mathbf{N}, t) \\
&= \frac{d}{dt} \sum_{\mathbf{N}} P(\mathbf{N}, t) \qquad (24) \\
&= \frac{d}{dt}(1) \\
&= 0.
\end{aligned}
$$

The penultimate equality in Eq. (24) holds because the probability of all states adds up to one, which is the normalization condition for any probability distribution. Eq. (24) holds for any time t, and of course for the steady state, too. Therefore, we find that the sum of all the steady state equations represented by Eq. (23) is zero:

$$0 = \sum_{\mathbf{N}} \sum_k \{-P_{SS}(\mathbf{N})a_k(\mathbf{N}) + P_{SS}(\mathbf{N} - \mathbf{s}_k)a_k(\mathbf{N} - \mathbf{s}_k)\}. \tag{25}$$

According to Eq. (25), if any $N-1$ equations out of the entire set of N equations represented by Eq. (23) are satisfied, the remaining one equation is automatically satisfied. In other words, we are trying to solve for N unknowns with only $N-1$ independent equations, which will not give a unique solution. (Many stochastic reaction systems have infinitely many states. The simple mRNA model, for instance, can have infinitely many values for m. But the relation between redundancy of equations and uniqueness of solution remains true for infinitely large systems of algebraic equations.)

Interestingly, the problem is rescued by its own culprit, the normalization condition, expressed in terms of the steady state probability distribution as

$$\sum_N P_{SS}(N) = 1. \tag{26}$$

Equation (26) supplies an additional algebraic equation to complete a system of N equations for N unknowns and makes it uniquely solvable.

Steady State Probability Distribution of mRNA Number

Next, we will present the steady state probability distribution for the basic stochastic gene expression model, and show some interesting predictions from the distribution.
For the model with the mRNA alone, the steady state equations read

$$0 = -P_{SS}(m)k_{sM} + P_{SS}(m-1)k_{sM} - P_{SS}(m)k_{dM}m \\ + P_{SS}(m+1)k_{dM}(m+1), \tag{27}$$

and

$$\sum_m P_{SS}(m) = 1. \tag{28}$$

The solution to Eqs. (27)-(28) reads[2]:

$$P_{SS}(m) = \frac{\mu^m e^{-\mu}}{m!}, \text{ where } \mu = \frac{k_{sM}}{k_{dM}}. \tag{29}$$

The solution is exactly a **Poisson distribution**!
For the sake of mathematical clarity in the following, please note that $P_{SS}(m)$ is a shorthand notation for $P(M(t = \infty) = m)$ or $P(M_{SS} = m)$, i.e., the probability that the random copy number of the mRNA at the steady state equals m. Mathematically speaking, M_{SS} is the random variable, while m is a generic possible value of the random variable.
With a Poisson distribution, the expectation, or mean value, of the steady state mRNA number is

---

[2]Interested readers can verify if Eq. (29) indeed satisfies Eqs. (27) and (28). To check if Eq. (29) satisfies Eq. (28), consider Taylor expansion of the exponential function, $e^{-\mu}$.

$$E(M_{SS}) = \sum_m P_{SS}(m)m = \mu = \frac{k_{sM}}{k_{dM}}. \tag{30}$$

This mean number is identical to the steady state mRNA number predicted by the deterministic model for mRNA (Eq. (5)). This finding is not surprising, because the deterministic model is supposed to delineate the average behavior of a system (this is why deterministic models are often called bulk average models!). In more complex systems the steady state mean molecule numbers predicted by stochastic models are not necessarily identical to the steady state numbers predicted by the corresponding deterministic models. In some cases, the stochastic model even brings about entirely new dynamics not existing in the deterministic model, such as oscillation driven by noise [3, 4].

What can be further predicted by a stochastic model, but not a deterministic model, is the variation of the system from its mean. With a Poisson distribution, **the variance of the mRNA number is equal to its mean:**

$$Var(M_{SS}) = \sum_m P_{SS}(m)(m - E(M_{SS}))^2 = \mu = \frac{k_{sM}}{k_{dM}} = E(M_{SS}). \tag{31}$$

Because of this interesting property of the Poisson distribution, a statistical quantity called **Fano factor**, defined as the ratio between the variance and mean of a molecule number, e.g.,

$$F(M_{SS}) = \frac{Var(M_{SS})}{E(M_{SS})}, \tag{32}$$

is often used to quantify the level of noise in gene expression. The Fano factor of experimentally measured mRNA counts in single cells suggests existence of additional regulation of gene expression noise. Data with Fano factor < 1 is called sub-Poissonian and implies noise suppression mechanisms (e.g., negative feedback); vice versa, data with Fano factor > 1 is called super-Poissonian and implies mechanisms that magnify noise (e.g., bursting transcription).

Another statistical quantity commonly used to evaluate the relative level of noise is the **coefficient of variation (CV)**, defined as the ratio between the standard deviation (square root of variance) and mean. Given the mean and variance predicted above, the CV of mRNA number is

$$CV(M_{SS}) = \frac{\sqrt{Var(M_{SS})}}{E(M_{SS})} = \frac{1}{\sqrt{\mu}} = \frac{1}{\sqrt{E(M_{SS})}}. \tag{33}$$

Equation (33) corroborates the statement posited at the beginning of this chapter: "*Generally speaking, **the relative level of randomness in a system scales with the inverse square root of the size of the system** (~ number of molecules).*" This prediction tells us why stochastic effect is prominent in systems with low molecule numbers.

Steady State Probability Distribution of Protein Number

For the mRNA-protein model, the steady state equations read

$$
\begin{aligned}
0 = \quad & - \ P_{SS}(m,p)k_{sM} + P_{SS}(m-1,p)k_{sM} \\
& - \ P_{SS}(m,p)k_{dM}m + P_{SS}(m+1,p)k_{dM}(m+1) \\
& - \ P_{SS}(m,p)k_{sP}m + P_{SS}(m,p-1)k_{sP}m \\
& - \ P_{SS}(m,p)k_{dP}p + P_{SS}(m,p+1)k_{dP}(p+1),
\end{aligned}
\tag{34}
$$

and

$$
\sum_{m}\sum_{n} P_{SS}(m,p) = 1.
\tag{35}
$$

Equations (34) and (35) are difficult to solve analytically, but the mean and stochastic variation of the steady state mRNA and protein numbers can be derived using some mathematical techniques beyond the scope of this chapter. We will just show the results below.

The mean, variance and CV of mRNA number are identical to those given in Eqs. (30), (31) and (33). This is totally expected, because in the model the mRNA dynamics does not depend on the protein dynamics and is thus identical to that in the model with the mRNA alone.

The predicted mean of protein number is [4–14]

$$
E(P_{SS}) = \frac{k_{sM}k_{sP}}{k_{dM}k_{dP}},
\tag{36}
$$

where P on the left-hand side is italicized to notate the random copy number of the protein, and should not be confused with the non-italicized P that represents probability.

Like the mean mRNA number, the mean protein number is identical to the steady state predicted by the deterministic model (Eqs. (5) and (6)). This is another example where the steady state mean of the stochastic model exactly matches the steady state of the deterministic model. Again, this is not generally true.

The predicted CV of protein number is [14]

$$
CV(P_{SS}) = \sqrt{\frac{1}{E(P_{SS})} + \frac{1}{E(M_{SS})}\frac{k_{dP}}{k_{dM}+k_{dP}}}.
\tag{37}
$$

The two terms inside the square root of Eq. (37) reveal two sources of noise for the protein number. The first term represents the Poissonian noise caused by stochastic translation and protein degradation, alike the Poissonian noise in the mRNA number caused by stochastic transcription and mRNA degradation (Eq. (33)). The second term represents an **additional noise relayed from the noise**

in mRNA number. If one replaces the degradation rate constants by the inverse of the average life times of the mRNA and protein, i.e., $k_{dM} = 1/\tau_M$, $k_{dP} = 1/\tau_P$, Eq. (37) becomes

$$\mathrm{CV}(P_{SS}) = \sqrt{\frac{1}{\mathrm{E}(P_{SS})} + \frac{1}{\mathrm{E}(M_{SS})}\frac{\tau_M}{\tau_M + \tau_P}}. \tag{38}$$

The second term inside the square root of Eq. (38) suggests that relay of the mRNA noise depends on the relative life times of the mRNA and protein: long-lived mRNAs contribute more significantly to the protein noise than short-lived mRNAs. This can be understood as a "time averaging" effect: random fluctuations in the number of a short-lived mRNA would be averaged out over a much longer life time of the protein. Because the copy number of a protein in a cell (typically thousands to millions according to BioNumbers [5]) is generally much higher than that of the mRNA (typically a few to hundreds according to BioNumbers [5]), the Poissonian noise of protein is generally lower than that of mRNA, $1/\mathrm{E}(P_{SS}) < 1/\mathrm{E}(M_{SS})$. Therefore, for a gene with long-lived mRNA, noise in the mRNA dynamics can become the dominant source of noise for the protein dynamics.

Simulated Stochastic Dynamics

Using Gillespie's stochastic simulation algorithm introduced above, we can simulate stochastic time trajectories of the mRNA and protein numbers.

Figure 6 shows simulated trajectories of mRNA decay alone, i.e., the mRNA model with $k_{sM} = 0$. The average of multiple stochastic trajectories approaches the exponential decay curve predicted by the deterministic model (Eq. (7)). Moreover, the more mRNA molecules the system starts with, the closer are single stochastic trajectories to the deterministic exponential decay. These results show that **the deterministic model depicts the bulk average behavior of the system and larger systems are subject to lower noise**.

Figure 7 shows the simulated stochastic trajectories of mRNA and protein numbers and the corresponding deterministic trajectories, using a physiologically realistic set of parameters for low-expression genes. Both mRNA and protein numbers fluctuate around their deterministic values, as expected. The mRNA trajectory exhibits strong noise. The protein trajectory apparently exhibits two types of noises: a weak noise superimposed upon a strong, but slow, fluctuation. The former reflects the Poissonian noise of the protein (the first term inside the square root of Eq. (38)) and the latter reflects the noise relayed from the mRNA (the second term inside the square root of Eq. (38)). For example, the strong increase of mRNA number around $t = 20$ h results in a protein peak around $t = 40$ h. This is followed by a fluctuating, but overall decreasing, leg of the mRNA trajectory that troughs around $t = 50$ h. Correspondingly, the protein number experiences a strong dip that troughs

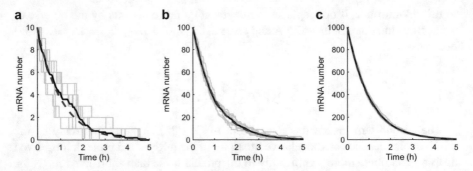

Fig. 6 Simulated stochastic dynamics of mRNA decay. Gray solid lines represent 10 simulated stochastic decay of individual mRNA molecules, with 10 (**a**), 100 (**b**) and 1,000 (**c**) mRNA molecules at the initial time. Black solid lines represent the average of the 10 simulated trajectories. Red dashed lines represent the deterministic trajectories given by Eq. (7). As the size of the system increases, the stochastic trajectories approach the deterministic trajectory. The average of the stochastic trajectories is also closer to the deterministic trajectory than single stochastic trajectories. In all cases, $k_{sM} = 0$, $k_{dM} = 1 \ h^{-1}$

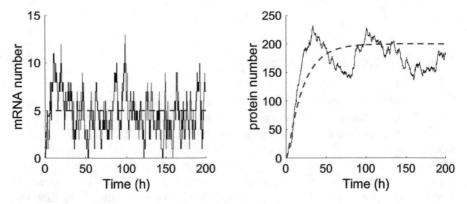

Fig. 7 Simulated stochastic dynamics (solid black lines) vs the corresponding deterministic dynamics (red dashed lines) of the mRNA and protein numbers. Parameter used: $k_{sM} = 2.5 \ h^{-1}$, $k_{dM} = 0.5 \ h^{-1}$, $k_{sP} = 2 \ h^{-1}$, $k_{dP} = 0.05 \ h^{-1}$. MATLAB code is given in Appendix D

around $t = 75$ h. Due to the "time averaging" effect, the protein trajectory reflects mRNA fluctuations in a coarse-grained manner, i.e., fast fluctuations diminished and only relatively long-term fluctuations displayed. Additionally, the 20-hour average life time of the protein ($k_{dP} = 0.05 \ h^{-1}$) causes a ~20 h delay in the protein peaks or troughs following the corresponding mRNA fluctuations.

With a large number of simulated stochastic trajectories, one can estimate the statistical distribution of the mRNA and protein numbers at any given time t (Fig. 8a–f) (MATLAB code in Appendix E). According to Fig. 7, the deterministic trajectory has reached the steady state at $t = 100$ h. (Strictly speaking, a dynamical

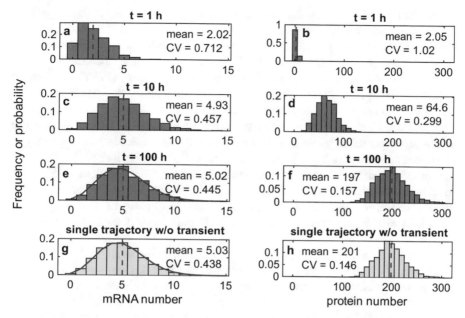

Fig. 8 Histograms of the mRNA and protein numbers. Histograms for 1,000 Gillespie simulation results at $t = 1$ h (**a, b**), $t = 10$ h (**c, d**), $t = 100$ h (**e, f**) and for data in a single trajectory between $t = 100$ h and $t = 4,100$ h (**g, h**). Red dashed lines in all panels represent the molecule numbers predicted by the deterministic model. Red solid lines in **e, g** represent the theoretical steady state (Poisson) distribution for the mRNA number (Eq. 29). The estimated mean and CV of the mRNA and protein numbers in each histogram are labeled on the plots. Note that due to randomness, statistical quantities estimated from stochastic simulations are expected to be close, but not identical, to the theoretical values, and are expected to vary each time the simulations are rerun. Model parameters are identical to those used in Fig. 7. MATLAB code is given in Appendix E

system can become infinitely close to its steady state as time goes, but can never reach it unless it starts at the steady state. In practice, when a system is sufficiently close to its steady state such that the difference is not noticeable on the plot (or detectable by experiments), we can consider that it has "reached" the steady state.) Consistently, the simulated distribution of the mRNA number at $t = 100$ h matches the Poisson distribution predicted by Eq. (29) (Fig. 8e). Moreover, plugging the parameter values used in Figs. 7 and 8, i.e., $k_{sM} = 2.5$ h^{-1}, $k_{dM} = 0.5$ h^{-1}, $k_{sP} = 2$ h^{-1}, $k_{dP} = 0.05$ h^{-1}, into Eqs. (30), (33), (36) and (38), we obtain the theoretical steady state mean and CV of the mRNA and protein numbers: $E(M_{SS}) = 5$, $CV(M_{SS}) = 0.447$, $E(P_{SS}) = 200$, $CV(P_{SS}) = 0.152$. Indeed, the estimated mean and CV at $t = 100$ h are close to the theoretical values. If we examine the values of the two terms inside the square root of Eq. (38) with this parameter set, we find that the Poissonian noise in the protein itself ($1/E(P_{SS}) = 0.005$) is several times lower than the noise relayed from the mRNA ($= 0.018$). Hence, this parameter set gives an example in which the relayed noise from mRNA is the major source of noise for the protein.

In practice, Gillespie simulations are often computationally expensive, especially as the stochastic reaction system gets more complex and includes more chemical species and reactions. Repeating the Gillespie simulation for many times may be computationally formidable. Alternatively, one can use a single stochastic trajectory to estimate the steady state distribution of molecule numbers (MATLAB code in Appendix E). Fig. 7 shows that after the system reaches the steady state, the stochastic trajectory fluctuates around the mean value. Hence, data in **a single stochastic trajectory less its initial transient would give a faithful estimation of the steady state distribution**. Obviously, the longer the steady state trajectory is, the better the steady state distribution can be estimated. Fig. 8g, h show the mean and CV of mRNA and protein numbers estimated from a single trajectory of 4,000 h long with the initial transient ($t < 100$ h) removed; the estimated values are indeed close to the theoretical values calculated above. Compared to running 1,000 Gillespie simulation of ~100 h long, this alternative method improves computational efficiency by 25 times.

Since the data in a steady state trajectory recapitulate the steady state distribution of molecule numbers, conversely, can we simply draw random samples from the steady state probability distribution (e.g., the Poisson distribution for the mRNA number) to generate a stochastic time trajectory? The answer is no. This is because the data generated by random sampling are completely independent of each other at different time points. In a stochastic time trajectory, however, the probability distribution of the data at time $t+\Delta t$ depends on the data at time t. In Fig. 7, for example, starting from zero, the mRNA number is very unlikely to change significantly in a very short time, say, rising to 5 in 0.1 h. In other words, the probability distribution at time 0 is not independent of that at 0.1 h. Hence, independent random sampling from the steady state probability distribution cannot be used to generate a stochastic trajectory for the system.

Discussion

In this chapter we learned from the example of gene expression how to build a stochastic model for a reaction system. Particularly, we learned how to write and analyze chemical master equations and how to simulate the stochastic dynamics using the Gillespie algorithm. These methods are applicable for stochastic modeling of any reaction systems, including cellular regulatory networks.

From the example model for stochastic gene expression, we learned that stochastic effect is prominent in a system with low molecule numbers. Eq. (33) gives a rule-of-thumb estimate for the level of noise in a system: the relative noise roughly scales with the inverse square root of the molecule number. Accordingly, the behavior of a reaction system in a test tube with gazillions of molecules of each species would stay close to the average and can be well described by a deterministic model. But molecular processes in a cell, such as gene expression, involve molecules with low copy numbers and experience strong random noise.

We also learned that noise can propagate in a coupled system. For example, we learned that the mRNA noise is relayed to the protein dynamics and may even be the major source of noise for the protein (Eq. (38)). Similarly, noise propagates in a cellular regulatory network. How it propagates depends on the topology of the network and dynamic properties of each component. Therefore, even if some components in a regulatory network may have millions of copies in a cell and generate low Poissonian noise by themselves, the overall network could be subject to significant noise arising from other low-copy molecular players. The bottom line is, cellular processes are generally noisy.

The example model discussed in this chapter is based on a simplistic schematization of the gene expression process. The model is built upon the following assumptions. First, obviously, the gene expression process has been significantly simplified as four reactions, transcription, mRNA degradation, translation, and protein degradation. Intermediate processes, such as post-transcriptional mRNA processing, nuclear export of the mRNA and ubiquitylation of the protein, are ignored. Second, the four resolved processes are assumed to have exponentially distributed waiting times. Note that this assumption can be derived from the mass action rate law commonly used in ODE models (see derivation in 'Deterministic model implies property of stochastic dynamics'). The mass action rate law is only strictly true for elementary, one-step chemical reactions and approximately true for complex reactions with a distinct rate-limiting substep. The four reactions in the simple gene expression model all comprise complex subprocesses, and therefore their waiting times may not follow the exponential distribution. Third, the model assumes constant reaction rate constants (which is assumed in the ODE models as well). This assumption actually implies two hidden assumptions about the cell, including (i) that the cell is a spatially homogeneous (i.e., well-mixed) system, and (ii) that the activity of the enzymes or molecular machineries that carry out the reactions stays constant. In reality, the intracellular environment is highly inhomogeneous, with a large number of subcellular organelles which are themselves dynamic. The concentration and activity of RNA polymerases and ribosomes are also not necessarily steady. In fact, fluctuations in these factors is believed to generate the **extrinsic noise** in gene expression, referring to the part of the noise that are observed to correlate across different genes, as opposed to the **intrinsic noise** which is independent across different genes and is the type of noise discussed in this chapter (see examples distinguishing extrinsic and intrinsic noise in [6]). Furthermore, the DNA is replicated over the cell cycle, which can double the gene dosage and transcription rate. Realizing the potential problems with these simple assumptions, a lot of researchers are trying to build more realistic models to investigate in more depth how noise is generated and regulated in gene expression. Finally, all models need to be ascertained by experiments. Genome-wide quantification of mRNA and protein noise from bacterial to eukaryotic cells shows that both mRNA and protein noise can span a wide range for different genes [7–9], suggesting gene-specific influences by more complex expression mechanisms and regulations. Nevertheless, the simplest models have provided critical insights to the basic principles of stochastic gene expression and stochastic reaction dynamics.

After all, what does noise do to a cell? Intracellular noise can interfere with cellular functions, such as making signals inaccurate or triggering cellular events at wrong times. To cope with these issues, the cell has evolved various strategies, such as negative feedback, to reduce noise in key cellular components or mechanisms. On the other hand, noise can play beneficial and even essential roles in some biological functions [10]. For example, genetic evolution in all organisms hinges on noise-induced errors in DNA replication [10]. Noise is also critical for force generation by actin filaments [11], probabilistic cell differentiation for hedge-betting against environmental fluctuations [10], stochastic resonance to enhance signal-to-noise ratio for oscillatory signals [12], etc. The study of noise and noise control in biological systems is an active field of research, where researchers continue to make discoveries about the mechanisms by which the cell avoids harms of noise and harnesses noise to its own benefit.

Exercises

Exercise 1. A cell contains a single copy of Gene X. Transcription activator A randomly binds to and unbinds from the promoter of Gene X, controlling its transcription. Suppose the binding rate constant between transcription activator A and the promoter of Gene X is k_b, and the unbinding rate constant is k_u. There are N copies of transcription factor A in the cell.

Build a stochastic model to describe the time evolution of probability distribution for the number of active Gene X in the cell.

(1) Let the state vector of the system be (x, a, c), where the three entries stand for the molecule numbers of free X, free A and A-X complexes, respectively. How many possible states does the system have and what are their corresponding state vectors?
(2) Sketch the probability flux diagram for the model.
(3) Write the chemical master equations based on the probability flux diagram.
(4) Find the steady state probability for all the possible states.
(5) Using the steady state probability distribution found in (4), compute the mean, variance and CV of the number of active Gene X (i.e., A-X complex) in the cell.
(6) Build an ODE model for the number of active Gene X, and compute its deterministic steady state when $N \gg 1$. (Hint: note the conservation laws, $x + c = 1$ and $a + c = N$.) How does the deterministic steady state of active Gene X compare to the mean number found in the stochastic model?

Exercise 2. Explore the effect of random fluctuation in gene activity on the noise of gene expression. The transcriptional activity of a gene can be randomly switched on and off, e.g., due to binding/unbinding of transcription factors with the gene, random folding/unfolding of the chromatin near the gene locus. Please revise the stochastic mRNA and protein expression model to include the following additional reactions for activation and inactivation of the gene, together with the two additional chemical species, G_a for the active form of the gene, and G_i for the inactive form of the gene.

$$G_a \xrightarrow{k_{iG}} G_i \qquad (39)$$

$$G_i \xrightarrow{k_{aG}} G_a \qquad (40)$$

Furthermore, the transcription reaction should now explicitly include the active form of the gene:

$$G_a \xrightarrow{k_{sM}} G_a + M \qquad (41)$$

(1) Run Gillespie simulation for the revised model using the same parameter values in Fig. 7 and $k_{iG} = k_{aG} = 0.1 \text{ h}^{-1}$. Assume the cell has 2 copies of the gene and both start with the inactive form at time 0. The numbers of the mRNA and protein also start from zero. Plot the time trajectories for the numbers of active gene, mRNA and protein.

(2) The total number of the gene, $g_t = g_a + g_i$, should stay constant at 2 throughout the time. Is this true in your simulated trajectory (plot $g_a + g_i$ vs. t)? Can you think of the reason why the gene number stays constant in the simulated trajectory?

(3) Estimate the steady state mean and CV of the numbers of active gene, mRNA and protein from the simulation results. Compared to the results from the original model (with constant gene activity), how does random fluctuation in gene activity affect the noise in the mRNA and protein dynamics?

(4) Explore the effect of fast versus slow fluctuation in gene activity. Repeat (1) and (2) for higher and lower rate constants, $k_{iG} = k_{aG} = 0.01, 1 \text{ h}^{-1}$. How do the simulated time trajectories differ under the influence of fast versus slow fluctuation in gene activity and why?

(5) Estimate the steady state mean and CV of the numbers of active gene, mRNA and protein from the simulation results in (4). Compare the statistical quantities in all three parameter sets, i.e., $k_{iG} = k_{aG} = 0.01, 0.1, 1 \text{ h}^{-1}$, to the theoretical results [13] given in Eqs. (42)–(47). How are mRNA and protein noise affected by fast versus slow fluctuation in gene activity and why? (Hint: examine and interpret the terms in the theoretical results.)

$$E(G_{a,SS}) = \frac{g_t k_{aG}}{k_{aG} + k_{iG}} \qquad (42)$$

$$CV(G_{a,SS}) = \sqrt{\frac{k_{iG}}{g_t k_{aG}}} \qquad (43)$$

$$E(M_{SS}) = \frac{k_{sM}}{k_{dM}} \frac{g_t k_{aG}}{k_{aG} + k_{iG}} \qquad (44)$$

$$\mathrm{CV}(M_{\mathrm{SS}}) = \sqrt{\frac{1}{\mathrm{E}(M_{\mathrm{SS}})} + \frac{k_{\mathrm{iG}}}{g_{\mathrm{t}}k_{\mathrm{aG}}} \frac{k_{\mathrm{dM}}}{k_{\mathrm{aG}} + k_{\mathrm{iG}} + k_{\mathrm{dM}}}} \tag{45}$$

$$\mathrm{E}(P_{\mathrm{SS}}) = \frac{k_{\mathrm{sP}}}{k_{\mathrm{dP}}} \frac{k_{\mathrm{sM}}}{k_{\mathrm{dM}}} \frac{g_{\mathrm{t}}k_{\mathrm{aG}}}{k_{\mathrm{aG}} + k_{\mathrm{iG}}} \tag{46}$$

$$\mathrm{CV}(P_{\mathrm{SS}}) = \sqrt{\begin{array}{l} \dfrac{1}{\mathrm{E}(P_{\mathrm{SS}})} + \dfrac{1}{\mathrm{E}(M_{\mathrm{SS}})} \dfrac{k_{\mathrm{dP}}}{k_{\mathrm{dM}} + k_{\mathrm{dP}}} \\[2ex] + \dfrac{k_{\mathrm{iG}}}{g_{\mathrm{t}}k_{\mathrm{aG}}} \dfrac{k_{\mathrm{dM}}}{k_{\mathrm{aG}} + k_{\mathrm{iG}} + k_{\mathrm{dM}}} \dfrac{k_{\mathrm{dP}}}{k_{\mathrm{dM}} + k_{\mathrm{dP}}} \left(1 + \dfrac{k_{\mathrm{dM}}}{k_{\mathrm{aG}} + k_{\mathrm{iG}} + k_{\mathrm{dP}}}\right) \end{array}} \tag{47}$$

Exercise 3. Explore the effect of autorepression on the noise of gene expression. Suppose the protein product of a gene serves as its own transcriptional inhibitor. To model such transcriptional repression, one can modify the gene activation and deactivation reactions in Exercise 2 as Reactions (48)–(49), such that binding between the gene and the protein inactivates the gene, and the unbound form of the gene is active.

$$G_{\mathrm{a}} + P \xrightarrow{k_{\mathrm{iG}}} G_{\mathrm{i}} \tag{48}$$

$$G_{\mathrm{i}} \xrightarrow{k_{\mathrm{aG}}} G_{\mathrm{a}} + P \tag{49}$$

Because Reaction (48) has two components on the reactant side, the reaction propensity assumes the form of a second-order mass action law: $a = k_{\mathrm{iG}}g_{\mathrm{a}}P$.

(1) Run Gillespie simulation for the revised model using the same parameter values in Fig. 7 and $k_{\mathrm{iG}} = 0.0005\ \mathrm{h}^{-1}$, $k_{\mathrm{aG}} = 0.1\ \mathrm{h}^{-1}$ (these values maintain the same mean numbers of the active gene, mRNA and protein such that the intrinsic noise of each molecular species remains unchanged). Assume the cell has 2 copies of the gene and both start with the inactive form at time 0. The numbers of the mRNA and protein also start from zero. Plot the time trajectories for the numbers of active gene, mRNA and protein.

(2) Estimate the steady state mean and CV of the numbers of active gene, mRNA and protein from the simulation results with fast versus slow fluctuation in gene activity (try multiplying k_{iG} and k_{aG} by the same factor, 0.1, 1, or 10), and compare them to the results in Exercise 2. (Tips: To improve accuracy of the estimation and allow more reliable comparison, you can run the simulation and statistics for multiple times and plot the results as mean $\pm$ SD. This will visualize whether the results are significantly different or not.)

Based on the model results, how do you think autorepression impact gene expression noise? Is the impact more significant in the presence of fast or slow fluctuation in gene activity?

(3) Now let us modify the autorepression mechanism in the model such that inactivation of the gene requires binding with two molecules of P. (Many

transcription factors indeed function in the dimeric form.) Reactions (48) and (49) are then modified to Reactions (50) and (51), respectively.

$$G_a + 2P \overset{k_{iG}}{\rightarrow} G_i \tag{50}$$

$$G_i \overset{k_{aG}}{\rightarrow} G_a + 2P \tag{51}$$

Note that Reaction (50) has three components on the reactant side, two of the same species. Its _reaction propensity_ is written as $a = k_{iG}g_a P(p-1)/2$; the term, $p(p-1)/2$, is the combinatorial number for selecting 2 copies of the protein out of p total copies of the protein in the system. Also note that the _stoichiometry vectors_ for Reactions (50) and (51) need to be modified correspondingly to reflect the stoichiometry of P in the reactions. (Hint: How many copies of P are consumed or produced in these reactions?)

Redo question (2) based on the new model and compare the results to those for (2). What do you conclude from the results about the effect of autorepression via monomeric vs. dimeric binding of the gene by its own protein product?

Open question: Exercises 2 and 3 give examples in which the stochastic gene expression model is extended to explore the potential impact of further mechanistic details of gene expression on gene expression noise. Gene expression is a highly complex process. How the complex details affect gene expression noise awaits integrated efforts from modeling and experimental studies. Have you learned any subprocess of gene expression that interests you? How would you design a model to test the effect of the subprocess on gene expression noise?

Appendix A: Sampling for Continuous Random Variables

Claim: If r is a uniformly distributed random variable, then $\tau = -\frac{1}{\lambda}\ln r$ is an exponentially distributed random variable with parameter λ.

To prove the claim above, let us examine the cumulative distribution function (CDF) of τ:

$$F_\tau(t) = P(\tau \le t) = P(e^{-\lambda\tau} \ge e^{-\lambda t}) = P(r \ge e^{-\lambda t}) = 1 - e^{-\lambda t}. \tag{52}$$

Equation (52) shows that the CDF of the random variable τ is exactly the CDF of the exponential distribution with parameter λ. Hence, τ follows the exponential distribution with parameter λ.

Appendix B: Probability Distribution for the Waiting Time of Next Reaction

Claim: Suppose a reaction system has N reactions with reaction propensities a_1, a_2, ..., a_N, respectively. The waiting time, T, for the next reaction in the system (regardless of which reaction) follows the exponential distribution with parameter $\lambda = \sum_{k=1}^{N} a_k$.

To prove the claim above, let us examine the cumulative distribution function (CDF) of T. Recall that the waiting time for each individual reaction, T_k, follows the exponential distribution with parameter equal to the reaction propensity, a_k. Hence,

$$
\begin{aligned}
F_T(t) &= P(T \le t) \\
&= 1 - P(T > t) \\
&= 1 - P(\text{None of the reactions occurs in } t) \\
&= 1 - \prod_{k=1}^{N} P(T_k > t) \\
&= 1 - \prod_{k=1}^{N} \exp(-a_k t) \\
&= 1 - \exp\left(-\left(\sum_{k=1}^{N} a_k\right) t\right).
\end{aligned}
\tag{53}
$$

Equation (53) shows that the CDF of the random waiting time, T, is exactly the CDF of the exponential distribution with parameter $\lambda = \sum_{k=1}^{N} a_k$. Hence, T follows the exponential distribution with parameter $\lambda = \sum_{k=1}^{N} a_k$.

Appendix C: Probability Distribution for the Index of Next Reaction

Claim: Suppose a reaction system has N reactions with reaction propensities a_1, a_2, ..., a_N, respectively. The index of the next reaction, I, follows the discrete distribution,

$$
P(I = i) = \frac{a_i}{\sum_{k} a_k}.
\tag{54}
$$

Recall that the stochastic waiting time for each individual reaction, T_k, follows the exponential distribution with parameter equal to the reaction propensity, a_k. Strictly speaking, the probability that the i-th reaction occurs next in the Gillespie algorithm is the conditional probability that the i-th reaction occurs next given that the next reaction occurs at time τ (the waiting time for the next reaction sampled in the previous step). Because the waiting time is a continuous random variable, its probability must be defined on a time interval (otherwise, the probability of hitting a number exactly is zero for a continuous random variable). For the convenience of the derivation below, we consider that the next reaction occurs at time τ if the waiting time for it falls in the time interval $(\tau - \Delta t, \tau]$. The length of the time interval, Δt, is chosen to be sufficiently small to allow an approximation shown below. Then, the conditional probability mentioned above is written as $P(I = i|\tau - \Delta t < T \le \tau)$.

Now let us evaluate the conditional probability that the i-th reaction occurs next given that the next reaction occurs at time τ. Note that the i-th reaction occurs next at time τ implies that the waiting times for the other reactions (with indices $k \ne i$) are all longer than τ. Hence,

$$
\begin{aligned}
P(I = i|\tau - \Delta t < T \le \tau) &= \frac{P(\tau - \Delta t < T_i \le \tau \cap T_k > \tau, \text{for all } k \ne i)}{P(\tau - \Delta t < T \le \tau)} \\[2mm]
&= \frac{P(\tau - \Delta t < T_i \le \tau)\prod_{k \ne i}P(T_k > \tau)}{P(\tau - \Delta t < T \le \tau)} \\[2mm]
&= \frac{\int_{\tau-\Delta t}^{\tau} a_i\exp(-a_i\sigma)d\sigma\prod_{k \ne i}\exp(-a_k\tau)}{\int_{\tau-\Delta t}^{\tau}(\sum_{k=1}^{N} a_k)\exp(-(\sum_{k=1}^{N} a_k)\sigma)d\sigma} \\[2mm]
&\approx \frac{a_i\exp(-a_i\tau)\Delta t\prod_{k \ne i}\exp(-a_k\tau)}{(\sum_{k=1}^{N} a_k)\exp(-(\sum_{k=1}^{N} a_k)\tau)\Delta t} \\[2mm]
&= \frac{a_i}{\sum_{k=1}^{N} a_k}.
\end{aligned}
\tag{55}
$$

The first equality in Eq. (55) is the definition of conditional probability. The second equality is based on that all reactions are independent of each other and thus the probability of the intersectional event equals the product of the probability of individual events. The third equality is based on that T_k and T are exponentially distributed random variables with parameters a_k and $\sum_{k=1}^{N} a_k$, respectively. The approximate equality on the fourth line holds for sufficiently small Δt.

Equation (55) proves Eq. (54).

Appendix D: MATLAB Code for Gillespie Simulation of Stochastic mRNA and Protein Dynamics

```
function [T, N] = Gillespie_mRNA_protein
% Gillespie simulation of mRNA and protein dynamics
%
% State vector = [nM, nP]
% Rxn:
% --> M, a1 = ksM, s1 = [1, 0]
% M -->, a2 = kdM * nM, s2 = [-1, 0]
% M --> M+P, a3 = ksP * nM, s3 = [0, 1]
% P --> , a4 = kdP * nP, s4 = [0, -1]
% Reseed the random number generator
rng('shuffle');
% Simulation parameters
qFinal = 5000; % total number of simulation steps
% Rate constants
ksM = 2.5; % --> M
kdM = 0.5; % M -->
ksP = 2; % M --> M + P
kdP = 0.05; % P -->
% Stoichiometry vectors
Nspecies = 2;
s1 = [1, 0]; % --> M
s2 = [-1, 0]; % M -->
s3 = [0, 1]; % M --> M+P
s4 = [0, -1]; % P -->
% Initialization (reserve space for T, N)
T = zeros(qFinal, 1);
N = zeros(qFinal, Nspecies);
N(1,:) = [0 0]; % initial molecule numbers at time 0
% Gillespie cycle
for q = 1:qFinal-1
 % Current number of molecules
nM = N(q,1); % # of mRNA
nP = N(q,2); % # of protein
% Calculate rxn propensities
a(1) = ksM;
a(2) = kdM * nM;
a(3) = ksP * nM;
a(4) = kdP * nP;
% Sum of rxn propensities
asum = sum(a);
% Sample for next reaction time
r1 = rand;
tau = (-1/asum)*log(r1);
% Sample for which reaction, update molecule number
r2 = rand;
if r2 < a(1)/asum
N(q+1,:) = N(q,:) + s1;
elseif r2 &lt; (a(1)+a(2))/asum
```

```
N(q+1,:) = N(q,:) + s2;
elseif r2 &lt; (a(1)+a(2)+a(3))/asum
N(q+1,:) = N(q,:) + s3;
else
N(q+1,:) = N(q,:) + s4;
end
% Update time
T(q+1) = T(q) + tau;
end
% Plot
subplot(1,2,1);
stairs(T,N(:,1),'k'); % mRNA trajectory
xlabel('Time (h)');
ylabel('mRNA number');
box off;
xlim([0 200]);
subplot(1,2,2);
stairs(T,N(:,2),'k'); % protein trajectory
xlabel('Time (h)');
ylabel('protein number');
box off;
xlim([0 200]);
```

Appendix E: MATLAB Code for Statistical Analysis of Results from Stochastic Simulation

The following code runs the stochastic simulation 1000 times, plots the histograms and calculates the mean, variance and CV for data at $t = 1, 10, 100$ h. To use the code, please save it as a separate file in the same folder where you save the code given in Appendix D.

```
% Repeat simulation 1000 times
Nsimu = 1000;
Tsample = [1;10;100]; % sampling times for averaging trajectories
Nspecies = 2;
Nsample = zeros(length(Tsample),Nspecies,Nsimu);
for i = 1:Nsimu
 [T, N] = Gillespie_mRNA_protein;
 % Sample at Tsample times
 [~,~,iTimeInterval] = histcounts(Tsample,T);
 Nsample(:,:,i) = N(iTimeInterval,:);
end
% Plot histograms
nedges_NM = -0.5:1:14.5; % edge points for the mRNA histogram
nedges_NP = -5:10:305; % edge points for the protein histogram
for i = 1:length(Tsample)
 subplot(length(Tsample), 2, 2*i-1);
 histogram(Nsample(i,1,:),nedges_NM,
 'Normalization','probability','FaceColor','k');
```

```
xlabel('mRNA number');
ylabel('Frequency');
title(['t = ' num2str(Tsample(i)) 'h']);
subplot(length(Tsample), 2, 2*i);
histogram(Nsample(i,2,:),nedges_NP,
'Normalization','probability','FaceColor','k');
xlabel('protein number');
ylabel('Frequency');
title(['t = ' num2str(Tsample(i)) 'h']);
end
% Calculate mean, variance and CV
for i = 1:length(Tsample)
mean_mRNA(i) = mean(Nsample(i,1,:));
var_mRNA(i) = var(Nsample(i,1,:));
CV_mRNA(i) = sqrt(var_mRNA(i))/mean_mRNA(i);
mean_protein(i) = mean(Nsample(i,2,:));
var_protein(i) = var(Nsample(i,2,:));
CV_protein(i) =
sqrt(var_protein(i))/mean_protein(i);
end
mean_mRNA
CV_mRNA
mean_protein
CV_protein
```

The following code computes the mean, variance and CV from a single trajectory. To use the code, please save it as a separate file in the same folder where you save the code given in Appendix D. Additionally, to simulate a long single trajectory, please increase the number of simulation steps in the code in Appendix D by 20 times, i.e., qFinal = 100000.

```
% Simulation
[T, N] = Gillespie_mRNA_protein;
qFinal = length(T);
% Remove initial transient
[~, qTransient] = min(abs(T-100)); % initial transient ~ t<100
NM = N(qTransient:qFinal-1, 1);
NP = N(qTransient:qFinal-1, 2);
Tau = T(qTransient+1:qFinal) - T(qTransient:qFinal-1); % time interval
between neighboring time steps ~ residence time of each state of molecule
counts
mean_mRNA = sum(NM.*Tau)/sum(Tau)
var_mRNA = sum(NM.^2.*Tau)/sum(Tau)-mean_mRNA^2;
CV_mRNA = sqrt(var_mRNA)/mean_mRNA
mean_protein = sum(NP.*Tau)/sum(Tau)
var_protein = sum(NP.^2.*Tau)/sum(Tau)-mean_protein^2;
CV_protein = sqrt(var_protein)/mean_protein
```

The following code plots histogram and computes the mean, variance and CV from a single trajectory, based on sampled data at regular time intervals. To use the code, please save it as a separate file in the same folder where you save the code given in Appendix D. Additionally, to simulate a long single trajectory, please increase the number of simulation steps in the code in Appendix D by 20 times, i.e., qFinal = 100000.

```
% Simulation
[T, N] = Gillespie_mRNA_protein;
qFinal = length(T);
% Remove initial transient
[~, qTransient] = min(abs(T-100)); % initial transient ~ t<100
NSS = N(qTransient:qFinal-1, :);
TSS = T(qTransient:qFinal-1);
% Sample the time trajectory at equally spaced time points over 90% of the
% whole time range (cutting the beginning and ending to avoid biased
% sampling)
Nsample = 10000;
Tsample_start = TSS(1)+(TSS(end)-TSS(1))*0.05;
Tsample_end = TSS(end)-(TSS(end)-TSS(1))*0.05;
Tsample = linspace(Tsample_start, Tsample_end,
Nsample);
% Get data values for each sample time point
[~,~,iTsample] = histcounts(Tsample,TSS);
Xsample = NSS(iTsample,:);
% Calculate mean, variance and CV
Average = mean(Xsample, 1);
Variance = var(Xsample, [], 1);
CV = sqrt(Variance) ./ Average;
% Plot histogram
nedges_NM = -0.5:1:14.5;
nedges_NP = -5:10:305;
 subplot(1, 2, 1);
 histogram(Xsample(:,1), nedges_NM, 'Normalization', 'probability',
 'FaceColor', 'k');
xlabel('mRNA number');
ylabel('Frequency');
title('single trajectory w/o transient'); subplot(1, 2, 2);
histogram(Xsample(:,2), nedges_NP, 'Normalization', 'probability',
 'FaceColor', 'k');
xlabel('protein number');
ylabel('Frequency');
title('single trajectory w/o transient');
end
```

References

1. Ingalls BP (2013) Mathematical modeling in systems biology, The MIT Press, p 408
2. Gillespie DT (1977) Exact stochastic simulation of coupled chemical reactions. J Phys Chem 81 (25):2340–2361
3. Ko CH et al (2010) Emergence of noise-induced oscillations in the central circadian pacemaker. PLoS Biol 8(10):e1000513

4. Vilar JM et al (2002) Mechanisms of noise-resistance in genetic oscillators. Proc Natl Acad Sci U S A 99(9):5988–5992
5. *BioNumbers*. Available from: https://bionumbers.hms.harvard.edu/search.aspx.
6. Elowitz MB et al (2002) Stochastic gene expression in a single cell. Science 297 (5584):1183–1186
7. Bar-Even A et al (2006) Noise in protein expression scales with natural protein abundance. Nat Genet 38(6):636–643
8. Newman JR et al (2006) Single-cell proteomic analysis of S. cerevisiae reveals the architecture of biological noise. Nature 441(7095):840–846
9. Taniguchi Y et al (2010) Quantifying E. coli proteome and transcriptome with single-molecule sensitivity in single cells. Science 329(5991):533–538
10. Eldar A, Elowitz MB (2010) Functional roles for noise in genetic circuits. Nature 467 (7312):167–173
11. Shaevitz JW, Fletcher DA (2007) Load fluctuations drive actin network growth. Proc Natl Acad Sci U S A 104(40):15688–15692
12. Dykman MI, McClintock PVE (1998) What can stochastic resonance do? Nature 391 (6665):344–344
13. Paulsson J (2005) Models of stochastic gene expression. Phys Life Rev 2(2):157–175
14. Thattai M, van Oudenaarden A (2001) Intrinsic noise in gene regulatory networks. Proc Natl Acad Sci U S A 98(15):8614–8619

Collective Molecular Motor Transport

Christopher Miles and Alex Mogilner

Introduction

Cells constantly need to move molecules and organelles throughout their interior. RNA produced in the nucleus must be exported to then be translated into proteins. Some of these proteins must then be imported back into the nucleus. Others are packaged into membrane vesicles, which in turn must be trafficked to the plasma membrane and exocytosed from the cell. The list of such examples has hundreds of entries [1]. An energetically 'cheap' way to transport microscopic particles is through diffusion [2]. The trick is done by Brownian movements: all particles in cytoplasm are bombarded by countless water molecules, and if these particles are smaller than a micron in size, momentous imbalances in hits by these smaller molecules results in random movement. This paradigm is surprisingly effective when particles only must spread over short distances. However, there are three major caveats that in many cases make diffusion ineffective. (1) Diffusion is direction-less, so if the cell need transport from a source to a well-defined target (e.g. from ER to Golgi), diffusion alone does not work. (2) Distance travelled by diffusion increases as a square root of time, while directional movement with constant speed generates displacement that grows linearly with time. Therefore, long-distance diffusive movements are slow. (3) Objects that are larger typically diffuse slower, so membrane vesicles, for example, diffuse far too slowly for cellular functions.

For all these reasons, evolution has provided cells with remarkable molecular machines called molecular motors, some of the most ancient complex proteins [3]. These proteins couple changes in shape to hydrolysis of ATP, hereby transducing chemical energy into mechanical work, much like a car burns gasoline to convert

C. Miles (✉) · A. Mogilner (✉)
Courant Institute and Department of Biology, New York University, New York, NY, USA
e-mail: Christopher.Miles@cims.nyu.edu; mogilner@cims.nyu.edu

© Springer Nature Switzerland AG 2021
P. Kraikivski (ed.), *Case Studies in Systems Biology*,
https://doi.org/10.1007/978-3-030-67742-8_13

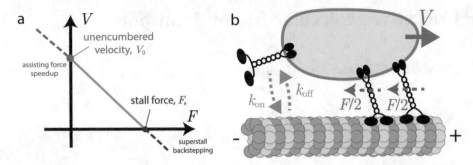

Fig. 1 Left: force-velocity curve for individual motors. Right: model diagram demonstrating the key components of the transport model: motors share force equally and bind and unbind stochastically

produced heat into force and movement. Thousands of studies resulted not only in atomic resolution of the motors' structures, but also in detailed mechanistic understanding of their mechanochemical cycles.

Here is a very crude and naïve caricature of how a generic motor works: imagine a molecule made of a 'body' (helix in Fig. 1) with two legs. The top of the body is attached to a cargo to be transported, like the big membrane vesicle depicted in Fig. 1. The legs can attach to a polar track such as a microtubule fiber shown in Fig. 1. Such tracks are laid down across the cell in an intricate pattern, much like a railroad network, so the motors can transport cargo from pretty much any place to any target. In the beginning of the cycle, both legs are bound to the same place on the track. Then, an ATP molecule binds to the motor, which triggers detachment of one leg. This leg, using energy of ATP, extends in the 'plus' direction and hovers above the track. Then, ATP is hydrolyzed, which brings the front leg down binding it firmly to the track. Now, the motor is like a person standing firmly on the ground with legs stretched apart. Finally, when the products of hydrolysis (ADP and phosphate group) are released, this induces the last shape change—the rear leg is pulled forward to the front leg. In the end of the cycle, the motor is in the same chemical and mechanical state as in the beginning, but it advanced one step forward, in the 'plus' direction. Then another step, then another, etc. Of course, real motors are not that simple, but they use similar principles.

There are two important characteristics of the motor's performance in moving cargo: rate of detachment from the track and the stepping rate. The former we need to know because occasionally, when motor is standing on one leg waiting for the other to reattach, this one attached leg detaches, and the motor floats away from the track. This is unavoidable—if the leg was attached too firmly, it would be impossible to drag it forward at the end of the mechanochemical cycle, so there is a trade-off between losing the cargo and ability to move at all. The distance a cargo moves before detachment is typically called the *run length*. The role of the stepping rate is more intuitive: while moving, we want to know how fast or slow the movement is. Both detachment and stepping are known to depend on the mechanical load applied to the motor—viscous resistance from dragging large cargo through the cytoplasm. Thus, we think of these rates as *functions* that characterize the motor: force-velocity relation, $F_{mot}(V)$, and force-dissociation relation, $k_{off}(F_{mot})$. The first tells us what force is acting against the motor moving at a given velocity (inversely, what velocity is against a given load force). The second tells us what is the rate with which the motor detaches providing a certain load force. Almost without exceptions, the velocity decreases with load force, and detachment rate increases with the force, which is obvious: one moves slower against a heavier load, and it is easier to rip a particle off by a stronger force.

It is very rare for a single motor to move cargo across the cell, and it is easy to understand why: one motor would have to bare the entire load of a very large cargo and move very slowly and detach often [4]. The solution to this problem is rather obvious—use teams of motors that operate together on the same cargo (Fig. 1), which is exactly what the evolution came up with [4]. Indeed, even if one motors detaches, another is still attached and can hold the cargo until the detached motor reattaches. Besides, two motors take on half of the load each, and thus could in principle move faster than one. But is it this simple? To check, we need a mathematical model of collective motor transport.

Transport by motors has other exciting twists. There are tens of different types of molecular motors, some transporting cargo in the plus direction on microtubules (like kinesin), others—in the opposite, minus, direction (like dynein). For example, large teams of kinesins and dyneins associate with the so-called IFT particles (platforms to which protein cargo binds) that are moving along super-long axons in nerve cells [5]. The purpose of this is rather clear: kinesins move one cargo in one direction—from the cell body, where the minus ends of most microtubules are, and then dyneins move another cargo from axonal terminals to the cell body. But what happens if both types of motors engage together, as actually happens often [6]? Intuition tells us that the cargo with opposing motors would get stuck. Is that so? We need a model to answer this question.

In this case study, we introduce, analyze, and simulate two closely related models—one describing a few motors of the same kind moving a cargo collectively, another—opposing motors engaged in a tug-of-war. Many such models were proposed; here we follow methods introduced in the pioneering paper [7]. What we want to focus on, and what the motors' problems are great example of, is *stochastic simulations*. Why?—because motors make steps and detach at random times, and because the motor numbers are usually small, and so the resulting collective behavior is very noisy. But there is a system to this noise, and the stochastic simulations, besides shedding light on this system, are also great fun.

Model

The force-velocity (F-V) relation, as mentioned above, describes the velocity of motor marching along the track as a function of the force opposing the motor. Without an opposing force, motors walk at an unencumbered velocity, V_0. With an opposing force, motors slow eventually stopping at a stall force, F_s. These force-velocity relations have been measured for a wide variety of motors and in many cases can be taken in the linear form:

$$F_{\text{mot}} = F_s\left(1 - \frac{V}{V_0}\right).$$

Note that this recovers the described behavior: when $F_{\text{mot}} = 0$, $V = V_0$ and when $F_{\text{mot}} = F_s$, $V = 0$. (Motors also exhibit exotic behavior when they are pushed ($F_{\text{mot}} < 0$) or feel superstall forces ($F_{\text{mot}} > F_s$) causing them to backstep, but we will not be exploring these behaviors further here.)

The cargo moves at velocity V due to motor stepping, as seen in Fig. 1. How do we determine this velocity? Let the number of motors currently driving the cargo to be n. Then, we assume that they each contribute an equal force. The intuition behind this is as follows: imagine the cargo as a chariot. The velocity of individual motors is determined by the force exerted on them via the F-V relation, so if one motor was pulling harder, it would be stepping at a different velocity and therefore not be in equilibrium with the others. Consequently, we can treat all the motors as identical and moving with the same velocity against the same force, F/n. Thus, the total force F is just the number of currently bound motors multiplied by the force F_{mot} which each of these motors exerts:

$$F = nF_{\text{mot}} = nF_s\left(1 - \frac{V}{V_0}\right).$$

We can rearrange this to be more useful for simulations:

$$V = V_0\left(1 - \frac{F}{F_s n}\right). \tag{1}$$

In other words, if we know the total load force F and number of currently bound motors n, we know the resulting velocity V.

The behavior of individual motors is stochastic. As they are stepping, motors randomly detach from and reattach to the track. The detachment rate is also measured and known to usually increase with the force exerted on the motor. We take the simplest form for the rate of detachment (there are deep physical arguments why such force-dissociation relation is realistic [7]):

$$k_{\text{off}} = \omega_0 \exp\left\{\frac{F_{\text{mot}}}{F_d}\right\}.$$

Here, ω_0 is the rate (in units of time^{-1}) of detachment from the track for an unencumbered motor. As the force on the motor increases, a detachment is more likely to occur. Another parameter here is the characteristic *detachment force*, F_d;

when the force on the motor is on the order of F_d, the motor detaches much faster than the freely moving motor.

After detachment, the leg of the molecular motor diffuses until a rebinding event, a complex process depending on geometry and other factors. For simplicity, we take the effective rebinding rate of a single motor to be a constant, $k_{on} = \omega_{on}$.

From our F-V relation, we know the velocity of the cargo for a given number of attached motors n, but this number evolves stochastically due to binding and unbinding at the described rates. The standard method for simulating such stochastic events is the **Gillespie algorithm**.

Gillespie Algorithm

Suppose we have a list of N possible types of stochastic events, each taking place with rates $r_1, \ldots, r_N$. These rates may change as time or the system evolves. We want an algorithm for choosing a stochastic sequence of events from this list, as well as the times at which they occur.

The main idea behind the Gillespie method [8] is that we need only choose the *next* event and time appropriately, and then repeat this process over and over. To do this, call τ_j the time at which reaction j occurs, ignoring all others. We are therefore looking for the time at which the first of these happens:

$$\tau_{next} = \min\{\tau_1, \ldots, \tau_N\}$$

and we must of course keep track of which type of event this corresponds to. The Gillespie method hinges on the fact that if these reaction times are modeled as exponential random variables (this is very common) with the prescribed rates, so $\tau_j \sim \exp(-r_j)$, then it can be shown that $\tau_{next} \sim \exp\left(-\sum_{j=1}^{N} r_j\right)$. That is, the time until the next reaction is also an exponential random variable with rate $r = \sum_{j=1}^{r} r_j$.

Thus, to simulate this time, we just need to generate a sample of an exponential random variable with this (known) parameter. This can be done easily in MATLAB by first generating a random uniform number w via w = rand (in Matlab, commands rand or randu generate a random number uniformly distributed on the interval $[0, 1]$) and transforming it $\tau_{next} = \left(\frac{1}{r}\right) \ln\left(\frac{1}{w}\right)$. (See details in [8].)

How do we decide which reaction occurred? Not all reactions should be equal. Intuitively, we might expect fast reactions should be more likely to occur and slow reactions less likely. To choose the right reaction appropriately, the probability of choosing reaction j is simply the weighted average of that rate compared to the others. That is, the probability of choosing reaction j is

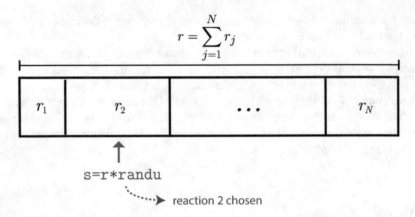

Fig. 2 Illustration of choosing a reaction in the Gillespie algorithm

$$p_j = \frac{r_j}{r} = \frac{r_j}{\sum\limits_{k=1}^{N} r_r}.$$

To implement this, we can generate a random (uniformly distributed on [0, 1]) number s and find the j such that $\sum\limits_{k=1}^{j-1} r_k < sr < \sum\limits_{k=1}^{j} r_k$. Visually, this corresponds to choosing at random a number between $[0, r]$ and choosing the reaction corresponding to the region this number falls into, as shown in Fig. 2. Note that this means we need *two* random numbers at each step: one for the time at which the reaction occurs and one to decide which reaction.

In summary, the Gillespie method is:

```
At each step:
    Compute r = r1+... rN
    Generate random number w = randu
    Compute next time tau_next = (1/r) log(1/w)
    Generate random number s = randu
    Find event j such that r1+...+rj-1 < r*s < r1+...+rj
    Perform event j, update r1,...,rN if necessary
```

Model 1: Transport by a Team

Equipped the Gillespie algorithm, we are now nearly ready to start modeling collective transport by molecular motors. The stochastic events deal with binding and unbinding of motors, but what is the velocity of the cargo between these events? Suppose we apply an external force F, then we know the velocity with n motors bound from F-V relation (1).

The binding and unbinding reactions occur at respective rates.

$$r_{n \to n+1} = (N - n)\omega_{\text{on}},$$

$$r_{n \to n-1} = nk_{\text{off}} = n\omega_0 \exp\left\{\frac{F}{n F_d}\right\}.$$

Why do both rates have multiplicative factors depending on n? The explanation in the Gillespie algorithm arises again here: if there are n motors that can unbind at exponentially distributed times with rate ω_{off}, then the rate at which the first one unbinds is $n\omega_{\text{off}}$. Similarly, there are $N-n$ unbound motors that each can bind with rate ω_{on}.

There are many parameters in the model. We can simplify our lives a bit by scaling forces, velocities, and time to become unitless quantities. Specifically, if we take $f = \frac{F}{F_s}$ and $v = \frac{V}{V_0}$, we have

$$v = \frac{V}{V_0} = \left(1 - \frac{F}{F_s n}\right) = \left(1 - \frac{f}{n}\right)$$

We scale all the times by k_{on} so

$$r_{n \to n+1} = (N - n), \quad r_{n \to n-1} = n\omega \exp\{f\alpha/n\},$$

where $\omega = \frac{\omega_0}{\omega_{\text{on}}}$ and $\alpha = \frac{F_s}{F_d}$. Now, we have reduced our simulation to just a few parameters: N, α, ω, f. The stall force of Kinesin-1 is $F_s = 6$ pN, the detachment force is $F_d = 3$ pN, and load forces are within this same range making $\alpha \approx 1 - 2$ *and* $f \approx 1$. The unbinding and binding rates are roughly $1 - 5$ s^{-1} [3], so we take their ratio $\omega = 1$ for simplicity.

The MATLAB code below simulates the first model with N motors of the same type. Initially, there is no external force $f = 0$ and at $t = 50$ an opposing force of $f = 0.5$ is exerted.

```
% set model/simulation parameters
omega=1, N=10, alpha=1, T=5000; f=0;

% initialize
n=N; time=zeros(size(T));
nn=N*ones(size(T)); v=ones(size(T));

for t=1:T % number of events
  rplus = (N-n); %binding rate
  if n==0 rminus=0; % unbinding rate
  else rminus = n*omega*exp(alpha*f/n); end
  r=rplus+rminus; % total rate
  T=-(1/r)*log(rand); % time until next reaction
  time(t+1)=time(t)+T; % update time
  if time(t)>50
```

```
    f = .5; % change external force at t=50
  end
  % choose reaction
  s=rand; if s<rplus/r n=n+1; % motor attaches
  else n=n-1; end % motor detaches
  nn(t+1)=n; % update motors
  if n == 0 v(t+1)=0; % update velocity
  else v(t+1)=(1-f/n); end %
end

subplot(2,3,[1 2]), stairs(time,v), % plot
subplot(2,3,3),
histogram(v, 'orientation','horizontal')
subplot(2,3,[4 5]), stairs(time,nn/N);
subplot(2,3,6),
histogram(nn/N, 'orientation','horizontal')
```

Model 2: Tug-of-War

Let the cargo be affiliated with N plus-end directed motors but also M minus-end directed motors. These two types of motors oppose each other in their preferred direction, but it is not clear what the resulting behavior of the cargo is.

As before, the individual motors can bind and unbind from the track, so let n and m be the currently bound numbers in each population. For simplicity, we assume them to have identical properties aside from their preferred direction.

Fig. 3 Model of molecular
motor tug-of-war

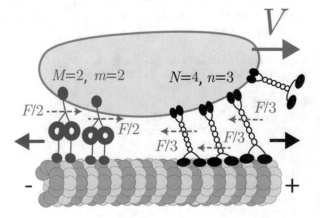

The net force between two groups of motors is split between the motors equally, meaning that

$$nF_+ = mF_- = F.$$

Note, that now F is not the load force on the cargo—such force is assumed to be insignificant—but rather the force stretching the cargo by the two motor groups. Again, we appeal to the argument that since the cargo is moving at some velocity V, all the associated motors must also be doing the same, otherwise an equilibrium would not be achieved. This means that

$$V = v_+\left(\frac{F}{n}\right) = v_-\left(\frac{F}{m}\right).$$

Plugging in the F-V relation for each of these, remembering that they only differ by their preferred direction (so $-V_0$ *for* the minus-directed population), we have

$$V = V_0\left(1 - \frac{F}{F_s n}\right) = -V_0\left(1 - \frac{F}{F_s m}\right).$$

This can be rearranged, and then plugged back in to solve for the force and velocity for a given number of attached motors

$$F = \frac{2F_s mn}{m+n}, \quad V = V_0\frac{n-m}{n+m}.$$

As we did before, we can reduce the number of parameters by scaling $f = \frac{F}{F_s}$ and $v = \frac{V}{V_0}$, which yields

$$f = \frac{2mn}{m+n}, \quad v = \frac{n-m}{n+m}.$$

Now, we just need the transition rates for n, m, which we can assume follow the same form, scaled to be unitless in the same manner as the previous model ($\omega = \frac{\omega_0}{\omega_{\text{on}}}$ and $\alpha = \frac{F_s}{F_d}$), yielding

$$r_{n,m\to n+1,m} = (N - n), \quad r_{n,m\to n-1,m} = n\omega\exp\left\{\frac{F/n}{F_d}\right\} = n\omega\exp\left\{\frac{2\alpha m}{m+n}\right\},$$

$$r_{n,m\to n,m+1} = (M - m), \quad r_{n,m\to n,m-1} = m\omega\exp\left\{\frac{F/m}{F_d}\right\} = m\omega\exp\left\{\frac{2\alpha n}{m+n}\right\}.$$

The MATLAB code for the tug-of-war scenario looks similar, just with four possible reactions now: binding and unbinding for both the + and − populations of motors.

```
% set model parameters
M=30; N=30; alpha=1.5; omega=1; n=N; m=M; T=5000;

% initialize
n=N; m=M; time=zeros(size(T)); nn=N*ones(size(T));
mm=M*ones(size(T)); v=ones(size(T));

for t=1:T
  Mon = (M-m)*omega; % bind rate -
  Non = (N-n)*omega; % bind rate +
  if m==0 f=0; Moff=0; Noff=n; % if only + motors
  elseif n==0 f=0; Noff=0; Moff=m; % if only -
  % else a mix
  else f=2*m*n/(m+n);
  Moff = m*exp(2*alpha*n/(m+n));
  Noff = n*exp(2*alpha*m/(m+n)); end
  % choose reaction time
  k=Mon+Non+Moff+Noff;
  T=-(1/k)*log(rand);
  time(t+1)=time(t)+T;
  % choose event
  r=rand; if r<Mon/k m=m+1; % - binds
  elseif r>Mon/k & r<(Mon+Non)/k n=n+1; % + binds
  elseif r>(Mon+Non)/k & r<(Mon+Non+Moff)/k
  m=m-1; % - unbinds
  else n=n-1; end % + binds
  nn(t+1)=n; mm(t+1)=m; v(t+1)=(n-m)/(n+m); % update
end

subplot(2,3,[1 2]), plot(time,v), %plot
subplot(2,3,3),
histogram(v, 'orientation','horizontal')
subplot(2,3,[4 5]), plot(time,nn/N,time,mm/M);
subplot(2,3,6),
histogram(nn/N, 'orientation','horizontal'); hold on;
histogram(mm/M, 'orientation','horizontal')
```

Results

Using the listed MATLAB codes, we simulated both models and show the results in Figs. 4 and 5. In the first model with just one population of motors, we see that with just a single motor, transport is not very robust. The motor detaches often and when faced with an opposing force, the net velocity decreases significantly. However, the team of 10 motors behaves cooperatively. Total detachments are rare and correspond to transportation across long distances. When faced with the same opposing force, the team's velocity changes very little. These simulations demonstrate the effective cooperativity of teams of motor transport.

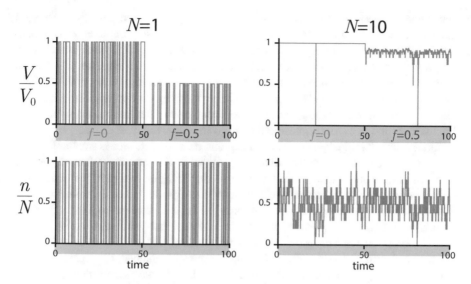

Fig. 4 Simulation results for model 1 for varying maximal number of motors, N. Top panels: the cargo velocity. Bottom panel: attached motor number, n. At $t = 50$, an external force in the opposing direction $f = 0.5$ is applied. For one motor, detachments occur frequently, and the cargo slows significantly to opposing loads. In contrast, teams of motors persist for long periods of time and are relatively unaffected by opposing forces

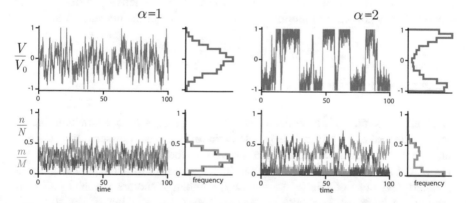

Fig. 5 Simulation results for model 2 for varying α. Top panels: the cargo velocity. Bottom panel: numbers of attached + motors n, and − motors, m. When α is low, the velocity fluctuates around 0 with motors stuck in a tug of war. For larger α, the cargo stochastically alternates moving in the + and − directions, with that corresponding family of motors dominating

Transport by opposing motor groups is quite a bit more random than that produced by model 1. This can be explained by model 2. Here, the question is: do the motors just get stuck opposing each other or does one team win? It turns out, both behaviors are possible! We see that for certain parameter values, the motors get stuck in a tug-of-war and the cargo moves nowhere. However, for larger values of α, we see much more exciting behavior: the cargo switches between periods of forward and backward transport. Here is the explanation: when parameter α is small, detachment force is much greater than the stall force. So, all motors detach with roughly the same, approximately force-independent and constant rate, at random, without correlation with one another. In this case, there is, on average, equal number of opposing motors attached at any time, pulling the cargo in opposite directions and almost balancing each other. There are, of course, random misbalances, but they fluctuate around zero force, causing the cargo to effectively just diffuse [9].

On the other hand, when parameter α is great, detachment force is smaller than the stall force—the motors detach relatively rapidly while still moving. What happens in this case is the winner-takes-all situation. Imagine that due to a rare fluctuation, the number of attached plus motors became much greater than that of the minus motors. The total load on *all* attached plus motors is the same as that on *all* attached minus motors, so the force per plus motor is much less than that per minus motor. So, the minus motors detach very rapidly one after another, and each next one detaches faster than the previous one—the fewer motors are still attached, the greater the load per motor, the greater the detachment rate is. The minus motors detach like an avalanche. The plus motors win and move the cargo collectively in the plus direction for a while; the minus motors keep attaching but keep being ripped off quickly by the overwhelming majority of the plus motors. This behavior persists until a rare fluctuation when a significant number of the minus motors attach at once and reverse the situation. As a result, the cargo makes long excursions back and forth. There are several reasons this might be beneficial for the cell's cargo delivery system, but this emergent behavior was not intuitively understood without mathematical modeling.

Discussion and Exercises

The ability to simulate is essential in understanding stochastic phenomenon like the ones studied here. However, it is problematic to try to learn lessons from a single simulation, since each should behave differently (they are random by definition!). The MATLAB code provided here generates single trajectories, but *statistics* (mean, standard deviation) of things like velocity, number of motors attached, are more informative in reporting results in stochastic systems. The fun of the stochastic model is that it is easy to explore them—we can just put the MATLAB file from the above into nested loops to repeat simulations hundreds of times, vary parameters and gather statistics of the multi-motor systems behaviors to understand patterns.

Exercise 1
Use Model 1 to find how the average run length before cargo detachment depends on each of the parameters: (a) N, (b) α, (c) F. Explain the results in words, qualitatively, as if you are trying to explain them to a collaborator with only biology background.

Exercise 2

Use Model 2 to find how the average cargo's run length before reversal and the average velocity of the cargo during this run depend on: (a) motor number N, M, (b) parameter α. Also, compute mean square displacement of cargo as function of time and calculate effective cargo's diffusion coefficient over a long time. What happens when the motors are not evenly matched? Again, explain the results in words, qualitatively.

The models considered here are very approximate, consisting of a motor in the model gliding on the track deterministically with a constant velocity before detachment. A more realistic description would respect the stochastic nature of individual steps. The model is a so-called 'Poisson stepper': the motor makes discrete steps of length δ along the track; the step is made after a random time τ, which is distributed exponentially. The stepping rate increases with the load force. At each step, there is a probability that the motor detaches before making the step; the detachment probability is an exponentially increasing function of force. Studies using probability theory have computed the effective velocity of such models [10] but again, we can resort to the fun and simplicity of stochastic simulations with the Gillespie algorithm.

Exercise 3

Write a code for a single Poisson stepper, then couple N such steppers to a cargo, and finally write a code to simulate equal numbers of opposing steppers coupled to the same cargo. Are there any qualitative differences in collective motor behavior between the stepper model and those with a deterministic velocity?

We conclude with noting that there are many more complexities that real motors demonstrate; the models we explored here are but caricatures. The force-velocity curves are not really linear [11]; the unbinding rates are not simple exponential functions of the force [12]; the motors are chemically regulated in addition to mechanical interactions; elastic connectors to the cargo produce complex forces and make 3D geometries of the microtubules and cargo significant [13]. Including these features and discerning their importance is not a cause of despair, but rather exciting avenues of future research. This case study could pave the way for your own possible future explorations.

References

1. Alberts B, Johnson AD, Morgan D, Lewis J, Roberts K, Raff M, Walter P (2014) Molecular biology of the cell, 6th edn. W. W. Norton & Company, New York
2. Berg HC (1993) Random walks in biology, Rev edn. Princeton University Press, Princeton
3. Howard J (2001) Mechanics of motor proteins and the cytoskeleton, 2nd edn. Sinauer Associates—Oxford University Press, Oxford
4. Gross SP, Vershinin M, Shubeita GT (2007) Cargo transport: two motors are sometimes better than one. Curr Biol 17:R478–R486
5. Signor D, Scholey JM (2000) Microtubule-based transport along axons, dendrites and axonemes. Essays Biochem 35:89–102

6. Gross SP (2004) Hither and yon: a review of bi-directional microtubule-based transport. Phys Biol 1:R1–R11
7. Müller MJ, Klumpp S, Lipowsky R (2008) Tug-of-war as a cooperative mechanism for bidirectional cargo transport by molecular motors. Proc Natl Acad Sci U S A 105:4609–4614
8. Mogilner A, Elston T, Wang H-Y, Oster G (2002) Molecular motors: theory. In: Fall CP, Marland E, Tyson J, Wagner J (eds) Joel Keizer's computational cell biology. Springer, New York, pp 321–355
9. Allard J, Doumic M, Mogilner A, Oelz D (2019) Bidirectional sliding of two parallel microtubules generated by multiple identical motors. J Math Biol 79(2):571–594
10. Miles CE, Lawley SD, Keener JP (2018) Analysis of nonprocessive molecular motor transport using renewal reward theory. SIAM J Appl Math 78(5):2511–2532
11. Kunwar A, Mogilner A (2010) Robust transport by multiple motors with nonlinear force-velocity relations and stochastic load sharing. Phys Biol 7(1):16012
12. Kunwar A, Tripathy SK, Xu J, Mattson MK, Anand P, Sigua R, Vershinin M, McKenney RJ, Yu CC, Mogilner A, Gross SP (2011) Mechanical stochastic tug-of-war models cannot explain bidirectional lipid-droplet transport. Proc Natl Acad Sci U S A 108(47):18960–18965
13. Bergman JP, Bovyn MJ, Doval FF, Sharma A, Gudheti MV, Gross SP et al (2018) Cargo navigation across 3D microtubule intersections. Proc Natl Acad Sci U S A:201707936

Principle of Cooperativity in Olfactory Receptor Selection

Jianhua Xing and Hang Zhang

Introduction

In this case study we will discuss how mathematical modeling provides mechanistic insight on an intriguing problem of olfactory receptor selection. Olfaction, or the sense of smell, is essential for the survival of most living organisms. Therefore, many organisms have developed sophisticated olfactory sensing systems. For example, millions of olfactory sensory neurons (OSNs) reside at the back the nasal epithelium of a human nose. Each neuron has many copies of transmembrane G-protein coupled olfactory receptors (ORs). Odorants bind to the receptors, trigger some conformational change, and the neurons send electric signals to the brain so we can smell different odors. Given that genes encoding ORs form the largest super-family of vertebrate genomes, an amazing observation is that each neuron only expresses one type of the receptors, actually one allele of the OR gene [1]. Richard Axel and Linda Buck received the 2004 Nobel Prize in Physiology or Medicine for discovering the receptors and this monoallelic expression phenomenon. But the mechanism for monoallelic OR expression remains unknown for decades. We will see that this seemingly "mission impossible" can be achieved from some very simple physics principles that we are familiar with from everyday life.

J. Xing (✉)
Department of Computational and Systems Biology, School of Medicine, Pittsburgh, PA, USA

Department of Physics and Astronomy, University of Pittsburgh, Pittsburgh, PA, USA

UPMC-Hillman Cancer Center, University of Pittsburgh, Pittsburgh, PA, USA
e-mail: Xing1@pitt.edu

H. Zhang (✉)
Department of Physics, Northeastern University, Boston, MA, USA

© Springer Nature Switzerland AG 2021
P. Kraikivski (ed.), *Case Studies in Systems Biology*,
https://doi.org/10.1007/978-3-030-67742-8_14

Physiology and Molecular Mechanisms

Physiology Background

Consider rat as an example. There are more than 1200 OR genes and over 500 pseudogenes. Pseudogenes can be transcribed to mRNAs, but the latter cannot be translated to proteins. That is, an OSN cell needs to select one and only one functional allele from thousands of possible candidates. It turns out that within the nasal epithelium the OSNs segregate into a number of regions. The OSNs within each region express only a subset of OR genes, while there is no spatial correlation on the OSNs expressing the same OR gene, i.e., their spatial distribution is random. Physiologically it is crucial that each OSN only expresses one type of the OR genes. OSNs expressing the same OR type bundle their axons together and form spherical olfactory glomeruli to send integrated electric signals to the brain, so the total amount of detected odorant molecules can be perceived. That is why each OSN needs to express only one type of the OR genes, otherwise the brain would receive ambiguous information about the odorant identity. In addition, all the OR genes should have approximately equal probability to be expressed within OSNs as a whole population so a variety of odors can be detected. Therefore the puzzle is, how can an OSN "count" and select one and only one receptor allele out of hundreds to thousands of possible choices, while at the population level OR expression can achieve diversity?

Summary of Known Experimental Information

Since the discovery of "one neuron, one type of receptor" phenomenon in 1990s, continuous efforts have been made to understand the molecular mechanism governing OR selection, and numerous studies converge onto two categories of regulation mechanisms, DNA cis-regulatory elements and histone modifications. Below we will only summarize some results closely relevant to construct the model discussed here.

Serizawa et al. identified a 2-kb sequence that acts as a cis-regulatory element for a cluster of downstream OR genes [2]. These researchers named the sequence H enhancer. Relocating the H enhancer to one OR gene leads to increased frequency of selecting this gene at the expense of decreased frequency of other genes in the OR cluster. Lomvardas et al. showed that the H enhancer in chromosome 14 actually can colocalize with the promoter of an OR gene in another chromosome with active transcription [3]. After adding an extra copy of the H enhancer in transgenic mice, they observed coexpression of an OR and a pseudogene. In a later study Lomvardas and coworkers further showed that multiple enhancer elements from different chromosomes colocalize with the promoter of the selected active OR allele and form an interacting network [4].

Markenscoff-Papadimitriou et al. found that the OR genes in an undifferentiated mouse OSN bear H3K9me3, a signature histone covalent modification mark of heterochromatin. After differentiation, one allele becomes H3K4m3, a mark typically activating gene transcription [5]. Similar epigenetic mark switch also takes place in zebrafish and *Drosophila* [6, 7]. One can view this epigenetic mark switch as a race among OR alleles with only one winner allowed. Participation of epigenetic modification in OR selection has been further confirmed with the observation that disruption of either histone methyltransferases or demethylases leads to violations of the rule of one-allele-activation and/or loss of diversity [7–9]. Lyons et al. noticed that during OSN differentiation a histone demethylase LSD1 is transiently expressed, and deletion of LSD1 before OSN maturation results in widespread loss of OR expression [8]. LSD1 removes both H3K4 and H3K9 methylation groups. Given that H3K4me3 is the mark to be added for the epigenetic switch, it seems inefficient to increase the expression of this bifunctional LSD1.

To confirm that the H3K9me3-to-H3K4me3 switch is necessary for OR selection, Lyon et al. performed an experiment using mice with one or more alleles of H3K9 methyltransferases (G9a and GLP) knocked out [9]. One would expect that such knockout tilts the tug-of-war between adding H3K9me3 and H3K4me3 onto nucleosomes towards the latter, thus facilitates OR activation. Surprisingly what they observed is that most OR genes were downregulated, and only a handful were upregulated; the more of the four copies of enzyme alleles were knocked out, the less OR genes were upregulated. Lyon et al. could not identify any significant difference between the promoters of the most upregulated ORs and the remaining ones in predicting the transcription-factor-binding-motifs [9]. This counterintuitive result complicates the simple picture of an H3K9me3-to-H3K4me3 switch in OSN maturation.

Besides enhancer elements and histone modifications, it is well recognized that a feedback loop must exists, which should be elicited by expression of the chosen functional OR gene, and functions to suppress further activation of other OR genes [2]. Expression of pseudogenes does not activate such negative feedback mechanism [2]. Recent studies reveal that expression of the winning allele causes endoplasmic reticulum stress and expression of enzyme Adcy3, which then down-regulates LSD1, completing a negative feedback loop and leading to an epigenetic trap that stabilizes the OR choice [8].

Formulation of the Problem

With all the above physiological considerations and known information, it is constructive to formulate the problem of OR selection as an engineering problem of multi-objective optimization that each olfactory system has evolved through natural selection,

1. Monolallelic expression:

    (a) Before differentiation, all OR genes should remain transcriptionally silent.
    (b) One and only one allele is stochastically selected to become transcriptionally active within a biologically relevant period of time (5–10 days for mice).
    (c) After differentiation the selected allele should be kept transcriptionally active while others remain inactive for the life time of an OSN, which is about 100 days for mice.
    (d) Existence of OR genes in hybrid (epigenetic) states should be discouraged to reduce potentially detrimental disruption of chromatin structure and dynamics. Notice that each OR allele has a number of nucleosomes. Before differentiation most nucleosomes of an allele bear the repressive mark. A selected OR allele has most nucleosomes bearing the active mark. A hybrid state refers to an intermediate stage between these two states. Intuitively this requirement is similar to request that for a race when the only winner reaches the finish line, all others should stay at the starting line.
    (e) If an OR pseudogene is selected it should be recognized to allow reselection of a functional OR gene.

2. Maximum diversity: Diversity of activated OR genes should be maximized so each gene has approximately equal probability of being expressed.

Mathematical Model

Here we will discuss a model developed by Tian et al. [10]. Most of the model description and results below are reproduced or adapted from this paper, and one can find more complete studies including a full list of model parameters in the original paper.

Figure 1 summarizes the mathematical model for the OR activation problem within one cell. We model an OSN cell with 100 alleles to recapitulate the selection process within a single zone of olfactory epithelium. Existing information suggests that an allele competes with its sister allele about equally as with other OR alleles. Therefore for simplicity we do not distinguish the gene identity of an allele, and just treat the OR genes within a cell as a number of individual alleles (Fig. 1a). To model the effect of pseudogenes, we assign N_p (=30) pseudo OR alleles and N_f (=70) functional OR alleles. The model can be divided into three parts: epigenetic (Fig. 1b), enhancer (Fig. 1c), and transcriptional regulation (Fig. 1d), as described in detail below.

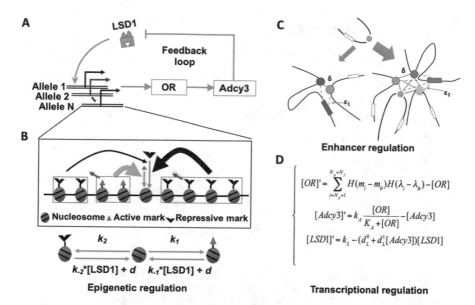

Fig. 1 Mathematical model of the experimentally revealed regulatory system of olfactory receptor activation. (**a**) Feedback regulated OR allele epigenetic activation. Each OSN contains N_p (=30) pseudo OR alleles and N_f (=70) functional OR alleles. Expression of an OR protein elicits a feedback to induce expression of enzyme $Adcy3$, which removes the demethylase LSD1. (**b**) Each allele is composed of a linear array of 41 nucleosomes. Each nucleosome bears active, no, or repressive mark, and a mark-bearing nucleosome facilitates an empty nucleosome to add the same mark in a distance dependent manner. A nucleosome changes its covalent modification state stochastically with the indicated rate constants. The methylation rate constants k_1 and k_2 are influenced by nearby nucleosomes. (**c**) Epigenetically active OR alleles compete for a finite number of enhancers, and the process is modeled through stochastic simulations. (**d**) Protein level changes are simulated by ordinary differential equations. See text for details

Epigenetic Dynamics

Each OR allele consists of a linear array of $N = 41$ nucleosomes, and each nucleosome can bear repressive H3K9 (R), no (E), or active H3K4 (A) methylations (Fig. 1b). Denote methylation state of a nucleosome R, E, and A as $s = -1$, 0, 1, respectively. Transitions between these states are governed by enzyme concentration dependent rates. Figure 2b gives forms of the overall transition rate constants for the four reactions. Specifically, demethylation steps R→E and A→E can take place either through stochastic exchange between nucleosome histones and the reservoir of unmarked histones with a turnover rate constant d, or through demethylation reactions with rates proportional to concentration of the catalyzing enzyme LSD1, which catalyzes both H3K4 and H3K9 demethylation. For simplicity we treat step-wise methylations/demethylations on a nucleosome as single steps, and treat participating enzymes other than LSD1 implicitly. One can propagate the nucleosome states with the tau-leap Gillespie algorithm [11]. An allele

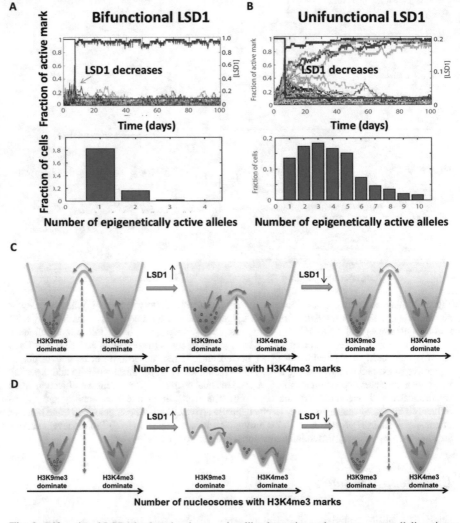

Fig. 2 Bifunctional LSD1 leads to barrier-crossing-like dynamics and ensures mono-allelic epige-netic activation. (**a**) Simulation results of the epigenetic dynamics. Top: Typical trajectories of the fraction of nucleosomes with active marks on each allele for 100 alleles (represented by different colors) within a cell. The temporal change of LSD1 level (blue curve, in relative unit) is also indicated. Bottom: Fraction of cells with various numbers of epigenetically active alleles at day 100. Sampled from 1000 independent simulations. (**b**) Same as A except for a hypothetical case that LSD1 only removes H3K9 methylation. (**c**) An analogous potential system during activation for the case in panel (**a**). (**d**) An analogous potential system during activation for the case in panel (**b**). In all simulations a cell has 100 OR alleles, and at time 0 the LSD1 level is elevated tenfolds from its basal value to simulate the onset of differentiation

(e.g., allele i) is epigenetically active if the fraction of active mark bearing nucleosomes in allele i, $\lambda_i \geq \lambda_\theta$, λ_θ being a cutoff fraction.

A key ingredient of epigenetic dynamics is that the methylation state change dynamics is cooperative between neighboring nucleosomes. That is, the rate of adding a specific methylation mark on a nucleosome is positively influenced by other nucleosomes bearing the same mark [12, 13]. Therefore, let us set the methylation rate constants k_1 and k_2 as functions of methylation states of other nucleosomes: k_1 (k_2) is promoted by H3K4 (H3K9) methylation in other nucleosomes, and the influence decreases with the nucleosome spatial separation. Mathematically we set the methylation rate constants for an empty nucleosome i as,

$$k_1^i = k_1^0 \exp \left(\sum_{j \neq i} \frac{\mu}{|i-j|} \delta_{s_j, -1} \right), k_2^i = k_2^0 \exp \left(\sum_{j \neq i} \frac{\mu}{|i-j|} \delta_{s_j, 1} \right),$$ where the sum is over all

other nucleosomes, and δ is a Kroneck-delta function. That is, each of the other nucleosomes influences the nucleosome to add the same mark of the latter, and the influence decreases with the nucleosome spatial separation. An insulating boundary is assumed. To model mutants with different degrees of H3K9 methyltransferase knockout (E_{K9M}^R and E_{K9M}^{RR}), we can reduce the value of k_2^0 for the wild type (WT) OSN to 90% and 80%, respectively.

Some DNA sequence specific molecular species, such as transcription factors and noncoding RNAs, help on recruiting histone modification enzymes, and form a nucleation region with higher enzymatic rate constants than other nucleosomes have. To model such effect, one can further choose some nucleosomes (e.g. the three located at the center of the nucleosome array) as the nucleation region, and assign higher enzymatic rate constants. One can use simulations to show that whether or not including this piece of biological detail in the model does not alter the qualitative picture of the OR selection mechanisms discussed below.

Enhancer Binding Dynamics

For simplicity let us assume that there are M enhancers available for an OR genomic cluster (see Fig. 1c), and we treat them equally. Each enhancer can bind to the epigenetically active i-th OR allele with a free energy of binding $\varepsilon_i < 0$, and can interact with any other enhancer bound to the same allele with energy $\delta < 0$. Enhancer binding to alleles with repressive marks is weak and can be neglected [4]. Let us further assume that there is no unbound enhancer. This assumption is not essential for the present discussions and can be easily removed at the expense of a few additional parameters.

Because of the term δ, enhancer binding to active alleles is cooperative: when two or more epigenetically active alleles compete for the enhancers, an enhancer preferentially binds to the allele that already has more enhancers bound since more enhancer-enhancer interactions can form. Actually a recent experimental study confirmed this predicted cooperative binding of multiple enhancers, and identified it as a super enhancer structure formed by multiple chromosomes [14].

If there is only one allele (e.g., allele i) being epigenetically active, then all the M enhancers bind to this allele to render the latter transcriptionally active. If the number of epigenetically active alleles >1, we model the enhancer binding dynamics explicitly. Consider a pair of epigenetically active alleles i and j. An enhancer can jump from allele i to j with rate $k_{i \to j} = v \exp \left\{ \frac{0.5 \left[\varepsilon_i - \varepsilon_j + \left(m_i - 1 - m_j \right) \delta \right]}{k_B T} \right\}$ to satisfy the detailed balance requirement, where v is a prefactor, m_i and m_j are the number of enhancers bound to allele i and j before the jump, respectively, $\sum_i m_i = M$, k_B is the Boltzmann constant, and T temperature. Here we simply choose the factor 0.5 to satisfy the detailed balance requirement, i.e., $k_{i \to j}/k_{j \to i}$ equals to the Boltzmann factor corresponding to the system free energy after the transition divided by that prior to the transition. One can simulate enhancer binding dynamics with the tau-leap Gillespie simulations.

Transcriptional Dynamics

We use three ordinary differential equations (ODEs) to model the expression dynamics of OR gene, Adcy3, and LSD1 (Fig. 1d). $H(x)$ is a Heaviside function, which assumes value 0 for $x < 0$, and 1 otherwise. The terms $H(\lambda_i - \lambda_\theta)$ and $H (m_i - m_\theta)$ ensure that an allele is transcriptionally active only if it is epigenetically active ($\lambda_i \geq \lambda_\theta$) and the number of bound enhancers is larger than a threshold value m_θ. The equation of Adcy3 reflects OR activation on Adcy3 synthesis, and the equation of LSD1 reflects Adcy3 facilitated LSD1 degradation. One can propagate the equations with an ODE solver simultaneously with the stochastic simulations on the epigenetic and enhancer binding dynamics.

Results

The top of Fig. 2a shows a typical simulated trajectory for differentiation of one OSN. Notice before differentiation, all the alleles are in a repressive mark dominated state (Objective 1a). With elevated LSD1, all alleles remain fluctuating in this state until one switches to an active mark dominated state, elicits the negative feedback to reduce LSD1 (Objective 1b). Then this active allele remains active while others stay inactive for ~100 days (Objective 1c). A remarkable observation is that except the allele being selected, all other alleles remain inactive throughout the period (Objective 1d). Sampling over 1000 cells reveals that throughout their lifespan most of the OSNs only have one allele epigenetically activated, while a small fraction has two and rarely three alleles epigenetically activated (Fig. 2a bottom). In comparison, the top of Fig. 2b shows a representative trajectory in a hypothetical but intuitively seemingly attractive case that LSD1 only removes the repressive mark. Compared to the model used in Fig. 1a, the only difference here is that the rate constant for active mark removal (see Fig. 1b) is independent of LSD1. Notice with elevated LSD1, all alleles start to increase the fraction of active marks albeit at different rates. After one allele becomes active to elicit degradation of LSD1, a number of alleles are in the middle of transition to the active mark dominated state, and relax either to the active or repressive mark dominated state slowly due to decreased LSD1 level. Repeated simulations reveal that in this case it is difficult to achieve monoallelic epigenetic activation (Fig. 2b bottom).

One can provide an intuitive mechanistic explanation for the above results by making analogy to particles moving under thermal fluctuations along a potential. Before differentiation, an OR allele is at the repressive mark dominated state, and transition to the active mark dominated state is prohibited due to lack of demethylases, analogous to a system trapped in a double-well shaped potential with a very high barrier (Fig. 2c left). Starting with the repressive mark dominated state, after initiation of OSN differentiation transient increase of LSD1 demethylates nucleosomes, and allows changing of methylation states in the nucleosomes. As a consequence, small patches of active mark nucleosomes may form, but are flanked by extended regions of repressive mark nucleosomes. Consider a newly formed unmarked nucleosome. It may add back active or repressive mark, but with a larger probability for the latter because of the cooperativity of methylation among nucleosomes and the dominance of repressive marks at the current stage (Fig. 1b). Consequently, the active mark patches grow and shrink stochastically, and are hard to expand. Nevertheless, when an active patch reaches a critical size - as a rare event, it is able to propagate spontaneously and generate an epigenetic conversion of the OR gene into the active mark dominated state. That is, LSD1 increase resembles lowering the transition barrier between the double-well shaped potential shown in the previous section, and allows rare transition to happen (Fig. 2c middle). Once one allele converts to the active mark dominated state, and triggers the negative feedback loop to remove LSD1, the system is kinetically trapped again with high "transition barrier" (Fig. 2c right). The converted allele is kept active with active marks, while the remaining alleles bear repressive marks. A prominent feature of this barrier-crossing-like dynamics is that throughout the process the probability of having an allele with hybrid pattern of epigenetic marks is low, and most alleles only fluctuate around the repressive mark dominated state (see also Movie at http://movie-usa.glencoesoftware.com/video/10.1073/pnas.1601722113/video-1).

The scenario with a unifunctional LSD1 shows qualitatively different ratchet-like dynamics. Again for an allele that is originally at a repressive mark dominated state (Fig. 2d left), an empty nucleosome has higher probability to gain back a repressive mark, and a low but finite probability to gain an active mark. Different from the bifunctional LSD1 case, since here the enzymatic activity of removing the active mark is low, the patches of nucleosomes with active marks increase size unidirectionally—a ratchet dynamics. It is analogous of having particles jumping a generally downhill potential (Fig. 2d middle). All particles move downhill, while some move faster than others do. After one particle reaches the destination and the negative feedback takes effect, other particles are likely in the middle and relax to either side of the double well potential (Fig. 2d right), so it is hard to achieve single allele activation.

If a pseudogene allele becomes epigenetically active, it cannot activate the negative feedback loop, so the selection process continues until a functional OR gene allele is selected (Fig. 3a) (Objective 1e). Occasionally multiple functional OR gene alleles also become epigenetically active, then they compete for the enhancers. Because of cooperative enhancer binding, only one allele having sufficient bound enhancers to be transcriptionally active (Fig. 3b). The enhancers can collectively shift from binding to one allele to another one due to thermal fluctuations (Fig. 3b left and middle), which explains the observed switching of OR gene selection. Suppose one of the alleles binds to an enhancer just slightly stronger than others do, say less than thermal energy of one $k_B T$, due to for example slight differences on sequences or proximity of the enhancers to an allele. This small difference is hard to detect experimentally. Then multi-enhancer binding can amplify the difference to a few $k_B T$, and this allele dominates the competition (Fig. 3b middle and right). This unequal competition leads to a seemingly astonishing prediction with H3K9 methyltransferase reduction. While the knockout leads to expected increase on cells with multiple alleles becoming epigenetically active alleles (Fig. 3c), most OR genes are transcriptionally down regulated (Fig. 3d). The more H3K9 methyltransferase enzymatic activity is reduced, the more OR genes down regulated and the less up regulated (Fig. 3e). These predictions are exactly what observed experimentally by Lyon et al. (Fig. 3f) [9]. Also Lyons et al. could out detect noticeable differences on the sequences of the upregulated OR genes from the downregulated ones, implying that the differences are subtle as we discussed above.

The model not only predicts the puzzling experimental observations, but also provides a simple mechanistic explanation. Here we use a toy system to illustrate why the enhancer-mediated regulation, when combined with weak or no epigenetic selection, causes reduced diversity of OR expression (Fig. 3g). Suppose L (=4) OR alleles exist in a zone, and these alleles have strong (allele 1), medium (allele 2), and weak (alleles 3 and 4) binding strength to the enhancers, respectively. The epigenetic activation step is stochastic and each allele has roughly equal probability $1/L$ to be chosen. For WT OSNs, most cells have only one epigenetically active allele, and the allele binds to the enhancers and becomes transcriptionally active as well. Therefore the overall transcriptional probability of each allele in the zone is $\sim \frac{1}{4}$. On the other hand, with reduced H3K9 methyltransferase level (E_{K9M}^{R} and E_{K9M}^{RR}, or G9a KO and G9a/GLP dKO, respectively), an OSN may have multiple epigenetically active OR alleles. For simplicity of argument let us assume that in a cell three alleles compete for enhancers. Since each allele has approximately the same probability of becoming epigenetically active, there are four possible combinations with equal probability, (123), (124), (134) and (234). As an allele with stronger enhancer binding dominates transcription, one expects that the first three combinations mainly express allele 1, and the last one expresses allele 2. That is, the expression of allele 1 is upregulated

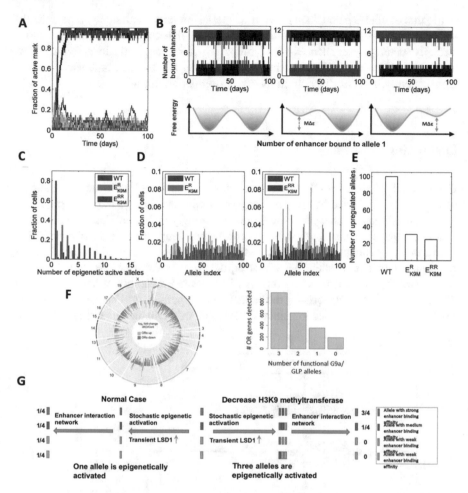

Fig. 3 Competition of cooperatively bound enhancers further reduces co-expression of multi-allele ORs. (**a**) Simulated allele trajectories of one cell with two epigenetically active alleles. (**b**) Simulated dynamics of enhancers binding to two epigenetic active alleles corresponding to the cell in panel A with the same (left) or different (by $\Delta\varepsilon = \pm 0.5\ k_B T$, middle and right) binding affinity. Also shown are schematic free energy profiles. (**c**) Simulated distribution of 1000 cells with various numbers of epigenetically active alleles under E_{K9M}^R, E_{K9M}^{RR} and WT on day 100. (**d**) Fractions of overall protein expression of each allele simulated with a population of 1000 cells under E_{K9M}^R and E_{K9M}^{RR}. E_{K9M}^R and E_{K9M}^{RR} refer to H3K9 methyltransferase level reduced and further reduced comparing to those with WT, respectively. (**e**) The number of transcriptionally upregulated alleles under E_{K9M}^R, E_{K9M}^{RR} and in WT. (**f**) Experimentally measured genome-wide OR gene expression under H3K9 methyltransferase knockout (left), and the number of transcriptionally upregulated OR genes under different levels of H3K9 methyltransferase knockout in mice (right) by Lyons et al. [9]. (**g**) Schematic illustration on the mechanism of reduced OR expression diversity with E_{K9M}^R and E_{K9M}^{RR} compared to that in WT

while that of alleles 3 and 4 are down regulated. Similarly, with more epigenetically active alleles coexisting in individual OSNs, the alleles that bind enhancers more strongly secure greater chances of outcompeting other epigenetically active allele in an OSN and getting expressed, whereas the weaker alleles have little such chance. Consequently, the OR diversity in the OSN population diminishes.

Discussion

Figure 4 summarizes the OR selection mechanism revealed by the model. A subset of the alleles is selected by the zonal segregation. Then they go through epigenetic barrier crossing and enhancer competition. Seemingly redundant, having both these two selection mechanisms in a definite order is necessary to solve the two-objective optimization challenge. The barrier crossing mechanism works well for the diversity requirement that each allele has about equal probability to be selected. It works pretty well for monoallelic selection but with a finite probability of having multiple alleles selected. In comparison, the enhancer competition mechanism works perfectly for monoallelic selection since the number of available enhancers is limited,

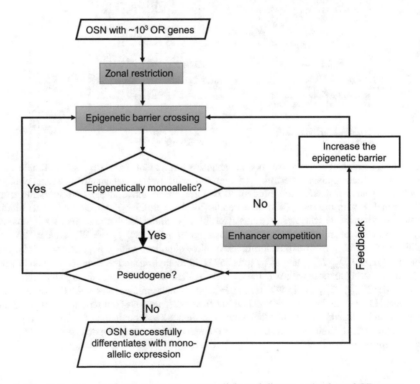

Fig. 4 The three-layer mechanism ensures mono-allele and diverse activation of OR genes

but works badly for diversity since it strongly biases towards a small subset of OR genes. During the selection process, most of the cells only have one epigenetically active and thus transcriptional active allele. In rare cases more than one allele is epigenetically activated, the enhancer competition mechanism ensures that only one epigenetically and transcriptional active allele exists, at the expense of some diversity but not much. That is, the selection procedure prioritizes monoallelic selection to be rigorously satisfied, and then maximize the diversity.

The model discussed here shows how Nature completes a seemingly impossible mission of diverse and monoallelic OR selection using some simple physics, namely, the principle of cooperativity, both in the epigenetic barrier crossing and enhancer competition steps. Cooperativity means that individual elements are dependent on others, and this dependence results in emergent collective phenomena absent from a system with only independent elements. For example, water molecules interact with each other that lead to the abrupt transition between liquid and solid phases. After a great performance the audience may spontaneously clap hands from random to synchronized rhythmic patterns through listening to others and adjusting accordingly. Another well-known example is binding of oxygen molecules to hemoglobin. A hemoglobin molecule has four oxygen binding sites. Binding of the first oxygen molecule can induce allosteric conformational change of the protein and facilitates binding of the second oxygen molecule, which then induces further protein conformational change. Consequently the more oxygen molecules already bound, the stronger binding of the next oxygen molecule. This is an example of positive cooperativity, and results in a sigmoidal shaped binding curve relative to the partial pressure of oxygen. That is, it is more likely to observe a hemoglobin molecule either with no or with four oxygen molecules bound, than in intermediate binding states. The existence of populated collective states and having depleted population of intermediate states are characteristics of positive cooperativity, which are exploited in the OR selection mechanism revealed by the model examined in this chapter. This mechanism is an example that simple principles often underlie seemingly complex phenomena.

As we have mentioned, the model examined in this chapter is minimal and generic, with lots of abstraction and removal of details of the real OR system. Future studies can expand the model to incorporate more details, such as specificity of the enhancer elements, three-dimensional structures of the chromatins, and parameters for specific OR clusters, so it can make more specific predictions on individual OR genes. Also the model predicts that elevating the level of bifunctional LSD1 is necessary to generate the barrier-crossing-like dynamics in the epigenetic activation step. There is debate on whether LSD1 is bifunctional. If not, then the model predicts that at least two types of demethylases for removing H3K4 and H3K9 methylation should be elevated during differentiation, and the model can be slightly modified to incorporate such information. Given our goal here is for revealing the basic mechanism of OR selection, and these details do not qualitatively change the mechanism summarized in Fig. 4, they are not necessarily to be included in the model. This simplification exemplifies a rule of thumb in modeling studies called Occam's razor, or as the words put by Einstein, "everything should be made as simple as possible, but no simpler".

Exercise

In this exercise, you are asked to write a computer code to simulate the enhancer binding model described in the main text, and reproduce the results shown in Fig. 3b. Consider two alleles that are epigenetically active, and the number of enhancers $M = 12$. You will simulate four cases, (A) symmetric case: $\varepsilon_1 = \varepsilon_2 = \varepsilon$, $\delta = -0.5\,k_BT$; (B) asymmetric case: $\varepsilon_1 = \varepsilon$, $\varepsilon_2 = \varepsilon + 0.5\,k_BT$, $\delta = -0.5\,k_BT$; (C) symmetric without cooperativity: $\varepsilon_1 = \varepsilon_2 = \varepsilon$, $\delta = 0$; (D) asymmetric case without cooperativity: $\varepsilon_1 = \varepsilon$, $\varepsilon_2 = \varepsilon + 0.5\,k_BT$, $\delta = 0$.

1. For each case, simulate a long or the multiple trajectories, plot histograms of observing m_1 enhancers bound to allele 1 (so $m_2 = M - m_1$ enhancers bound to allele 2).
2. Convince yourself that the total free energy of a configuration with m_1 enhancers bound to allele 1 is $E_{m_1} = m_1\varepsilon_1 + m_2\varepsilon_2 + \frac{m_1(m_1-1)}{2}\delta + \frac{m_2(m_2-1)}{2}\delta$. Then the equilibrium probability of observing such configuration is given by the Boltzmann distribution, $P_{m_1} = \frac{\frac{M!}{m_1!m_2!}\exp\left(-E_{m_1}/k_BT\right)}{\sum_{m_1=0}^{m_1=M}\frac{M!}{m_1!m_2!}\exp\left(-E_{m_1}/k_BT\right)}$. Compare the Boltzmann distribution with what you obtained in question 1.
3. Compare the symmetric (asymmetric) case with and without cooperativity, and discuss how the qualitative shape of the distribution changes.

For reference a sample Matlab code is provided.

Matlab Code

We provide a main code and a subroutine function to simulate the enhancer binding model as described in the exercise. You will need to save the two codes as "OR_ST_traj_bookchapter.m", and "simulation.m".

1. main code

```
%this program will generate a trajectory for 2 allele competing for M
enhancers
clear;
tic % set up timer
%%%%%%%%%%%%%%%%%%%%%%%%
% these are the parameters one can modify
%parameter initialization
Mtotal = 12; % total enhancer number
delta = -0.5; % coopertivity energy, unit in kBT, this can be changed
according to the excises.
v = 1; % transition coefficient
N_init = 1e6; % total initialization step (to relax to equilibrium before
sampling)
Nstep = 1e6; % total simulation step
%%%%%%%%%%%%%%%%%%%%%%%%
```

```
M1 = randi([0 Mtotal],1,1); %number of enhancers bind to allele 1
M2 = Mtotal - M1; %number of enhancers bind to allele 2


% symmetric case, same binding energy towards two alleles
[epsilon1, epsilon2] = deal(0, 0);


%--------------------------------------------------------%
%run the simulation N_init steps, so that the system can relax to the
%equilibrium state (to avoid the initial condition being far from
equilibrium state)
%--------------------------------------------------------%
dt_init = zeros(N_init,1); % time array for each simulation step
M1_record_init = zeros(N_init,1); % enhancer number which binds to
allele 1
% call the gillespie simulation sub-routine
[dt_init, M1_record_init] = simulation(N_init, v, epsilon1, epsilon2,
delta, M1, M2, Mtotal, dt_init, M1_record_init);


%--------------------------------------------------------%
% symmetric simulation begin
%--------------------------------------------------------%

dt = zeros(Nstep,1); % time array for each simulation step
M1_record = zeros(Nstep, 1); % enhancer number which binds to allele 1

% call the Gillespie simulation sub-routine
[dt, M1_record] = simulation(Nstep, v, epsilon1, epsilon2, delta, M1,
M2, Mtotal, dt, M1_record);


%--------------------------------------------------------%
% Here comes the asymmetric case simulation
%--------------------------------------------------------%

% re-initialize M1, or you can use M1 from the previous case
M1 = randi([0 Mtotal], 1, 1);
M2 = Mtotal-M1;
dt2 = zeros(Nstep,1);
M1_record2 = zeros(Nstep,1);

% asymetric case
[epsilon1, epsilon2] = deal(0, 0.5);

%--------------------------------------------------------%
% run the simulation N_init steps, so that the system can relax to the
% equilibrium state (to avoid the initial condition being far from
equilibrium state)
%--------------------------------------------------------%
dt2_init = zeros(N_init,1); % time array for each simulation step
M1_record2_init = zeros(N_init,1); % enhancer number bound to allele 1
% call the Gillespie simulation sub-routine
[dt2_init, M1_record2_init]=simulation(N_init, v, epsilon1,
epsilon2, delta, M1, M2, Mtotal, dt2_init, M1_record2_init);
```

```
%----------------------------------------------------------%
% Asymmetric simulation begin
%----------------------------------------------------------%

M1_record2 = zeros(Nstep,1);
[dt2, M1_record2]=simulation(Nstep, v, epsilon1, epsilon2, delta, M1,
M2, Mtotal, dt2, M1_record2);

%%
%----------------------------------------------%
% simulation result trajectory plot
%----------------------------------------------%

set(gca,'DefaultTextFontSize',32)

%find the xrange:

xmax = min(dt(end), dt2(end));

if isempty(min(find(dt > xmax)))
 dt_idx = length(dt)
else
 dt_idx = min(find(dt > xmax))
end

if isempty(min(find(dt2 > xmax)))
 dt_idx = length(dt2)
else
 dt2_idx = min(find(dt2 > xmax))
end

%Symmetric case
subplot(2,1,1);
plot(dt(1:dt_idx), M1_record(1:dt_idx))
ylim([0 Mtotal])
xlim([0 xmax])

ylabel('Number of Enhancers bound to allele 1', 'FontSize', 12)
xlabel('Simulation time (a.u.)', 'FontSize', 12);
title_str = sprintf('Symmetric case: \crdelta = %1.1f \critk_BT',
delta);
title(title_str,'FontSize', 14)

%Asymmetric case
subplot(2,1,2);
plot(dt2(1:dt2_idx), M1_record2(1:dt2_idx))
ylim([0 Mtotal])
xlim([0 xmax])

ylabel('Number of Enhancers bound to allele 1', 'FontSize', 12)
xlabel('Simulation time (a.u.)', 'FontSize', 12);
```

```
title_str=sprintf('Asymmetric case: \crdelta = %1.1f \critk_BT,
\crDelta\crepsilon = %1.1f \critk_BT', delta, epsilon2);
title(title_str, 'FontSize', 14);


%%

%**********************************************
% Here comes the dwelling time statistics
%**********************************************

% dwelling time statistics
M1_record_t = zeros(Mtotal+1, 1);
for i = 1:length(M1_record) %length(M1_record)
 idx = M1_record(i)+1;
 M1_record_t(idx) = M1_record_t(idx) + dt(i); %dt(idx)
end
Normalized_M1_record_t = M1_record_t/sum(M1_record_t);

M1_record2_t = zeros(Mtotal+1, 1);
for i = 1:length(M1_record2) %length(M1_record)
 idx = M1_record2(i)+1;
 M1_record2_t(idx) = M1_record2_t(idx) + dt2(i);
end
Normalized_M1_record2_t = M1_record2_t/sum(M1_record2_t);

ymax = max(max(Normalized_M1_record_t), max
(Normalized_M1_record2_t));
ymax_lim = ymax + 0.3*ymax;

% Symmetric case
subplot(2,1,1)
bar([0:Mtotal],Normalized_M1_record_t)
xlim([-1 Mtotal+1])
ylim([0 ymax_lim])
xlabel('Number of enhancers bound to Allele 1', 'FontSize', 12);
ylabel('Normalized dwelling time (Density)', 'FontSize', 12)

title_str=sprintf('Symmetric case: \crdelta = %1.1f \critk_BT, \n
dwelling time distribution', delta);
title(title_str,'FontSize', 14);

% Asymmetric case
subplot(2, 1, 2)
bar([0:Mtotal], Normalized_M1_record2_t)
xlim([-1 Mtotal+1])
ylim([0 ymax_lim])
xlabel('Number of enhancers bound to Allele 1','FontSize', 12);
ylabel('Normalized dwelling time (Density)', 'FontSize', 12)

title_str = sprintf('Asymmetric case: \crdelta = %1.1f \critk_BT,
\crDelta\crepsilon = %1.1f \critk_BT, \n dwelling time distribution',
```

```
delta, epsilon2);
title(title_str, 'FontSize', 14);


toc
```
2. subroutine for Gillespie simulations


```
% This function performs Gillespie simulations
function [dt, N1_record] = simulation(Nstep, v, epsilon1, epsilon2,
delta, N1, N2, Ntotal, dt, N1_record)

for step = 1:Nstep
%step
k1 = v*N1*exp(0.5*(epsilon1-epsilon2+(N1-N2-1)*delta)); %1-->2, 1--
k2 = v*N2*exp(-0.5*(epsilon1-epsilon2+(N1-N2+1)*delta)); %2-->1, 1++

alpha0 = k1 + k2;
alpha = [k1/alpha0 1];
r1 = rand;
r2 = rand;
dt_tempt = 1/alpha0*log(1/r1);

if step == 1
dt(step) = dt_tempt;
else
dt(step) = dt(step-1) + dt_tempt;
end


N1_record(step) = N1;

if alpha(1)>r2 %select k1
N1 = N1 - 1;
else
N1 = N1 + 1;
end
N2 = Ntotal - N1;
end
```


References

1. Rodriguez I (2013) Singular expression of olfactory receptor genes. Cell 155(2):274–277
2. Serizawa S et al (2003) Negative feedback regulation ensures the one receptor-one olfactory neuron rule in mouse. Science 302(5653):2088–2094
3. Lomvardas S et al (2006) Interchromosomal interactions and olfactory receptor choice. Cell 126 (2):403–413
4. Markenscoff-Papadimitriou E et al (2014) Enhancer interaction networks as a means for singular olfactory receptor expression. Cell 159(3):543–557
5. Magklara A et al (2011) An epigenetic signature for monoallelic olfactory receptor expression. Cell 145(4):555–570

6. Sim CK et al (2012) Epigenetic regulation of olfactory receptor gene expression by the Myb-MuvB/dREAM complex. Genes Dev 26(22):2483–2498
7. Ferreira T et al (2014) Silencing of odorant receptor genes by G protein betagamma signaling ensures the expression of one odorant receptor per olfactory sensory neuron. Neuron 81 (4):847–859
8. Lyons DB et al (2013) An epigenetic trap stabilizes singular olfactory receptor expression. Cell 154(2):325–336
9. Lyons DB et al (2014) Heterochromatin-mediated gene silencing facilitates the diversification of olfactory neurons. Cell Rep 9(3):884–892
10. Tian XJ et al (2016) Achieving diverse and monoallelic olfactory receptor selection through dual-objective optimization design. Proc Natl Acad Sci U S A 113(21):E2889–E2898
11. Cao Y, Gillespie DT, Petzold LR (2006) Efficient step size selection for the tau-leaping simulation method. J Chem Phys 124(4):044109–044111
12. Dodd IB et al (2007) Theoretical analysis of epigenetic cell memory by nucleosome modification. Cell 129(4):813–822
13. Zhang H et al (2014) Statistical mechanics model for the dynamics of collective epigenetic histone modification. Phys Rev Lett 112(6):068101
14. Monahan K, Horta A, Lomvardas S (2019) LHX2- and LDB1-mediated trans interactions regulate olfactory receptor choice. Nature 565(7740):448–453

Applying Quantitative Systems Pharmacology (QSP) Modeling to Understand the Treatment of *Pneumocystis*

Tongli Zhang

Introduction

Pneumocystis, a yeast-like fungi, resides in lung alveoli and can cause a lethal infection known as *Pneumocystis pneumonia* (PCP) in hosts with impaired immune systems. In order to develop novel drugs or novel combinations of available drugs, quantitative Systems Pharmacological (QSP) models were built to help us understanding the complex pharmacokinetic (PK) and pharmacodynamic (PD) of multiple treatment regimens. In this chapter, we illustrated how to construct and independently validate PK modules of a number of drugs with available pharmacokinetic data. Characterized by simple structures and well constrained parameters, these PK modules could serve as a convenient tool to summarize and predict pharmacokinetic profiles. With the current hypotheses on the life stages of *Pneumocystis*, we also constructed a PD module to describe the proliferation, transformation, and death of *Pneumocystis*. By integrating the PK module and the PD module, the QSP model was further constrained with observed levels of asci and trophic forms following treatments with multiple drugs. Developed and validated in a data-driven manner, this QSP model integrated available data and promises to facilitate the design of future therapies against PCP.

T. Zhang (✉)
Department of Pharmacology and Systems Physiology, College of Medicine, University of Cincinnati, Cincinnati, OH, USA
e-mail: zhangtl@ucmail.uc.edu

© Springer Nature Switzerland AG 2021
P. Kraikivski (ed.), *Case Studies in Systems Biology*,
https://doi.org/10.1007/978-3-030-67742-8_15

Physiology and Molecular Mechanisms

Pneumocystis is a common opportunistic infection. In hosts with functional immune systems, the growth of these organisms is repressed and few pathological symptoms are observed. On the other hand, PCP is a cause of morbidity in HIV-positive patients as well as hosts with other immune defects, or in patients undergoing therapy with immunosuppressive agents [1–3].

The genus *Pneumocystis* is comprised of many species, including *P. carinii, P. jirovecii* [4], *P. wakefieldiae, P. murina* [5], and *P. oryctolagi* [6–8]. These different species are characterized by their ability to infect different hosts. For example, *P. jirovecii* resides in the human lung alveoli. Despite their differences in host preference, all *Pneumocystis* species are hypothesized to have a bi-phasic life cycle: a) an asexual phase of replication via the binary fission of the trophic forms; b) a sexual phase in which the conjugation of trophic forms results in formation of asci which contain eight ascospores, that are released and either continue in the sexual phase or enter the asexual phase [6].

Unlike mammalian cells, *Pneumocystis* is unable to harvest folate from the environment and must synthesize it *de novo* [9]. To take advantage of this weakness, the primary therapy for PCP is TMP-SMX, which inhibits dihydropteroate synthase and dihydrofolate reductase, the integral enzymes involved in folate synthesis in host cells and fungi [10–12]. Despite high success rates in treating PCP, TMP-SMX therapy leads to significant side effects, including neutropenia and serious allergic skin reactions that can result in death. It's estimated that between 25% and 50% of HIV-infected patients are unable to tolerate prolonged TMP-SMX treatment due to these harsh side effects and must seek other treatment options [13].

Currently, alternative medications include atovaquone, clindamycin-primaquine, echinocandins, and pentamidine isethionate. Atovaquone inhibits nucleic acid and adenosine triphosphate synthesis [14], thus disrupting DNA replication, energy production, and proliferation of the fungi. A combination of clindamycin and primaquine suppresses fungal protein synthesis and mitochondrial function [15]. The echinocandin family (i.e. anidulafungin, caspofungin, and micafungin) are β-1,3-D-glucan (BG) synthase inhibitors. Since BG is an essential component of the cellular wall that surrounds the asci of *Pneumocystis*, these drugs selectively target fungi in this phase [16–18]. When compared to TMP-SMX, these alternative therapies suffer from high rates of relapse and recurrence [19, 20]. Development of new drugs to treat PCP promises to deliver effective treatment with reduced side effects.

In comparison to other pathogens, the study of *Pneumocystis* is particularly challenged by the fact that these fungi cannot be reliably cultured in vitro for any significant length of time, nor continuously passaged to identify whether drugs are pneumocysticidal or pneumocystistatic. Due to this limitation, preclinical drug efficacy studies are carried out in animal models of *Pneumocystis* infection, typically in mice or rats [21]. Such reliance on animal studies significantly increases both the time and costs associated with the development of treatments to combat PCP. To alleviate this, it will be beneficial to integrate currently available knowledge on the treatment of PCP and our current knowledge of the *Pneumocystis* lifecycle into a QSP model to facilitate the drug development process. By combining traditional PK and PD analysis with systems biology modeling, QSP can summarize available information into a convenient framework, which can then be used to rigorously test different hypotheses, and scan through treatment regimens in an efficient and cost-effective manner [22, 23]. QSP modeling has been useful in the treatment of infectious diseases, such as Tuberculosis, where it has been used for dose optimization of anti-Tuberculosis drugs [24–26]. In addition, QSP models have shown great promise as powerful quantitative tools to study the dosing regimens for novel compounds [27].

A QSP model for the treatment of *Pneumocystis* is not yet available, and the scarcity of data from human patients makes the development of a human model difficult. With available data in mice, we constructed and validated a QSP model of PCP. This model includes both a PK module and a PD module. The PK module describes the distribution and decay of an applied drug, with different drugs characterized by their respective rate constants. This module was parametrized using independent construction and validation datasets. Following validation, the model can then be used to predict the temporal PK profiles of standard dosing regimens used in mice.

The PD module specifies the proliferation, transformation, and death of *Pneumocystis in* infected mice. The PK module and PD modules were then integrated into a population of QSP models. The parameters of this integrated model were estimated using a population of models that recapture the steady state distributions of the trophic forms and asci following drug treatment. The temporal dynamics generated by these QSP models were further validated with the observed dynamics of *Pneumocystis* following these same drug treatments.

As result, the QSP models developed in this work promise to serve as a solid first step towards understanding the temporal dynamics of *Pneumocystis* infection and facilitating the design of novel therapies. In the future, this model can be further improved and projected to a human version.

Mathematical Model

The overall QSP model, as illustrated in Fig. 1, was built in a step-by-step manner. After a PK module and a PD module were constructed, they were integrated into a comprehensive QSP model.

The structure of the systems pharmacology model

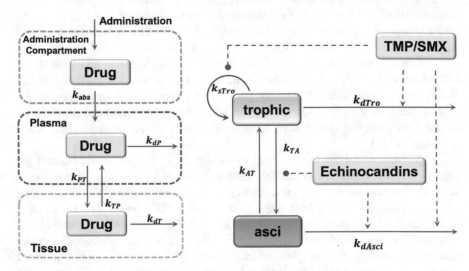

Fig. 1 The structure of the QSP model. Left panel: A three-compartment PK module was used to describe the reported pharmacokinetic data. The first compartment was the AC, the second compartment was plasma, and the third was 'peripheral tissue'. Drug decay was assumed to occur in plasma and 'peripheral tissue' compartments. The rates of drug distribution and decay were described by the corresponding parameters. Right panel: The dynamics of *Pneumocystis* were described by a two-stage model which involves both trophic forms and asci. The temporal changes of trophic forms and asci were also controlled by the indicated parameters. The drug effects were indicated by arrows (promoting) and lines with solid circle heads (inhibiting)

Step One: Construction of the PK module

A three-compartment PK module was used to describe drug dynamics. Overall the module is comprised of the administration compartment (AC), a plasma compartment and a "*peripheral tissue*" compartment (combining all organs, muscles and fat *etc.*). The parameters that govern drug distribution and decay are labeled near the corresponding reaction arrows. Three ordinary differential equations (ODEs) with identical structures (detailed below), but different parameters, were used to describe all drugs. This is due to the assumption that all these drugs follow similar distribution and decay processes, but are characterized by different rates. The equations and parameters are available in Table 1, and it can be implemented in any mathematical modeling software, such as the free software XPPAUT (http://www.math.pitt.edu/~bard/xpp/xpp.html), or MATLAB from Mathworks (https://www.mathworks.com).

Table 1 The equations and parameters of the PK module

Absorptive compartment

$$\frac{dDrug_{AC}}{dt} = -RAP * K_{abs} * Drug_{AC}$$

Plasma compartment

$$\frac{dDrug_P}{dt} = K_{abs} * Drug_{AC} - (K_{PT} + K_{dP}) * Drug_P + K_{TP} * Drug_T$$

Peripheral tissue compartment

$$\frac{dDrug_T}{dt} = RTP * (-K_{TP} * Drug_T + K_{PT} * Drug_P) - K_d * Drug_T$$

PK parameters for each drug

	Anidulafungin	Caspofungin	Micafungin	TMP/SMX
K_{dP} (h^{-1})	0.035	0.18	0.06	0.2
K_{PT} (h^{-1})	1.5	5	0.6	0.17
K_{TP} (h^{-1})	5	2	1.8	5
RTP(dimensionless)	0.2	1	0.4	0.01
K_{dT} (h^{-1})	0.035	0.18	0.06	0.2
RAP(dimensionless)	3	0.1	1	3
K_{abs} (h^{-1})	5	5	0.75	5

Step Two: Construction of the PD module

The life cycle of *Pneumocystis*, including its proliferation, life cycle stage transformation, and death, was simplified into a two-stage model which included both trophic forms and asci. The simplified model was described using a pair of ODEs and 5 control parameters.

This model describes the transformation between trophic forms and asci following a similar multistate model of tuberculosis [28]. Following logistic growth models, the decay of the trophic form is a second order reaction since the trophic forms actively proliferate and compete for space and nutrients. On the contrary, the asci do not actively proliferate but rather result from the transformation of trophic forms. Hence, the decay of the asci is set to be a first order reaction.

The basal values of these control parameters were estimated on the basis of relevant experimental data (Table 2). In order to incorporate the experimentally observed heterogeneities of *Pneumocystis* levels, a population of PD models with parameter values selected from a uniform distribution that covers 70–130% of the basal values (Table 2) were also used.

Step Three: Integration of the PK Module and the PD Modules into QSP Models

Various drugs target *Pneumocystis* via diverse mechanisms, and such mechanistic diversities need to be incorporated into the QSP models. TMP-SMX represses folate synthesis which is essential for genome replication in the organism [10]. Therefore,

Table 2 The equations and parameters of the PD modules

Trophic form
$\frac{dTro}{dt} = K_{sTro} * Tro - K_{dTro} * Tro * Tro - K_{TA} * Tro + K_{AT} * Asci$

Asci
$\frac{dAsci}{dt} = K_{TA} * Tro - K_{AT} * Asci - K_{dAsci} * Asci$

Basal parameter values

Parameter	Value (unit)	Constraining data
K_{sTro}	1 day^{-1}	The observed accumulation of total Pneumocystis constrained the
K_{dTro}	1×10^{-7} day^{-1}	time scale The steady state values of trophic form constrained the $K_{sTro} : K_{dTro}$ ratio
K_{AT}	0.1 day^{-1}	The $K_{TA} : K_{AT}$ ratio was constrained with experimental observa-
K_{TA}	0.1 day^{-1}	tions [30]
K_{dAsci}	2×10^{-12} day^{-1}	Degradation rate of the asci is assumed to be small [33]

in our simplified model, TMP-SMX was assumed to inhibit the proliferation rate of the trophic forms and increase the death rates of both the trophic forms and asci. Echinocandins on the other hand, block the construction of the cellular wall of the asci. Therefore, this family of drugs were assumed to reduce the level of asci by promoting their death as well as inhibiting their formation. The equations for integration are available at Table 3.

Results

The Constructed PK and PD Modules Are Consistent to Multiple Experimental Data

As it is straightforward to justify the equations of the PK modules based on the drug distribution and decay, the parameters values need to be estimated using data reported in the literature. Here, we illustrate some of the data and the how the estimated PK modules compare to these data. Gumbo *et al.* measured the plasma concentration of anidulafungin following a single 10 mg/kg i.p. injection [29], which we used to estimate the PK parameters for a three-compartment pharmacokinetic model of anidulafungin. (Fig. 2a). After estimating the PK parameters, they were used to simulate further experimental scenarios with either *i.p.* or *i.v.* administration of anidulafungin and compared to their respective data sets (Fig. 2a). Because these additional data sources were not used for the initial parameter estimation, the consistency between the model simulation and these additional data sources served as a validation of the estimated parameters for the anidulafungin PK model. In a similar fashion, the parameters for PK models of caspofungin, micafungin, and SMX were estimated and validated with different literature sources (Figs. 2b–3d).

Table 3 Integrating the PK modules and PD modules into QSP models

Echinocandin effect on asci death			
$v_{dAsci} = k_{dAsci} + ME_{Echi} \frac{Echi_{pla}^n}{Echi_{pla}^n + Ec50_{Echi}^n}$			
Echinocandin effect on asci formation			
$v_{TA} = k_{TA}\left(1 - \frac{Echi_{pla}^n}{Echi_{pla}^n + Ec50_{Echi}^n}\right)$			
Parameters			
Echinocandin family member	Ec50 (μg ml^{-1})	ME	n
Anidulafungin	0.039	0.42	1
Caspofungin	0.0007	0.45	1
Micafungin	0.04	0.1	1
TMP/SMX effect on asci death			
$v_{dAsci} = k_{dAsci} + ME_{Asci} \frac{SMX_{eff}^n}{SMX_{eff}^n + Ec50_{SMX}^n}$			
TMP/SMX effect on trophic proliferation			
$v_{sTro} = k_s\left(1 - \frac{SMX_{eff}^n}{SMX_{eff}^n + Ec50_{SMX}^n}\right)$			
TMP/SMX effect on trophic death			
$v_{dTro} = k_{dTro}\left(1 + ME_{Tro} \frac{SMX_{eff}^n}{SMX_{eff}^n + Ec50_{SMX}^n}\right)$			
Delay in SMX effect			
$SMX_{eff}^n = SMX_{pla}^n(t - \tau)$			

Parameters				
$Ec50_{SMX}^n$	$MEAsci$	$METro$	N	τ
0.2 μg ml^{-1}	0.75	650	2	7 days

After the PD wiring diagram was converted into ODEs, the parameters of the module were estimated with currently available data. In order to check whether the estimated parameters are reasonable, the temporal simulations of the PD module need to be compared to these experimental observations. We encourage interested readers to simulate the PD module and compare it to the relevant sources as an exercise.

Quantitative Systems Pharmacology Model Construction and Validation

By integrating the PK modules and the PD module, the QSP model can describe the changes of asci and trophic forms following treatment for a population of models. Modules were integrated by adjusting the parameters that control: the growth and death of the cyst form (for the echinocandin family of drugs), or the death rates of the trophic and cyst forms along with the growth of the trophic form (for TMP/SMX treatment).

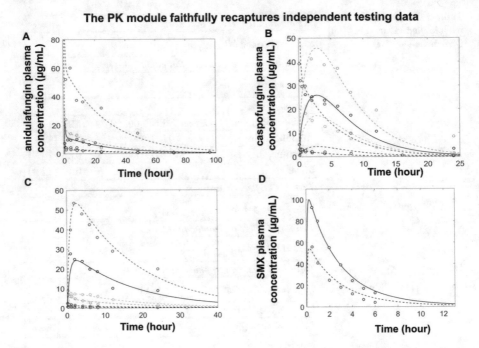

Fig. 2 The temporal simulations of the PK modules were consistent with diverse experimental data. The temporal simulations of the plasma concentrations of anidulafungin (**a**), caspofungin (**b**), micafungin (**c**) and SMX (**d**) were compared to relevant experimental data. The black dots and black solid curves represent the construction data and corresponding model simulations; the colored dots and colored dashed curves represent the validation data and corresponding simulations. The data sources were elaborated in **Remove**. The colors in each panel were used to indicate different administration methods and dosages. In (**a**), blue, *i.v.* of 1 mg/kg; magenta, green and red, *i.p.* of 80 mg/kg, 20 mg/kg and 5 mg/kg respectively. In (**b**), blue and magenta, *i.v.* of 0.5 mg/kg and 5 mg; red, cyan and green, *i.p.* of 1 mg/kg, 5 mg/kg and 80 mg/kg; In (**c**), blue, red and green, *i.v.* of 0.32 mg/kg, 1 mg/kg and 3.2 mg/kg; cyan and magenta, *i.p.* of 5 mg/kg and 80 mg/kg; In (**d**), blue, oral of 50 mg/kg

Following the experimental setting as reported by Cushion et al., each drug in the model was administrated three times per week for 3 weeks [30]. The simulated levels of asci at day 56 were then compared to the experimental observations from Cushion et al. (Fig. 3a). At a dose of 1 mg/kg, treatment with all three echinocandins (anidulafungin, caspofungin and micafungin) considerably reduced asci burdens. At lower doses (0.5 and 0.1 mg/kg), anidulafungin and caspofungin still decreased the number of asci, while micafungin caused no notable decrease in the levels of asci (Fig. 3a). In contrast to the dramatic reductions in asci, the simulated trophic forms were not meaningfully altered following treatment with any of the echinocandins (Fig. 3b). The model showed a marked decrease in both asci and trophic forms in response to TMP/SMX treatment (Fig. 3a and b). These simulated results were consistent with the experimental observations [30], indicating that our integrated

The integrated QSP models are further validated with experimental data

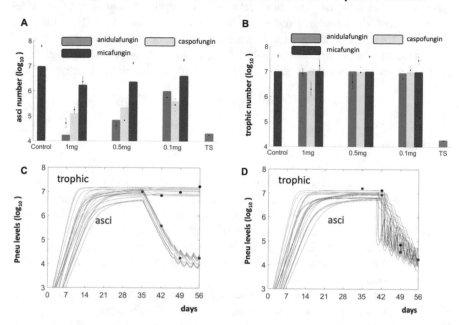

Fig. 3 The simulations of the QSP models were consistent to relevant data. (**a** and **b**). Bar plots of average simulated $\log_{10}$ levels: of asci (**a**) and trophic forms (**b**) at day 56 post-treatment of *Pneumocystis* from: untreated mice (Control), mice treated with varying doses of anidulafungin, caspofungin and micafungin; as well as mice treated with TMP-SMX. Corresponding experimental data are represented as dot plots with standard error. (**c**) The simulated dynamic changes of the trophic forms (black curves) and asci (red curves), on a $\log_{10}$ scale were consistent to the corresponding experimental data (black and red dots) following anidulafungin treatment. (**d**) The simulated dynamic changes of trophic forms (black curves) and asci (red curves) were consistent to the corresponding data (black dots and red dots) following TMP-SMX treatment

QSP models are reasonable in describing the therapeutic effects of the echinocandin family of drugs and those of TMP/SMX.

With the constructed QSP models, we then simulated the temporal changes of asci and trophic forms prior to and after anidulafungin treatment (Fig. 3c). Prior to drug administration, the simulated accumulation of both trophic forms and asci are consistent to experimental data collected in the absence of drugs, as elaborated in the description of the PD modules above. At about 35 days, the levels of both trophic forms and asci reached a steady state of about 10^7, in agreement to the experimental data (Fig. 3c). Following anidulafungin treatment (starting at day 35), the level of asci decreased dramatically while the level of trophic form remained constant. These simulated responses to anidulafungin were consistent with the corresponding experimental data from our lab (Fig. 3c).

When compared to anidulafungin, treatment with TMP-SMX decreased the levels of both asci and trophic forms. However, in comparison with the rapid antifungal effect of anidulafungin, the experimental evidence suggests that the effect of TMP-SMX was delayed. This time delay was incorporated into our QSP model (Table 3), and the simulated responses of trophic and asci levels (Fig. 3d) were consistent to corresponding experimental data (Fig. 3d). In summary, the QSP models serve as a reasonable tool to describe the temporal dynamics of Pneumocystis upon treatment with either the echinocandin class of antifungals or TMP/SMX.

Discussion

In this chapter, we have illustrated how to develop a QSP model to simulate how the numbers of *Pneumocystis* are altered by commercially available echinocandins and TMP-SMX. In addition to describing the temporal dynamics of these drugs, this novel QSP model also incorporated two different life cycle stages of the infecting fungi. Since the different life stages are presumably conserved in a broad range of hosts, the QSP model would be useful for studying *Pneumocystis* infections in a number of hosts including humans.

QSP modeling, which integrates knowledge from pharmacology and systems biology, is emerging as a powerful approach in pharmaceutical development [31, 32]. To the encouragement of the QSP community, QSP modeling aided in studying the dosing regimens of a new biologic, NATPARA, in the regulatory domain [27]. Particularly, QSP modeling has been useful in aiding the treatment of infectious diseases, such as tuberculosis, where it has been used for dose optimization of anti-Tuberculosis drugs [24–26]. Moreover, QSP models have shown great promise as powerful quantitative tools to study the dosing regimen for novel pharmaceutical compounds [27]. Thus, it is worthwhile to carefully evaluate the power as well as limitations of QSP modeling.

The benefits of QSP modelling originate from its ability to integrate all available knowledge and data to predict the effect of novel treatment regimens. In this way, the modeling provides some guidance for choosing effective strategies and avoiding plans that might have little chance for success. In this way, QSP combines traditional PK/PD modeling with systems biological modeling and provides a more comprehensive picture than single indices such as steady state or the AUC. In order to generate faithful predictions, both the PK and PD portions of the QSP models must be carefully constructed and independently validated. For the current QSP model, the PK module has been well constrained with the abundant data available in the literature, however the PD module needs to be further validated with additional dynamic data of the asci and trophic forms following treatment with different drugs as well as dynamic data of the growth of the organism prior to treatment. These additional data sources will either validate the model's current parameter settings or allow for further refinement of the parameters.

The complexity and scope of the current model aim to achieve a balance between incorporation of mechanistic details and constraint by currently available data. When additional details become available, the current PD module can be expanded to include a more detailed description of the Pneumocystis life stages, while the PK module can be expanded to incorporate additional compartments, such as a lung compartment. Furthermore, the model can be tailored to investigate additional drugs such as atovaquone or clindamycin-primaquine.

The current model, constrained with data collected in mice, promises to serve as a useful framework to understand and predict the growth, death and drug response of Pneumocystis in human patients, assuming the conservation of Pneumocystis life stages between species. Such predictions of *Pneuomocystis* levels in human, being orthogonal to the observed symptoms, will provide valuable insight for the clinicians to understand the progression of the infection as well as its response to treatment.

Exercises

Exercise 1. Once constructed and validated, the PK modules may serve as convenient tools to predict the plasma level of each drug following more than a single dose. With the PK modules, predict plasma levels of anidulafungin, caspofungin, micafungin and SMX when they are administrated three times/week. Use different dosages of anidulafungin, caspofungin, micafungin (in mg/kg) to see how the pharmacokinetic profiles are changed. Compare your simulation results with those reported in the original paper.

Exercise 2. With the PD module, simulate the dynamic changes of trophic form and asci. Start with differential initial levels of the trophic forms to see how the systems evolve in the absence of any drug treatment. Compare your simulation results with those reported in the original paper.

```
#XPPAUT code
### ODE file for QSP-Pneu
### SMX-SMX, 50/200 mg/kg, 3 times/ week

################################### the PK model
###################################\

dSMXAC/dt = -RAP*kab*SMXAC

dSMX_Pla/dt = kab*SMXAC - (kPT + kdSMX) * SMX_Pla + kTP * SMX_Ti

dSMX_Ti/dt = RTP*(-kTP*SMX_Ti + kPT*SMX_Pla) - kdSMX*SMX_Ti

### new parameters (jan 16 2018) par kOP=5, ROP=0.01, kd=0.2, kPO=0.17

### new par in days
par kTP=120, kPT=4.08, kdSMX=0.48
par RTP=0.01
par kab=120, RAP=3

################################### end of the PK model
###################################
```

```
################################### Randomized parameter
setting ###################################

par range1 = 0.7, range2 = 0.5

!kCT=0.1 * (range1 + ran(range2))
!kTC=0.1 * (range1 + ran(range2))
!kdCyst=0.000000000002 * (range1 + ran(range2))
!ks=1 * (range1 + ran(range2))
!kdTro=0.0000001 * (range1 + ran(range2))

################################### Drug effect
###################################

### Since SMXdulafungin blocks Cyst Wall, it is assumed to promote
killing Cyst forms and repress transformation to Cyst forms

SMX_Eff = delay(SMX_Pla,d)

!d=7*(range1 + ran(range2))

vdCyst = kdCyst + MECyst*SMX_Eff^n/(SMX_Eff^n+EC50SMX^n)

vsTro = ks * ( 1 - SMX_Eff^n/(SMX_Eff^n+EC50SMX^n))

vdTro = kdTro * ( 1 + METro*SMX_Eff^n/(SMX_Eff^n+EC50SMX^n))

par n=2, EC50SMX=40, MECyst=0.85, METro=700

################################### Drug effect
###################################

init Tro=100, Cyst=100

dTro/dt = vsTro*Tro - vdTro * Tro * Tro  - kTC * Tro + kCT * Cyst

dCyst/dt = kTC * Tro - kCT * Cyst - vdCyst * Cyst

################################### End of the PD model
###################################

################################### addition of Drugs
###################################

T_add'=0

init T_add=35.1311

par T_tem=56.1311, OPdelay=0
```

```
global 1 {t-(T_add+OPdelay)} {SMXAC=SMXAC+dose}

global 1 {t-(T_add+OPdelay)} {T_add=T_add+7/3}

global 1 {t-(T_tem+OPdelay)} {T_add=T_add+90000}

par dose=500

aux x=SMX_Eff

@ method=stiff, delay=20
@ xplot=t, yplot=Tro, xlo=0, xhi=200, ylo=0, yhi=2000000000000000
@ total=100, bound=100000000000000000, nmesh=400, dt=0.01
done
```

References[1]

1. Huang YS, Yang JJ, Lee NY, Chen GJ, Ko WC, Sun HY, Hung CC (2017) Treatment of Pneumocystis jirovecii pneumonia in HIV-infected patients: a review. Expert Rev Anti-Infect Ther 15(9):873–892
2. Liu Y, Su L, Jiang SJ, Qu H (2017) Risk factors for mortality from pneumocystis carinii pneumonia (PCP) in non-HIV patients: a meta-analysis. Oncotarget 8(35):59729–59739
3. Luraschi A, Cisse OH, Pagni M, Hauser PM (2017) Identification and functional ascertainment of the Pneumocystis jirovecii potential drug targets Gsc1 and Kre6 involved in glucan synthesis. J Eukaryot Microbiol 64(4):481–490
4. Nahimana A, Rabodonirina M, Bille J, Francioli P, Hauser PM (2004) Mutations of Pneumocystis jirovecii dihydrofolate reductase associated with failure of prophylaxis. Antimicrob Agents Chemother 48(11):4301–4305
5. Hauser PM, Macreadie IG (2006) Isolation of the Pneumocystis carinii dihydrofolate synthase gene and functional complementation in Saccharomyces cerevisiae. FEMS Microbiol Lett 256 (2):244–250
6. Beck JM, Cushion MT (2009) Pneumocystis workshop: 10th anniversary summary. Eukaryot Cell 8(4):446–460
7. Weiss LM, Cushion MT, Didier E, Xiao L, Marciano-Cabral F, Sinai AP, Matos O, Calderon EJ, Kaneshiro ES (2013) The 12th international workshops on opportunistic protists (IWOP-12). J Eukaryot Microbiol 60(3):298–308
8. Calderon EJ, Cushion MT, Xiao L, Lorenzo-Morales J, Matos O, Kaneshiro ES, Weiss LM (2015) The 13th international workshops on opportunistic protists (IWOP13). J Eukaryot Microbiol 62(5):701–709
9. Skold O (2000) Sulfonamide resistance: mechanisms and trends. Drug Resist Updat 3 (3):155–160
10. Huang L, Crothers K, Atzori C, Benfield T, Miller R, Rabodonirina M, Helweg-Larsen J (2004) Dihydropteroate synthase gene mutations in Pneumocystis and sulfa resistance. Emerg Infect Dis 10(10):1721–1728
11. Nahimana A, Rabodonirina M, Zanetti G, Meneau I, Francioli P, Bille J, Hauser PM (2003) Association between a specific Pneumocystis jiroveci dihydropteroate synthase mutation and

---

[1]This book chapter is based on a previous publication [34]. Published under a CC BY 4.0 license.

failure of pyrimethamine/sulfadoxine prophylaxis in human immunodeficiency virus-positive and -negative patients. J Infect Dis 188(7):1017–1023

12. Nahimana A, Rabodonirina M, Francioli P, Bille J, Hauser PM (2003) Pneumocystis jirovecii dihydrofolate reductase polymorphisms associated with failure of prophylaxis. J Eukaryot Microbiol 50(Suppl):656–657

13. Castro JG, Morrison-Bryant M (2010) Management of Pneumocystis Jirovecii pneumonia in HIV infected patients: current options, challenges and future directions. HIV AIDS (Auckl) 2:123–134

14. Artymowicz RJ, James VE (1993) Atovaquone: a new antipneumocystis agent. Clin Pharm 12 (8):563–570

15. Schlunzen F, Zarivach R, Harms J, Bashan A, Tocilj A, Albrecht R, Yonath A, Franceschi F (2001) Structural basis for the interaction of antibiotics with the peptidyl transferase centre in eubacteria. Nature 413(6858):814–821

16. Powles MA, Liberator P, Anderson J, Karkhanis Y, Dropinski JF, Bouffard FA, Balkovec JM, Fujioka H, Aikawa M, McFadden D et al (1998) Efficacy of MK-991 (L-743,872), a semisynthetic pneumocandin, in murine models of Pneumocystis carinii. Antimicrob Agents Chemother 42(8):1985–1989

17. Letscher-Bru V, Herbrecht R (2003) Caspofungin: the first representative of a new antifungal class. J Antimicrob Chemother 51(3):513–521

18. Espinel-Ingroff A (2009) Novel antifungal agents, targets or therapeutic strategies for the treatment of invasive fungal diseases: a review of the literature (2005–2009). Rev Iberoam Micol 26(1):15–22

19. Patel N, Koziel H (2004) Pneumocystis jiroveci pneumonia in adult patients with AIDS: treatment strategies and emerging challenges to antimicrobial therapy. Treat Respir Med 3 (6):381–397

20. Thomas M, Rupali P, Woodhouse A, Ellis-Pegler R (2009) Good outcome with trimethoprim 10 mg/kg/day-sulfamethoxazole 50 mg/kg/day for Pneumocystis jirovecii pneumonia in HIV infected patients. Scand J Infect Dis 41(11–12):862–868

21. Lobo ML, Esteves F, de Sousa B, Cardoso F, Cushion MT, Antunes F, Matos O (2013) Therapeutic potential of caspofungin combined with trimethoprim-sulfamethoxazole for pneumocystis pneumonia: a pilot study in mice. PLoS One 8(8):e70619

22. Agoram BM, Demin O (2011) Integration not isolation: arguing the case for quantitative and systems pharmacology in drug discovery and development. Drug Discov Today 16 (23–24):1031–1036

23. Knight-Schrijver VR, Chelliah V, Cucurull-Sanchez L, Le Novere N (2016) The promises of quantitative systems pharmacology modelling for drug development. Comput Struct Biotechnol J 14:363–370

24. Lyons MA, Reisfeld B, Yang RS, Lenaerts AJ (2013) A physiologically based pharmacokinetic model of rifampin in mice. Antimicrob Agents Chemother 57(4):1763–1771

25. Lyons MA (2014) Computational pharmacology of rifampin in mice: an application to dose optimization with conflicting objectives in tuberculosis treatment. J Pharmacokinet Pharmacodyn 41(6):613–623

26. Lyons MA, Lenaerts AJ (2015) Computational pharmacokinetics/pharmacodynamics of rifampin in a mouse tuberculosis infection model. J Pharmacokinet Pharmacodyn 42(4):375–389

27. Peterson MC, Riggs MM (2015) FDA advisory meeting clinical pharmacology review utilizes a quantitative systems pharmacology (QSP) model: a watershed moment? CPT Pharmacometrics Syst Pharmacol 4(3):e00020

28. Clewe O, Aulin L, Hu Y, Coates AR, Simonsson US (2016) A multistate tuberculosis pharmacometric model: a framework for studying anti-tubercular drug effects in vitro. J Antimicrob Chemother 71(4):964–974

29. Gumbo T, Drusano GL, Liu W, Ma L, Deziel MR, Drusano MF, Louie A (2006) Anidulafungin pharmacokinetics and microbial response in neutropenic mice with disseminated candidiasis. Antimicrob Agents Chemother 50(11):3695–3700

30. Cushion MT, Linke MJ, Ashbaugh A, Sesterhenn T, Collins MS, Lynch K, Brubaker R, Walzer PD (2010) Echinocandin treatment of pneumocystis pneumonia in rodent models depletes cysts leaving trophic burdens that cannot transmit the infection. PLoS One 5(1):e8524
31. Leil TA, Bertz R (2014) Quantitative systems pharmacology can reduce attrition and improve productivity in pharmaceutical research and development. Front Pharmacol 5:247
32. Leil TA, Ermakov S (2015) Editorial: the emerging discipline of quantitative systems pharmacology. Front Pharmacol 6:129
33. Icenhour CR, Kottom TJ, Limper AH (2003) Evidence for a melanin cell wall component in Pneumocystis carinii. Infect Immun 71(9):5360–5363
34. Liu GS, Ballweg R, Ashbaugh A, Zhang Y, Facciolo J, Cushion MT, Zhang T (2018) A quantitative systems pharmacology (QSP) model for Pneumocystis treatment in mice. BMC Syst Biol 12(1):77

Virus Dynamics

Stanca M. Ciupe and Jonathan E. Forde

Introduction

Viruses are organisms that infect bacterial, animal or human cells and use their machinery to replicate into new viruses. An individual virus particle, called a virion, is composed of genetic material, either RNA or DNA, surrounded by an outer capsid [1]. The first known virus to cause a human disease, yellow fever, was discovered in 1901 [2]. Since then hundreds of other harmful human viruses have been identified, most of which emerged through animal spillover [3]. Viruses differ in their routes of transmission, the types of cell they infect, and disease severity, with some, such as HIV, hepatitis B, hepatitis C, measles and SARS-CoV-2 [4] viruses, causing emerging and re-emerging global pandemics.

In an individual virus infection, the first line of defense is the innate immune system, which is non-specific and does not result in immune memory. The second line of defense is the adaptive immune system, which is virus-specific, results in immune memory, and reacts quickly to secondary viral challenge. Adaptive immunity is composed of two branches: humoral and cell-mediated. Humoral immunity involves B cells, which can produce high affinity neutralizing antibodies that facilitate virus removal and are capable of blocking virus entry into a target cell. Cell-mediated immunity involves T cells, which recognize viral fragments presented on the surface of infected cells and subsequently kill those cells. Ultimately,

S. M. Ciupe (✉)
Department of Mathematics, Virginia Polytechnic Institute & State University, Blacksburg, VA, USA
e-mail: stanca@vt.edu

J. E. Forde (✉)
Department of Mathematics and Computer Science, Hobart and William Smith Colleges, Geneva, NY, USA
e-mail: forde@hws.edu

© Springer Nature Switzerland AG 2021
P. Kraikivski (ed.), *Case Studies in Systems Biology*,
https://doi.org/10.1007/978-3-030-67742-8_16

successful adaptive immune responses result in the formation of memory B and T cells, which respond rapidly upon secondary challenge with the same virus variant [5, 6]. Immune memory of this sort can be acquired through natural infection or through vaccination, the most cost-effective public health intervention of the last century [6, 7]. Understanding the interactions between a virus and the immune system and quantifying the mechanisms of virus removal are fundamental necessities for vaccine development.

Over the last few decades, within-host mathematical models have been developed with the aim of gaining insight into disease progression in a single infected individual and the potential role of various immune interventions (following in the footsteps of the SIR-type models used to produce epidemic forecasts [8]). In order to draw inferences about immunological interventions, one must design reliable models that capture the characteristics of virus infection and transmission, while accurately describing the emerging immune responses. Mathematical modelers are often required to choose an appropriate balance between accurately representing biological complexity and ensuring robustness in the model's analytical and numerical predictions, with complexity being limited by the availability of the empirical data.

One of the first within-host models, used to describe the HIV infection within an infected individual, is called the *basic model*. This model does not explicitly incorporate any immune responses [9–12]. Instead, it focuses on the interactions between target cells, infected cells and HIV. While advancements in virology have added increasingly detailed knowledge of virus and immune response characteristics, alterations to the basic models have been more limited. It is still largely used in its original form, with a small number of adaptations, to predict the dynamics of HIV and other virus infections, such as influenza [13], dengue [14, 15], Zika [16] West Nile [17], hepatitis B [18–21], hepatitis C [22], and SARS-CoV-2 viruses. In this chapter, we describe the general principles of modeling within-host virus dynamics with the basic model as our primary example. First, we present the basic model and discuss its usefulness in predicting disease establishment inside an infected individual. We explain how such a model can be used to quantify virus and host characteristics that capture the kinetics apparent in empirical viral data. Second, we provide an overview of models incorporating additional biological processes, such as the explicit role of immune responses and drug therapy in blocking transmission and ensuring viral clearance. Lastly, we explain how empirical data can be used to select among models with various levels of biological complexity. For simplicity, we focus on HIV and briefly explain how models of HIV infection can be adapted for other endemic or emerging viruses.

Mathematical Model

The basic model of viral dynamics describes the interactions of three populations: target cells, T, infected cells, I, and virions, V. The target cell population includes cells that are susceptible to infection by virus, and will therefore represent different cell types in models of different viral infections. Infected cells are cells that have

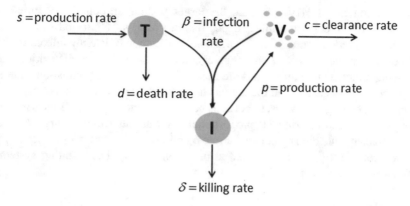

Fig. 1 Diagram for the basic virus dynamics model (Eq. 1)

been infected by the virus and begun producing new virions. The virion population consists of all those viral particles which have been released from infected cells. The relationships between these populations are represented in Fig. 1.

In order to translate these relationships into a concrete mathematical model, decisions/assumptions must be made about the specific nature of the interactions between the populations. These decisions determine the rates of change of the given population, and so the mathematical form of the differential equations composing the model. The most commonly used forms are constant, linear, Hill-type, and mass action. Given a population X,

- A constant term (a) describes a rate of change that is *independent of current population levels*;
- A linear term (aX for growth or $-aX$ for death) describes a rate of change that is *proportional to the current population level*. The parameter value a is called the *per capita* growth/decay rate;
- A Hill-type function $rX^h/(X^h + L^h)$ is commonly used to describe situations of resource or physical constraint, which may limit the rate at which an interaction occurs. In such cases, the rate of change saturates at some maximal rate. On the other hand, there is also sometimes a minimum (threshold) population level that is required for an interaction to be expected to take place. Parameter r is the maximum rate of change. Parameter L represents the population level at which the rate is half of r, and can be used as a proxy for a threshold interaction level. Parameter h is the Hill coefficient, and represents how rapidly the rate transitions from sub-threshold levels to maximal rates.

Given a second population Y,

- A mass action term (aXY) describes a rate of change that is *proportional to each of two interacting populations*. When used to describe the conversion of one cell type to another, mass action terms generally occur in pairs, with $-aXY$ representing the removal from one class and aXY representing addition to the other class.

In the context of HIV infection, the target population consists of uninfected activated CD4 T cells, the infected population consists of infected CD4 T cells that actively produce virus (note that a second population of latently infected CD4 T cells exist, but are ignored in this chapter), and the virus is free HIV virions. The following simplifying assumptions are made about how these populations interact. Target cells are released from the thymus at constant rate s and die at per capita rate d. Upon interacting with virus, target cells become infected at rate β. This is modeled through a mass action term βTV, proportional to the amounts of both target cells and virus. Infected cells die at a per capita rate δ and produce p new virions per day per infected cell. Lastly, virus is cleared at per capita rate c. The resulting system of differential equations is

$$
\begin{aligned}
\frac{dT}{dt} &= s - \beta TV - dT, \\
\frac{dI}{dt} &= \beta TV - \delta I, \\
\frac{dV}{dt} &= pI - cV.
\end{aligned}
\tag{1}
$$

The initial conditions are $T(0) = s/d$, $I(0) = 0$, $V(0) = v_0$.

Per Capita Rates and Average Lifespans

An important part of the modeling process is determining the meaning of the parameter values from a biological point of view, and how their values can be translated into useful predictions that can be tested in patients or in the lab. In the basic HIV model, the death rate of the infected population is described by the negative linear term $-\delta I$. One mathematical consequence of this assumption is that the average lifespan of an infected CD4 T cell is $1/\delta$. Similarly, the average life span of a free virion is $1/c$ and of an uninfected cell is $1/d$. This interpretation allows for bidirectional translation between the model and laboratory experiments. If the average lifespan of the infected cell or virion is known experimentally, then the parameter value (δ or c) can be determined. On the other hand, if the parameter value is known (for example by fitting the model to data), this provides information about the lifespan, which may not be observable experimentally.

Results

The usefulness of the basic model comes from its ability to describe virus and host characteristics that are not directly measurable in the lab. In HIV research, it is customary to collect virus titers over the course of an individual's infection. It is harder, however, to collect measurements of the CD4 T cell concentrations (uninfected or infected), since the majority of CD4 T cells are found in the lymph nodes, and reaching them requires invasive methods. The analytical and/or numerical solution of model (Eq. 1) can describe kinetics over time for both HIV and CD4 T cell populations. To ensure they correctly represent an individual infection, parameter values have to be chosen such that the virus population V matches empirical data. This is done through data fitting (see the subsection *Reflections on data fitting and model selection* below and reference [23]).

Before infection, system (Eq. 1) is in an uninfected state $(T_1, I_1, V_1) = (s/d, 0, 0)$. A small initial amount of virus v_0 results in an exponential increase of the virus population if the basic reproduction number R_0, defined as the average number of new virions resulting from a single virion in a susceptible population of target cells, is greater than 1. For the basic model, the basic reproduction number is given by the formula

$$R_0 = \frac{1}{c} \times \beta \times \frac{s}{d} \times \frac{p}{\delta}, \tag{2}$$

(see [24] for a derivation). Thus, the basic reproduction number is the product of the average lifespan of a virion $1/c$, the infection rate β, the number of target cells available for infection s/d and the number of virions produced by a cell once it is infected p/δ. If this product is greater than 1 then, on average, each virion leads to the production of more than one new virion, and the infection is sustained.

An example of CD4 T cells and HIV dynamics over time, for $R_0 > 1$, is given in Fig. 2, panels (a) and (b). Virus titers increase from 1 copy/ml to maximum values of 1.2×10^7 copies/ml in 3.3 days after infection. Following the initial period of viral increase, the system converges to an equilibrium infected state $(T_2, I_2, V_2) = \left(\frac{\delta c}{\beta p}, (R_0 - 1)\frac{cd}{\beta p}, (R_0 - 1)\frac{d}{\beta}\right)$. In the numerical example in Fig. 2, panel (b), the virus equilibrium value corresponds to 1.5×10^5 copies/ml. While these viral characteristics are chosen to match empirical data, the corresponding dynamics of the uninfected and infected CD4 T cell populations (T and I) are unknown empirically. Figure 2, panel (a), shows uninfected CD4 T cell population loss and replacement by the infected CD4 T cell population. Following maximal infection level, at the peak of virus infection, the uninfected CD4 T cells rebound to low values of 1.2×10^5 cells/ml, while the infected CD4 T cells level off at 8×10^3 cells/ml. The total CD4 T cell population after the rebound is less than 20% of the healthy level ($s/d = 10^6$ cells/ml). This information is particularly useful for the medical community, since a patient whose total CD4 T cells decay below 2×10^5 cells/ml is classified as having AIDS [12]. Since CD4 T cells are immune cells required for protection against other diseases, it is important to maintain the CD4 T cell levels of at least 50% normal values at all times.

The basic model has been used to suggest interventions that may prevent an infection or improve outcomes if an infection has already been established. An infection can be prevented if R_0 is reduced below 1. This can be accomplished by increasing viral clearance c, decreasing infectivity β, increasing the death rate of

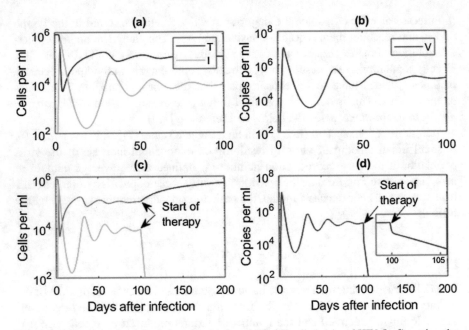

Fig. 2 Kinetics over time of uninfected and infected CD 4 T cells (**a, c**) and HIV (**b, d**) as given by model (Eq. 1) in: the absence of therapy (**a, b**) and when therapy is started 100 days after infection (**c, d**). Parameter values are $s = 10^4$, $\beta = 10^{-6}$, $p = 400$, $c = 23$, $d = 0.01$, $\delta = 1$, (**a, b**): $\epsilon = \eta = 0$ and (**c, d**): $\epsilon = 0.5$, $\eta = 0.9$. Initial conditions are $T(0) = 10^6$, $I(0) = 0$, $V(0) = 1$

infected cells δ or reducing the viral production rate of infected cells p. We present two strategies that can be used to alter these parameter values: induction of immune responses and administration of drug therapy.

Mathematical Models of Immune Responses

As mentioned in the introduction, two types of adaptive immune response can be induced through vaccination or natural infection: humoral (antibodies) and cell-mediated (T cells). The simplest way to model the effect of antibodies on HIV viral infection is by increasing or decreasing parameters in model (Eq. 1) corresponding to neutralizing and non-neutralizing processes [25–29]. For example, if the antibody population $A(t)$ is known, neutralization effects can be included by reducing the infection rate constant β by a factor $(1 + \alpha_1 A(t))$, therefore target cells would become infected at rate $\beta T V /(1 + \alpha_1 A(t))$. Increasing the number of antibodies $A(t)$ would reduce the efficiency of infection, β. The parameter α_1 corresponds to the quality of the neutralizing effect of the antibodies on the particular virus. Increased infected cell removal (and virus clearance) due to the non-neutralizing antibody effects can be included by increasing the infected cells death rate constant δ by a factor $(1 + \alpha_2 A(t))$ and virus clearance rate c by a factor $(1 + \alpha_3 A(t))$; therefore infected cells die at rate $\delta(1 + \alpha_2 A(t))I$ and virus is cleared at rate $c(1 + \alpha_3 A(t))V$.

Similarly, if the effector T cell population $E(t)$ is known, killing of infected cells can be included by increasing the infected cells death rate constant δ by a factor $(1 + \mu E(t))$ [30]. If any or all of these effects can be achieved, the basic reproduction number will be reduced, with a reduction below one leading to virus clearance.

Another approach is to include the antibody or effector cell population (or both) as an additional model state variable. The rates of change of these populations depend, via various biological mechanisms, on the levels of infected cells and virus present in the host. The inclusion of additional state variables results in a larger system of differential equations and a more complex model. Some examples of how HIV and CD4 T cell populations change when one or both of these interventions are considered can be found in Exercise 2.

Mathematical Models of Drug Therapy

Several drugs that act against HIV have been approved by the FDA [31–33]. These drugs have high efficacies in facilitating the removal of free virions and actively (but not latently) infected CD4 T cells. Here we present two such types of drug therapy, reverse transcriptase inhibitors and protease inhibitors. Reverse transcriptase inhibitors prevent cells from becoming productively infected. This can be included into model (Eq. 1) through reduction of the infectivity rate from β to $\beta(1 - \eta)$, where $0 \leq \eta \leq 1$ is the efficacy of the drug. Protease inhibitors block virus production. This can be included in model (Eq. 1) through reduction of the production rate, from p to $p(1 - \epsilon)$, where $0 \leq \epsilon \leq 1$ is the efficacy of the drug [34, 35]. An example of how the three populations of model (Eq. 1) change under combination therapy is given in Fig. 2, panels (c) and (d). When combination therapy with $\eta = 0.9$ and $\epsilon = 0.5$ starts 100 days post infection, free HIV virus decays below one copy/ml in 10 days. Most importantly, the total CD4 T cell level rebounds to 7×10^5 cells per ml (70% of the normal levels) in 100 days. This allows for the reconstitution of the immune system of the infected individual, which is needed to fight not only HIV, but other diseases as well.

Finding the Basic Reproduction Number

Since the quantification of the basic reproduction number (R_0 below or above one) is the simplest way to determine disease outcomes, it is useful to understand how one derives its formula [36]. We give a simple geometric method. For more advanced methods based on local linearization that can be applied in higher dimensional systems, see [24]. Over a short period of time at the beginning of infection, we make the simplifying assumption that the target cell population is constant $T(t) = s/$

d. This reduces the basic model to a two-dimensional system of linear differential equations:

$$\frac{dI}{dt} = \frac{s\beta V}{d} - \delta I,$$

$$\frac{dV}{dt} = pI - cV. \tag{3}$$

System (Eq. 3) has a single equilibrium solution $(I_1, V_1) = (0, 0)$, which corresponds to the infection-free state that must be approached to ensure virus clearance. We analyze the first quadrant of the $I - V$ phase-plane to determine the stability of the infection-free equilibrium. The I–nullcline of system (Eq. 3), along which $dI/dt = 0$, is given by the formula $V = \frac{d\delta}{s\beta}I$, while the V–nullcline, along which $dV/dt = 0$, is given by the $V = \frac{p}{c}I$. The graph of each nullcline is a line through the origin.

In general, there are three possible arrangements of these lines, as determined by the relative values of the slopes of the nullclines. In the first arrangement, the graph of the I–nullcline lies above that of the V–nullcline (see Fig. 3, panel (a)). This occurs if $d\delta/s\beta > p/c$, i.e. $R_0 = p\beta s/c\delta d < 1$. By testing points on the nullclines, we can see that trajectories move downward through the upper nullcline, and to the left through the lower nullcline. All trajectories passing into the middle region thus move down and to the left, approaching the origin. So, the infection-free equilibrium is locally asymptotically stable if $R_0 < 1$.

A second arrangement of the nullclines occurs if $d\delta/s\beta < p/c$, i.e. $R_0 = p\beta s/c\delta d > 1$. In this case, the relative position of the nullclines and the direction of trajectories passing through them are all reversed (see Fig. 3, panel (b)). So, trajectories passing through the upper nullcline move to the right, and those passing through the lower nullcline move upward. All trajectories passing into the middle

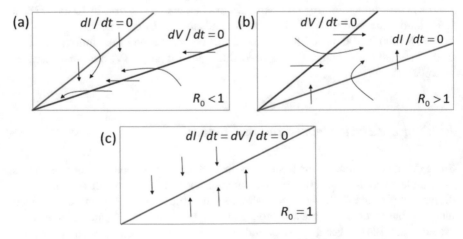

Fig. 3 Phase portrait for model (Eq. 1) for $T = s/d = \text{const.}$ and different R_0 values

region then move up and to the right, away from the origin. So, the infection-free equilibrium is unstable if $R_0 > 1$.

A degenerate case occurs when $R_0 = 1$, in which case the two nullclines are identical. This represents a transitional state between the two primary arrangements described above (see Fig. 3, panel c). In sum, this analysis shows that the value of R_0 relative to 1 is exactly the right threshold for determining whether a nascent infection will establish itself within the host.

Biphasic Decay

Following the initiation of drug therapy, there is a sharp decline in levels of infected cells (I) and free virions (V) (see Fig. 2, panels (c) and (d)). On closer inspection, the decline in viral load occurs in two distinct phases: a first sharp decline, followed by a second, more gradual decrease. By looking at the slopes associated with these phases, we can determine two essential viral parameters: δ and c.

Let us begin with the simplifying assumption that the reverse transcriptase inhibitor is completely effective ($\eta = 1$) and the protease inhibitor is not applied $\epsilon = 0$. Under these assumptions, the differential equation for I becomes the simple exponential decay $\frac{dI}{dt} = -\delta I$, with solution $I(t) = I_0 e^{-\delta t}$, where t is the time since the onset of treatment and I_0 is the number of infected CD4 T cells at the start of treatment. Hence, the value of δ is given by the slope of the downward line in the semilog plot of $I(t)$ (Fig. 2, panel (c)).

Clinically, free virus concentration is far easier to measure than the concentration of infected cells, so it is useful that we can determine both δ and c from the biphasic decline observed in the graph of $V(t)$ (Fig. 2, panel (d), inset). Some further calculations yield that $V(t) = \frac{V_0}{c-\delta}(ce^{-\delta t} - \delta e^{-ct})$ for quasi-equilibrium conditions before treatment $pI_0 = cV_0$. It is also reasonable to assume that c is significantly greater than δ, since the life span of an infected cell is longer than that of a free virus. It follows that the first term $\frac{V_0 \delta}{c-\delta}e^{-ct}$ goes to zero much faster than the second term $\frac{V_0 c}{c-\delta}e^{-\delta t}$. Therefore, in a semilog plot of $V(t)$ vs. t, the first slope of $V(t)$ decay is equal to c and the second slope is equal to δ. This means that therapy first leads to the clearance of free virus and then the loss of infected cells (and their associated viral production). This observation allows us to interpret patient data. If time-series data on viral titers from a patient at the beginning of drug therapy is collected, it can be used to determine c and δ. As seen in Fig. 2, panel (c), the first decay phase occurs on a very short time scale, so early measurements would be required. From these parameter values, we can infer the average lifespan of infected cells and of free virions. Since the system is in quasi-equilibrium before the onset of drug therapy, $p = cV_0/I_0$, the rate of viral production by infected cells can also be calculated. This rate is particularly important for assessing the risk to treatment effectiveness posed by viral mutations.

Reflections on Data Fitting and Model Selection

When monitoring an individual infection or response to treatment, experimentalists usually measure serum virus titers over time. Let us assume that the data is given by $\left(t_i, V_i^{data}\right)$, where t_i are the times post infection/treatment when virus titers V_i^{data} are collected $(1 \leq i \leq n)$, and that model (Eq. 1) is a good representation of virus dynamics. Then all of the model's parameters $\Theta = \{\beta, s, d, p, c, \delta\}$, or some subset of them, can be estimated by minimizing a least square functional $J = \min_{\Theta} \left(\sum_{i=1}^{n}\left(V_i^{data} - V(t_i, \Theta)\right)^2\right)^{1/2}$ over the parameter space, where $V(t_i, \Theta)$ is the viral population given by model (Eq. 1) with parameters Θ at time t_i. The minimization of J is implemented numerically, and a variety of algorithms and software packages can be used [37]. One example is the 'fminsearch' function in Matlab, see [38, 39]. A small J guarantees a close agreement between model predictions for V and the data.

The form of the data collected and the desired biological predictions play an important role in the construction of a model. One natural question is how to choose among different possible mathematical models that describe the same data. Let us look at a simple example where we compare two such models for the same data set $\left(t_i, V_i^{data}\right)$, $1 \leq i \leq n$. The first model uses k_1 unknown parameters, and the second model uses k_2 unknown parameters. From an information theoretic point of view, the model that best describes the data is the one with the smallest Akaike information criterion (AIC) [40], given by the formula

$$\text{AIC}_j = n\ln\left(\frac{1}{n}J_j\right) + 2(k_j + 1), \tag{4}$$

where $J_j = \min_{\Theta_j} \left(\sum_{i=1}^{n}\left(V_i^{data} - V^j(t_i, \Theta_j)\right)^2\right)^{1/2}$ is the least square estimate for model $j = \{1, 2\}$. It is often the case that increasing the number of fitted parameters in a model decreases the value of the least square functional J. However, in order to prevent overfitting, the AIC penalizes for including too many parameters. Thus, in some cases, simpler models with higher J may have a smaller AIC value and be considered superior.

While modelers may prefer the simplest model that sufficiently describes the data, the addition of complexity is usually driven by biological considerations. The biological question at hand may require the inclusion of new mechanistic interactions and the addition of more variables and parameters. A modeler should always consider multiple models and deem 'best' the one that satisfies the mathematical requirement (smallest AIC) while still meeting all biological requirements (creating testable predictions).

Discussion

The basic model of virus dynamics, developed almost 30 years ago, is still used to model individual infections with pathogens that cause persistent diseases. This model is easily adaptable to modeling the more rapid and short-lived infections with pathogens that cause acute infections, such as influenza, dengue and SARS-CoV-2 viruses. This can be accomplished by removing the production and death rates of the susceptible population. As a result, the initial viremia is followed by virus approaching zero over time [11].

As a group, these models are called 'target cell limitation' models, since virus decay past the initial peak is due to the infection of all available target cells. When no new infections can occur and infected cells die, virus is lost as well. Since it is known that virus loss is often immune mediated, more complex models that incorporate immune system dynamics have been proposed. While we discussed simple modeling inclusions that describe adaptive immune responses, many studies have adopted more sophisticated ways to account for not only innate and adaptive immunity development and function, but also immune memory [41–43]. Such models, of increased complexity, allow for a better description of the mechanisms of virus clearance and control.

The basic model used in this chapter describes the HIV infection of activated CD4 T cells that become productively infected. A second class of CD4 T cells, usually memory cells, revert to a latent stage upon HIV infection and only start producing virus upon activation, an event which is relatively rare, and therefore occurs on a longer time scale [44, 45]. The existence of this latent class, however, makes HIV a life-long disease where antiretroviral therapy is needed at all times to avoid virus rebound following the activation of the latent class. Several models have helped determine the relationship between this latent class and serum HIV dynamics, and have helped quantify the efficacy of drug therapy against the latent CD4 T cell population [44, 46–48].

An important feature of any mathematical model is its ability to suggest testable hypotheses and human interventions. In the case of the basic model, fitting to virus data under antiretroviral therapy led to the first characterization of important parameters such as virus lifespan, daily virus production and drug efficacy. One conclusion stemming from this was that the rate of viral production by infected cells is high (7×10^8 virions per day) [9]. Together with the high mutation rate in HIV replication, this indicates that this virus poses a high risk of developing resistance to any single-drug treatment. For this reason, combination drug therapy is recommended in all cases.

More recently, adaptations of the basic model have been used to determine the relationship between the size of virus inoculum, immune responses and disease outcome [39], the amount of antibody needed to block an infection [49], and the role of virus persistence in immune exhaustion [50]. They have also proven important in determining the size of the basic reproduction number in emerging and reemerging infections, such as Zika [16] and Ebola [51].

Mathematical models have opened avenues for interdisciplinary research and built bridges between the theoretical and experimental scientific communities. The modeling process is dynamic and iterative, requiring constant dialogue to improve model design, make predictions that can be validated in the lab, and incorporate new data and discoveries back into the models. Mathematical models have played a crucial role in advancing our understanding of viral dynamics over the past 30 years, and are now an essential tool, used routinely alongside experiments, in the study of viral infections.

Methods

We provide two Matlab functions (Matlab R2019b). To use these functions, you will need to save the two files with their function titles as the file names with a ".m" in the same directory. The two included functions are: HIV_ode and HIV_main_code.

To simulate System (Eq. 1), use the function HIV_main_code which calls the function HIV_ode.

HIV_main_code.m

```
clear all
axis manual;

% defining all parameters as global variables
global s d c delta beta p;

% parameter values
s=1e+4; % production rate
d=0.01; % death rate
delta=1; % killing rate
c=23; % clearance rate
p=400; % production rate
beta=1e-6; % infection rate

% solving the ODE @HIV_ode for time span [0, 300] and initial conditions
% T(0)=10^6; I(0)=0; V(0)=1;
[t,x] = ode45(@HIV_ode, [0 300], [1e+6; 0; 1]);

% plotting the T and I on logarithmic scale for 30 days
subplot(1,2,1)
semilogy(t, x(:,1), 'b', 'LineWidth',1);
hold on
semilogy(t, x(:,2), 'c', 'LineWidth',1);
axis([0 30 1e+2 1.2e+6]);
xlabel('Days after infection');
ylabel('Cells per ml')
hold on
```

```
% plotting V on logarithmic scale for 50 days.
subplot(1,2,2)
semilogy(t, x(:,3), 'r', 'LineWidth',1);
axis([0 50 1e+2 1e+8]);
xlabel('Days after infection');
ylabel('Copies per ml')
hold on
```

HIV_ode.m

```
% ode model: y(1)=T, y(2)=I; y(3)=V
% yprime(1)=dT/dt, yprime(2)=dI/dt; yprime(3)=dV/dt
function yprime=HIV_ode(t,y)

% defining all parameters as global variables
global s d c delta beta p;

% defining the differential equations
yprime=zeros(3,1); % matrix initialization

yprime(1)=s-d*y(1)-beta*y(1)*y(3); % T
yprime(2)=beta*y(1)*y(3)-delta*y(2); % I
yprime(3)=p*y(2)-c*y(3); % V
```

Exercises

Exercise 1. Each of the systems of differential equations below is a mathematical model of a viral infection within a human host.

System I

$$\frac{dT}{dt} = s - \beta TV - dT,$$

$$\frac{dE}{dt} = \beta TV - kE - \mu E,$$

$$\frac{dI}{dt} = kE - \delta I,$$

$$\frac{dV}{dt} = pI - cV.$$

System II

$$\frac{dT}{dt} = s - \beta TV_1 - dT,$$

$$\frac{dI}{dt} = \beta TV_1 - \delta I,$$

$$\frac{dV_1}{dt} = \epsilon pI - cV_1,$$

$$\frac{dV_2}{dt} = (1 - \epsilon)pI - cV_2.$$

$$\frac{dT_1}{dt} = s_1 - \beta_1 T_1 V - dT_1$$

$$\frac{dT_2}{dt} = s_2 - \beta_2 T_2 V - dT_2,$$

System III

$$\frac{dI}{dt} = \beta_1 T_1 V + \beta_2 T_2 V - \delta I,$$

$$\frac{dV}{dt} = \epsilon pI - cV.$$

For each system, complete the following tasks.

(a) Draw a diagram representing the interactions of the given populations. Refer to Fig. 1 as a sample.
(b) On the basis of your diagram, write a paragraph describing the viral infection that the system represented.
(c) Describe the ways in which this model differs from the basic model, both mathematically and biologically.

Exercise 2. Consider an extension of model (Eq. 1) that includes constant adaptive immune responses: T cell immune responses E and antibody immune responses A. The model diagram is given in Fig. 4 and the corresponding ODE system is:

$$\frac{dT}{dt} = s - \frac{\beta TV}{1 + \alpha A} - dT,$$

$$\frac{dI}{dt} = \frac{\beta TV}{1 + \alpha A} - \delta I - \mu IE, \qquad (5)$$

$$\frac{dV}{dt} = pI - cV.$$

The parameters values are $s = 10^4$, $\beta = 10^{-6}$, $d = 0.01$, $p = 400$, $\delta = 1$, $c = 23$, $\mu = 10^{-4}$, $\alpha = 1$ and the initial conditions are $T(0) = 10^6$, $I(0) = 0$, $V(0) = 1$.
Consider the following scenarios.

  I. There are no immune responses, i.e. $E = 0$ and $A = 0$.
 II. The immune responses are $E(t) = 5 \times 10^3$ and $A(t) = 5$.
III. The T cell immune response varies over time $E(t) = 5 \times 10^3 \frac{t^2}{t^2 + 100}$ and the antibody is constant $A(t) = 5$.
IV. The T cell immune response is given by the additional ODE

$$\frac{dE}{dt} = s_E + \alpha_E EI - d_E E, \qquad (6)$$

with $s_E = 5 \times 10^2$, $\alpha_E = 5 \times 10^{-4}$, $d_E = 0.1$ and initial condition $E(0) = 10$. The antibody immune response decreases exponentially $A(t) = 5e^{-0.5t}$.

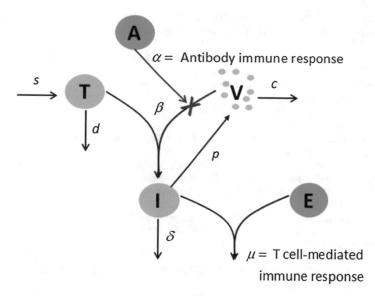

Fig. 4 Diagram for virus dynamics model that includes immune responses

(a) For scenarios I and II, determine the basic reproduction number.
(b) Interpret your answers for part (a).
(c) For each scenario, plot a numerical solution of T, I, and V and describe the observed dynamics of these populations. For scenario II, use the time interval $0 < t < 20$. For the other scenarios, use $0 < t < 300$.

References

1. Knipe DM, Howley PM (2013) Fields virology, 6th edn. Wolters Kluwer/Lippincott Williams & Wilkins Health, Philadelphia, PA. 2 volumes
2. Taylor LH, Latham SM, Woolhouse ME (2001) Risk factors for human disease emergence. Philos Trans R Soc Lond Ser B Biol Sci 356(1411):983–989
3. Woolhouse M, Gaunt E (2007) Ecological origins of novel human pathogens. Crit Rev Microbiol 33(4):231–242
4. Amanat F, Krammer F (2020) SARS-CoV-2 vaccines: status report. Immunity 52(4):583–589
5. Murphy K, Weaver C (2016) Janeway's immunobiology. Garland Science, New York
6. Plotkin SA (2010) Correlates of protection induced by vaccination. Clin Vaccine Immunol 17 (7):1055–1065
7. Nabel GJ (2013) Designing tomorrow's vaccines. N Engl J Med 368(6):551–560
8. Brauer F (2005) The Kermack–McKendrick epidemic model revisited. Math Biosci 198 (2):119–131
9. Ho DD et al (1995) Rapid turnover of plasma virions and CD4 lymphocytes in HIV-1 infection. Nature 373(6510):123–126
10. Perelson AS et al (1996) HIV-1 dynamics in vivo: virion clearance rate, infected cell life-span, and viral generation time. Science 271(5255):1582–1586
11. Ciupe SM, Heffernan JM (2017) In-host modeling. Infect Dis Model 2(2):188–202

12. Nowak M, May RM (2000) Virus dynamics: mathematical principles of immunology and virology: mathematical principles of immunology and virology. Oxford University Press, Oxford, UK
13. Baccam P et al (2006) Kinetics of influenza A virus infection in humans. J Virol 80 (15):7590–7599
14. Nikin-Beers R, Ciupe SM (2018) Modelling original antigenic sin in dengue viral infection. Math Med Biol 35(2):257–272
15. Nikin-Beers R, Ciupe SM (2015) The role of antibody in enhancing dengue virus infection. Math Biosci 263:83–92
16. Best K, Perelson AS (2018) Mathematical modeling of within-host Zika virus dynamics. Immunol Rev 285(1):81–96
17. Banerjee S et al (2016) Estimating biologically relevant parameters under uncertainty for experimental within-host murine West Nile virus infection. J R Soc Interface 13(117)
18. Ciupe SM et al (2007) Modeling the mechanisms of acute hepatitis B virus infection. J Theor Biol 247(1):23–35
19. Ciupe SM et al (2007) The role of cells refractory to productive infection in acute hepatitis B viral dynamics. Proc Natl Acad Sci 104(12):5050–5055
20. Ciupe SM et al (2011) Dynamics of hepatitis B virus infection: what causes viral clearance? Math Pop Stud 18(2):87–105
21. Ciupe SM (2018) Modeling the dynamics of hepatitis B infection, immunity, and drug therapy. Immunol Rev 285(1):38–54
22. Neumann AU et al (1998) Hepatitis C viral dynamics in vivo and the antiviral efficacy of interferon-α therapy. Science 282(5386):103–107
23. Durbin J (1960) The fitting of time-series models. Revue de l'Institut International de Statistique:233–244
24. Heffernan JM, Smith RJ, Wahl LM (2005) Perspectives on the basic reproductive ratio. J Roy Soc Int 2(4):281–293
25. Liu P et al (2011) Dynamic antibody specificities and virion concentrations in circulating immune complexes in acute to chronic HIV-1 infection. J Virol 85(21):11196–11207
26. Perelson AS (2002) Modelling viral and immune system dynamics. Nat Rev Immunol 2 (1):28–36
27. Ciupe SM, Ribeiro RM, Perelson AS (2014) Antibody responses during hepatitis B viral infection. PLoS Comput Biol 10(7):e1003730
28. Ciupe SM, Schwartz EJ (2014) Understanding virus–host dynamics following EIAV infection in SCID horses. J Theor Biol 343:1–8
29. Ciupe SM (2015) Mathematical model of multivalent virus–antibody complex formation in humans following acute and chronic HIV infections. J Math Biol 71(3):513–532
30. Ciupe M et al (2006) Estimating kinetic parameters from HIV primary infection data through the eyes of three different mathematical models. Math Biosci 200(1):1–27
31. Volberding PA (2017) HIV treatment and prevention: an overview of recommendations from the 2016 IAS–USA Antiretroviral Guidelines Panel. Top Antiv Med 25(1):17
32. Cihlar T, Fordyce M (2016) Current status and prospects of HIV treatment. Curr Opin Virol 18:50–56
33. Infection, P.o.C.P.f.T.o.H (1998) Guidelines for the use of antiretroviral agents in HIV-infected adults and adolescents. Afr J Med Pract 5(2):79
34. Perelson AS et al (1997) Decay characteristics of HIV-1-infected compartments during combination therapy. Nature 387(6629):188–191
35. Bonhoeffer S et al (1997) Virus dynamics and drug therapy. Proc Natl Acad Sci 94 (13):6971–6976
36. Perelson AS, Nelson PW (1999) Mathematical analysis of HIV-1 dynamics in vivo. SIAM Rev 41(1):3–44
37. Floudas CA, Pardalos PM, SpringerLink (Online service). Encyclopedia of optimization

38. Moré JJ (1978) The Levenberg-Marquardt algorithm: implementation and theory. In: Numerical analysis. Springer, New York, pp 105–116
39. Ciupe SM, Miller CJ, Forde JE (2018) A bistable switch in virus dynamics can explain the differences in disease outcome following SIV infections in rhesus macaques. Front Microbiol 9:1216
40. Sakamoto Y, Ishiguro M, Kitagawa G (1986) Akaike information criterion statistics. D. Reidel, Dordrecht, The Netherlands, p 81
41. Antia R, Ganusov VV, Ahmed R (2005) The role of models in understanding CD8+ T-cell memory. Nat Rev Immunol 5(2):101–111
42. Antia A et al (2018) Heterogeneity and longevity of antibody memory to viruses and vaccines. PLoS Biol 16(8):e2006601
43. Youngblood B et al (2017) Effector CD8 T cells dedifferentiate into long-lived memory cells. Nature 552(7685):404–409
44. Hill AL et al (2014) Predicting the outcomes of treatment to eradicate the latent reservoir for HIV-1. Proc Natl Acad Sci 111(37):13475–13480
45. Sengupta S, Siliciano RF (2018) Targeting the latent reservoir for HIV-1. Immunity 48 (5):872–895
46. Forde J, Volpe JM, Ciupe SM (2012) Latently infected cell activation: a way to reduce the size of the HIV reservoir? Bull Math Biol 74(7):1651–1672
47. Ke R et al (2015) Modeling the effects of vorinostat in vivo reveals both transient and delayed HIV transcriptional activation and minimal killing of latently infected cells. PLoS Pathog 11 (10):e1005237
48. Rong L, Perelson AS (2009) Modeling latently infected cell activation: viral and latent reservoir persistence, and viral blips in HIV-infected patients on potent therapy. PLoS Comput Biol 5 (10):e1000533
49. Cohen YZ et al (2018) Relationship between latent and rebound viruses in a clinical trial of anti-HIV-1 antibody 3BNC117. J Exp Med 215(9):2311–2324
50. Baral S, Antia R, Dixit NM (2019) A dynamical motif comprising the interactions between antigens and CD8 T cells may underlie the outcomes of viral infections. Proc Natl Acad Sci 116 (35):17393–17398
51. Guedj J et al (2018) Antiviral efficacy of favipiravir against Ebola virus: a translational study in cynomolgus macaques. PLoS Med 15(3):e1002535

Identifying Virus-Like Regions in Microbial Genomes Using Hidden Markov Models

Frank O. Aylward

Background

The Role of Viruses in Genome Evolution

Viruses represent an enormous reservoir of genetic diversity on Earth [1]. Although viruses that infect humans, plant crops, or domesticated livestock have been studied for over a century due to their medical or agricultural importance, there is growing appreciation for the large diversity of viruses that inhabit the biosphere and infect microorganisms such as bacteria, archaea, single-celled eukaryotes, and even other viruses [2–5]. All of these viruses collectively encode a vast diversity of genetic and genomic novelty that is only beginning to be explored, and it represents a major frontier for current and future investigations of biodiversity on the planet.

An intriguing characteristic of many viruses is their ability to integrate their genomes into those of their host. For some viruses, such as retroviruses and some dsDNA viruses, this is part of their normal life-cycle, and the endogenized "proviruses" can subsequently excise themselves at a later time [6]. For other viruses, endogenization appears to be accidental rather than the result of normal viral reproduction, and the exact mechanism is not clear. In either case, the introduction of new genetic material into the host genome results in a "genetic chimera", since the contents derive from multiple distinct sources with unique evolutionary histories. The term "endogenous viral element", or EVE, can be used as catch-all term for the myriad forms of integrated viruses, which can range in size from partial genomic fragments of only a few base-pairs to complete viral genomes encoding up to hundreds of genes. The implications for the host are sizeable, since viral genes in

F. O. Aylward (✉)
Department of Biological Sciences, Virginia Polytechnic Institute and State University, Blacksburg, VA, USA
e-mail: faylward@vt.edu

© Springer Nature Switzerland AG 2021
P. Kraikivski (ed.), *Case Studies in Systems Biology*,
https://doi.org/10.1007/978-3-030-67742-8_17

EVEs can potentially participate in multitudes of different metabolic pathways, and in turn shape the physiology of the host. In multicellular organisms, EVEs would not be inherited unless they endogenize into a germ-line cell, and we would therefore expect that the majority of endogenization events would not be passed on through generations. For unicellular lineages such as protists, bacteria, or archaea, however, the incidence of inherited EVEs is likely much higher.

The endogenization of viral genomes can sometimes have implications for host evolution that are not immediately evident. The introduction of viral genes into the host genome creates opportunities for adaptation at a later time (a phenomenon called "exaptation") because viral genes can be co-opted to participate in alternative processes that are distinct from their original functions. For example, retroviral-derived genes in mammals have been shown to participate in placental development, and it has been postulated that the initial introduction of these viral genes into ancestral mammals was a key even in the history of this group [7]. Moreover, encapsulin proteins involved in storage or transportation functions in some bacteria and archaea have been postulated to derive from viral capsid proteins, suggesting viral genes were co-opted for a different function after an integration event [8, 9]. As these examples illustrate, because evolution operates by "tinkering" with existing genetic material, the introduction of diverse viral genes therefore can have immense and often unforeseen consequences for the evolution of the host.

Integrated viruses have perhaps been studied the most in-depth in bacteria, where they have been shown to play key roles in their hosts [10, 11]. Toxin genes in many pathogenic bacteria are encoded within integrated proviruses, for example, demonstrating that the ability of these bacteria to cause disease is tightly linked to their endogenized viruses. Well known cases where this has been explored include *Vibrio cholerae*, *Corynebacterium diphtheriae*, *Clostridium perfringens*, and *Shigella dysenteriae* (the causative agents of cholera, diphtheriae, gangrene, and dysentery, respectively). In these cases, the ecological lifestyle of these pathogenic bacteria hinges on their encoded proviruses. Other proviruses of unknown function have been retained within some bacterial lineages for millions of years, suggesting they play important but cryptic roles [12].

Overview of Current Methods

Detection of EVEs

Methods for the identification of integrated viruses in the genomes of cellular lineages is of great importance due to their critical roles in the evolution of their hosts, but the accurate detection of EVEs in their host genome is complicated by a number of issues. Firstly, depending on the method of integration, some EVEs may consist of only short fragments of viral genomes that are not suitable for classification. An EVE that is only a few hundred base-pairs long may bear no obvious signature of its viral origin, for example, and it will be difficult to detect and definitively show that it is viral. Secondly, our knowledge of viral genomic diversity is still in its infancy, and EVEs that derive from novel lineages may not bear similarity to known viruses and may therefore elude detection. Thirdly, EVEs often degrade over time due to deletions, chromosomal rearrangements, and the accumulation of mutations, leading to genomic regions that, although originally derived from a virus, may not be recognized as such. Due to these caveats, methods for the detection of EVEs probably underestimate their occurrence and return only high-confidence cases where the viral signatures of an EVE are unambiguous. Even when EVEs can be identified with high-confidence, the exact boundaries of the EVE may remain unclear unless the integration event took place recently.

The detection of EVEs can rely on multiple lines of evidence, such as abnormal codon usage patterns or nucleotide signatures (e.g., % G+C content), but for EVEs that are short or derive from integration events in the distant past only the presence of virus-specific genes is likely to provide unambiguous detection of an EVE. The accurate classification of protein-coding genes into protein families based on sequence homology is therefore a critical step in the detection of EVEs. Genes that encode capsid proteins, virion structural proteins, or virus-specific transcription factors or DNA polymerase subunits are high-confidence signatures of EVEs, since it is highly unlikely that they would occur in the genome of a cellular organism independent of any viral integration. Some large DNA viruses have recently been found to encode a wide variety of metabolic genes heretofore only found in cellular lineages, however, such as fermentation genes [13], light-harvesting rhodopsins [14], and even citric acid cycle components [15], illustrating the difficulty in designating any protein family as "virus-specific" or "cellular-specific".

Homology Searches and Hidden Markov Models: Practical Considerations

Because gene classification is critical for evaluating whether or not a genomic region derives from a virus or a native host genome, I have provided a brief summary here of some practical considerations and popular methods for gene classification. In order to classify a gene, one typically searches the query gene's sequence against a database of known genes to identify sequence similarity. If sufficient sequence similarity is found, one can establish that the query gene and the reference are *homologous*, implying that the sequence similarity is due to common evolutionary origin and not random chance. In practice this can be done using the nucleotide sequence of the query gene or its translated amino acid sequence, but when comparing highly divergent genes (i.e., those for which abundant mutations would be expected to obscure the signal of sequence similarity), it is best to rely on protein-protein comparisons. This is because the amino acid sequence of a protein is more highly-conserved than its nucleic acid sequence; changes to the amino acid sequence of a protein would be expected to potentially deteriorate the function of a protein and therefore be removed by natural selection, whereas changes to the nucleic acid sequence of a gene may persist without necessarily inhibiting the function of the encoded protein. Because viral genes typically evolve very quickly, the detection of EVEs therefore relies on gene classification strategies that use the encoded amino acid sequence of a gene (i.e., the protein).

One of the most popular and earliest-developed strategies for classifying proteins is Basic Local Alignment Search Tool (BLAST, specifically BLASTp for proteins) [16]. Using this tool one can search a query protein sequence against a database of known proteins for sequence similarity. BLAST generates "local alignments", meaning that is does not always align the entire query sequence to reference sequences, and it reports various metrics that one can use to assess alignment quality. Perhaps the most popular is the e-value, which provides the number of hits that would be expected to be found by random chance (lower e-values imply the local alignment would be highly unlikely to be found by random chance, and is therefore a very good match).

Another popular and effective strategies for classifying protein-coding genes leverages Hidden Markov Models (HMMs) and the HMMER3 tool [17]. In this approach the reference database consists of HMMs rather than individual proteins. Each HMM represents a protein family, and is generated by aligning related proteins to each other in a multi-sequence alignment before generating the HMM. From the perspective of protein classification, one can think of HMMER3 as operating in a similar way to BLAST, with the exception that a query protein is compared to alignments of proteins with HMMER3, while with BLAST it is compared to individual reference proteins. HMMER3 is useful when classifying divergent proteins that bear only minor similarities to known protein families, since comparisons with whole protein families can often succeed in identifying similarities that are missed when evaluating individual protein-protein alignments independently.

Figure 1 shows an example of an amino acid multi-sequence alignment of the NifH protein, an enzyme involved in nitrogen fixation that is present in diverse microbial lineages separated by billions of years of evolution. The alignment has been colored to help identify highly conserved regions. Proteins from different

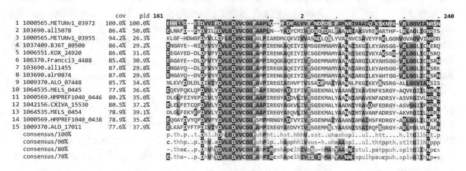

Fig. 1 Multi-sequence alignment at the amino acid level. The alignment was generated using the web interface of the MUSCLE tool [18] (https://www.ebi.ac.uk/Tools/msa/muscle/).

species are shown as rows of the alignment, and highly conserved regions can be seen as consistent vertical strips (the GDVVC region denotes a conserved Glycine-Aspartate-Valine-Valine- Cysteine sequence in these proteins, for example). The nucleic acid sequences homology of the corresponding genes would likely be undetectable in this example given the long amount of time separating the species encoding the NifH protein. Given conservation at the amino acid level, however, we can still detect invariant regions of the proteins (likely important for proper functioning of the enzyme) which reveal these proteins to be members of the same protein family with shared evolutionary origin.

ViralRecall: A Simple Approach for Identifying EVEs and Other Virus-Like Regions in Genomic Data

ViralRecall is a tool that illustrates a homology-based method for detecting virus-like regions in genomic data. The tool has a relatively simple approach and is therefore useful for examining a basic bioinformatic workflow. This tool takes a nucleotide sequence of a genome as input and leverages HMMs for the classification of protein-coding genes as likely-viral or likely-cellular in origin. The algorithm is divided into the following steps:

1. Protein coding gene prediction. Proteins are predicted from the input nucleic acid sequence using Prodigal.
2. Proteins are then searched against two HMM databases. One database consists of the Protein Families database (Pfam), but with viral HMMs removed. The other database consists of the Viral Orthologous Groups database (VOG). Proteins with similarity to HMMs in the Pfam database therefore provide signatures we would expect in a cellular genome while, those with similarity to HMMs in the VOG database provide evidence of viral provenance.
3. A cumulative alignment score (referred to as the ViralRecall score) is calculated by subtracting the Pfam score (if any) from the VOG score (if any), resulting in a net positive score for genes with higher scores to the VOG database.
4. A rolling average of all gene cumulative ViralRecall scores is calculated for each replicon in the input nucleotide sequence.
5. Regions that have a net positive ViralRecall score and >4 genes with hits to VOG HMMs are reported as putative viral regions.

A summary of an example ViralRecall output can be found in Fig. 2.

Note that in Fig. 2 there are a wide range of "virus-like" regions. Two regions have high peaks and encompass a relatively large region of the chromosome, two others have shorter peaks and encompass smaller regions, and there are some additional very small peaks scattered in between.

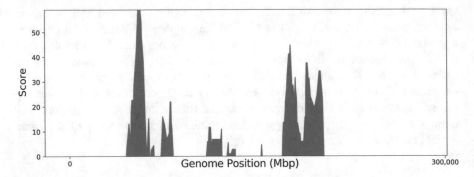

Fig. 2 Illustration of an example ViralRecall output. The x-axis shows the chromosome position, while the y-axis provides the ViralRecall score. This was run on a fragment of second chromsome of *Vibrio cholera* O1 biovar El Tor str. N16961. Blue areas denote elevated ViralRecall scores, which indicate genes encoded in these regions have homology to known viral genes. Higher scores denote higher bit scores to known viral protein families. The tight clustering of these regions suggests they correspond to bona fide endogenous viruses (prophage).

Conclusions

Through a variety of mechanisms viruses can integrate their genetic material into that of their host, thereby profoundly impacting the genome evolution of cellular life. Identifying virus-like regions in the genomes of cellular organisms is difficult, in part because of the vast diversity of viruses and the erosion of viral genetic material after integration. However, by examining homology to known viral genes, as is done by tools such as viralrecall, it is possible to identify virus-like regions in the genomes of cellular organisms, and thereby assess the overall impact viruses have had on the genome evolution of their hosts.

Exercise

For this exercise we will predict virus-like regions in the genome of *Escherichia coli* O157:H7 str. Sakai, a causative agent of foodborne illness. The commands below were run on a Ubuntu 16.04 command line, but they should run on a variety of Unix or Unix-like operating systems. For detailed information on how to install VIralRecall and its dependencies, please see the GitHub page: https://github.com/faylward/viralrecall.

First, we will download the latest version of ViralRecall using the "git clone" command, and then navigate into this directory:

```
git clone https://github.com/faylward/viralrecall

cd viralrecall
```

Then we will download the archived HMM files needed to run this tool, and unpack the data:

```
wget -O hmm.tar.gz https://data.lib.vt.edu/downloads/8k71nh28c
```

```
tar -xvzf hmm.tar.gz
```

To download the genome of *Escherichia coli* O157:H7 str. Sakai we can use the FTP server at the National Institute for Biotechnology Information (NCBI). It is possible to browse a wide variety of available genomes on this site (https://www.ncbi.nlm.nih.gov/genome).

```
wget -O o157.fna.gz https://ftp.ncbi.nlm.nih.gov/genomes/all/GCF/
000/008/865/GCF_000008865.2_ASM886v2/
GCF_000008865.2_ASM886v2_genomic.fna.gz
```

```
gunzip o157.fna.gz
```

Now that we have downloaded the tool and the data, we can run the analysis. We will just use default parameters here:

```
python viralrecall.py -i o157.fna -p o157_out -f
```

This creates an output file with several files. The *.pdf file shows a visual representation of the results, with blue peaks representing virus-like regions and red marks indicating chromosome boundaries (the same as in Fig. 2 above). The *.vogout and *.pfamout files have the raw output for the HMM search, and we will not discuss them further. The *.faa file has the predicted proteins that were used in the HMM search. The *.summary.tsv file has summarized statistics for the virus-like regions that were found, including the start and end of the virus-like regions, the length of the region, the overall viralrecall score (higher scores meaning more virus-like, negative scores indicating more cellular-like), and the presence of viral "hallmark" genes that are excellent indicators of integrated viral elements (i.e., capsid, tail, portal, and terminase genes, among others). In this case, viralrecall identified 18 virus-like regions, which is consistent with the prophage regions that were originally identified in this genome [19]. Note that most of these regions have bacteriophage hallmark genes, but there are some regions that lack genes with clear viral provenance. It is therefore necessary to manually inspect some regions in detail before deciding whether they are *bona fide* prophage or merely anomalous regions with some virus-like signatures. Manual inspection of the annotations is possible by looking at the *full_annot.tsv file. If we inspect Viral Region 5 (found on chromosome NC_002695.2 from bp 1,408,259-1,451,604) we see that this region was characterized as virus-like because of the presence of several transposases, insertion elements, tellurium resistance proteins that have homology to those encoded in viruses. It is therefore unclear if this region is derived from a viral integration event or through the proliferation of other selfish genetic elements, such as transposases. This degree of ambiguity is regrettable but unavoidable given the complex nature of bacterial genome evolution. Further analyses examining the phylogenetic relationships between the proteins encoded in this region and those present in known viruses may reveal additional insight into the potential viral provenance of this region, but these analyses are outside the scope of the current article.

References

1. Suttle CA (2016) Environmental microbiology: viral diversity on the global stage. Nat Microbiol 1(11):16205
2. Hatfull GF (2015) Dark matter of the biosphere: the amazing world of bacteriophage diversity. J Virol 89(16):8107–8110
3. Duponchel S, Fischer MG (2019) Viva lavidaviruses! Five features of virophages that parasitize giant DNA viruses. PLoS Pathog 15(3):e1007592
4. Coy SR, Gann ER, Pound HL, Short SM, Wilhelm SW (2018) Viruses of eukaryotic algae: diversity, methods for detection, and future directions. Viruses 10(9) https://doi.org/10.3390/v10090487
5. Prangishvili D (2013) The wonderful world of archaeal viruses. Annu Rev Microbiol 67:565–585
6. Holmes EC (2011) The evolution of endogenous viral elements. Cell Host Microbe 10(4):368–377
7. Mi S, Lee X, Li X, Veldman GM, Finnerty H, Racie L, LaVallie E et al (2000) Syncytin is a captive retroviral envelope protein involved in human placental morphogenesis. Nature 403(6771):785–789
8. Krupovic M, Koonin EV (2017) Multiple origins of viral capsid proteins from cellular ancestors. Proc Natl Acad Sci U S A 114(12):E2401–E2410
9. Koonin EV, Krupovic M (2018) The depths of virus exaptation. Curr Opin Virol 31(August):1–8
10. Howard-Varona C, Hargreaves KR, Abedon ST, Sullivan MB (2017) Lysogeny in nature: mechanisms, impact and ecology of temperate phages. ISME J. https://doi.org/10.1038/ismej.2017.16
11. Brüssow H, Canchaya C, Hardt W-D (2004) Phages and the evolution of bacterial pathogens: from genomic rearrangements to lysogenic conversion. Microbiol Mol Biol Review: MMBR 68(3):560–602, table of contents
12. Bobay L-M, Touchon M, Rocha EPC (2014) Pervasive domestication of defective prophages by bacteria. Proc Natl Acad Sci U S A 111(33):12127–12132
13. Schvarcz CR, Steward GF (2018) A giant virus infecting green algae encodes key fermentation genes. Virology 518(May):423–433
14. Yutin N, Koonin EV (2012) Proteorhodopsin genes in giant viruses. Biol Direct 7(October):34
15. Moniruzzaman M, Martinez-Gutierrez CA, Weinheimer AR, Aylward FO (n.d.) Dynamic genome evolution and blueprint of complex virocell metabolism in globally-distributed giant viruses. https://doi.org/10.1101/836445.
16. Altschul SF, Gish W, Miller W, Myers EW, Lipman DJ (1990) Basic local alignment search tool. J Mol Biol 215(3):403–410
17. Eddy SR (2011) Accelerated profile HMM searches. PLoS Comput Biol. https://doi.org/10.1371/journal.pcbi.1002195
18. Edgar RC (2004) MUSCLE: a multiple sequence alignment method with reduced time and space complexity. BMC Bioinform 5(August):113
19. Perna NT, Plunkett G 3rd, Burland V, Mau B, Glasner JD, Rose DJ, Mayhew GF et al (2001) Genome sequence of enterohaemorrhagic Escherichia coli O157:H7. Nature 409(6819):529–533

Computational Software

Pavel Kraikivski

The computational modeling in case studies is performed using various computational software packages, including routinely used XPPAUT [1], MATLAB (Octave is a free alternative to MATLAB), and Python. Also freely available is Capasi [2] and VCell [3] computational tools, and BioNetGen [4] and Kappa [5] for rule-based modeling. These tools are frequently used by the Systems Biology community.

XPPAUT is a free computational software package that works on Windows, Mac OS, and Linux operating systems. It is recommended for Computational and Systems Biology classes, computational labs, and research projects. XPPAUT is straightforward and an easy-to-use tool, even without any coding experience.

It is a very powerful environment, especially when it is used to analyze dynamical systems. XPPAUT contains two integrated programs XPP and AUTO. The first one is used to integrate differential equations, and perform stability and phase plane analyses, whereas the second (AUTO) is used to solve continuation and bifurcation problems in ordinary differential equations.

MATLAB is widely used in systems and computational biology fields to simulate and analyze dynamic biological systems. MATLAB programming language is designed for mathematical and technical computing, and, in general, it is easy to implement and simulate a mathematical model using MATLAB. In addition to numerous general mathematical tools, MATLAB also includes a systems biology toolbox that can be used to perform deterministic or stochastic simulations, steady-state and stability analyses, parameter estimation, sensitivity analysis, and many other computational modeling-related operations for analysis and simulation of biological systems.

Python is a general-purpose programming language that supports object-oriented and functional programming and provides numerous mathematical and scientific

P. Kraikivski (✉)
Academy of Integrated Science, Division of Systems Biology, Virginia Polytechnic Institute and State University, Blacksburg, VA, USA
e-mail: pavelkr@vt.edu

© Springer Nature Switzerland AG 2021
P. Kraikivski (ed.), *Case Studies in Systems Biology*,
https://doi.org/10.1007/978-3-030-67742-8_18

libraries. Python is widely used in the computational biology community for modeling complex biological networks and systems. Python also provides a PySB framework for building mathematical, rule-based models of biochemical systems as Python programs [6]. PySB inter-operates with BioNetGen and Kappa software tools. Python also supports PyDSTool package that allows users to analyze dynamical systems and perform phase plane analysis, continuation and bifurcation analyses [7]. Therefore, PyDSTool allows users to perform the same computational operations as the XPPAUT computational software.

COPASI (COmplex PAthway SImulator) is a computational simulator software that supports deterministic, stochastic, and hybrid deterministic–stochastic methods for the simulation and analysis of biochemical reaction networks. COPASI allows non-expert users to automatically convert reaction equations to the appropriate mathematical formalism (ordinary differential equations for deterministic simulation or reaction propensities for stochastic methods). COPASI's graphical interface is similar to that of Windows Explorer in its operation with a window containing a set of functions and model editing; for example, it is done through tables and specialized widgets. COPASI is available for all major operating systems (Linux, Mac OS X, Windows).

VCell (Virtual Cell) is an open-source software platform for modeling cell biological systems. VCell is available via the web, and the client-server implementation allows users to run simulations without requiring their own high-performance computer hardware. VCell is designed to be a computational modeling platform that is easily accessible to cell biologists who may not have sufficient training in mathematics and programming, and with limited access to sophisticated and high-performance computer hardware. VCell provides a graphical model building interface where a user can introduce and describe cellular compartments, molecular species, and interactions, after which VCell generates the appropriate mathematical encoding for running simulations.

BioNetGen and Kappa are software tools with languages developed for rule-based modeling. Using BioNetGen and Kappa languages, a user can describe molecules as structured objects and molecular interactions as rules. The rule-based model description is easier to understand and error-check than a large system of differential equations. Also, the rules are easier to reuse. Furthermore, a model that is described using the rule-based language can be translated into mathematical equations for simulation.

The following sections provide further essential information and user manuals for computational tools that are extensively used in this book.

XPPAUT

The XPPAUT software, installation instructions and documentation are available online [1]. XPPAUT is a tool for solving differential equations, difference equations, delay equations, functional equations, boundary value problems, and stochastic equations. The detailed guide to XPPAUT can be found in Ermentrout's book [8]. Here, a quick guide to XPPAUT is provided, which is just enough to reproduce all XPPAUT simulation results in this book.

XPPAUT can read a model code in a text file with the file extension .ode. The model description in the file should specify a set of differential equations, the model parameters, and initial conditions. It is optional to provide the initial conditions in the file; the default initial values are automatically set to zero and can be changed later in the "Initial condition window" of the XPP interface. The .ode file can also contain information on various parameters related to the solver (e.g., the total simulation time), axes and graphics window settings, auxiliary functions, a random number generator function, delay command, and other specific settings that can be set directly through the XPP graphical user interface menus.

For example, we consider the following mathematical model:

$$\frac{dx}{dt} = y^2 - ax,$$
$$\frac{dy}{dt} = -x + by. \tag{1}$$

where a, b are parameters.

The corresponding model description in the .ode file has the following form:

```
#Example 1

dx/dt=y^2-a*x
dy/dt=-x+b*y



par a=2.5, b=2
done
```

Lines that begin with a hash mark # are ignored by the XPPAUT program; thus # is used to write comments. XPP reads dx/dt as the derivative of a variable x with respect to time t. The asterisk symbol * is used for multiplication. XPP returns an error message if the expression containing multiplication is written without an asterisk symbol; for example, k(x+1) will generate an error message, because the correct expression is k*(x+1). The command "par" is used to indicate parameters; a = 2.5 and b = 2 are the starting parameter values that can be later changed in XPPAUT. "done" tells XPP that the model description is completed; thus, this command is always the last one. Parameters can be described before or after equations. By default, initial conditions are x = 0, y = 0. To define different initial conditions, the line that begins with the init command is used. For example, the following line in the .ode file:

```
init x=1, y=0.5
```

will set initial values for x and y variables to 1 and 0.5, respectively. Note, that this line should be introduced before the "done" command.

Other useful commands that are often used in XPP codes in this book are "aux" to define an auxiliary function, and @ total to specify total integration time. For example, the auxiliary function 2x + 3y can be defined by introducing the following line in the .ode file:

```
aux plotxy=2*x+3*y
```

This combination of variables then can be plotted in the graphics window of the XPP program.

The Heaviside function can be used to trigger events at certain times. For example, the following line in the .ode file:

```
event1=2.5*heav(t-t0)
```

defines a fixed variable event1 that equals zero when t is less than t0 (t0 is a parameter that should be defined in the par line of the code) and equals 2.5 when t is larger or equal to t0. The notation heav() is recognized by XPP as the function that returns zero for negative arguments and one for positive arguments.

XPP has a default total integration time, and the default time equals 20. Such small integration times often must be adjusted. The integration time can be adjusted in the program after opening the code, by going to nUmerics, then Total, and entering a new number, or the integration time can be set in the code using the following command line in the .ode file:

```
@ total=1000
```

This line will set the total integration time to be equal to 1000. Lines that begin with @ are used to set computational parameters. A space should always be included after the @ symbol. For example, to set the bound on the variable values the following command can be used:

```
@ bound=500
```

This will change the default bound of ±100 to ±500. This can also be done by using the XPP menu items that are described in the XPP graphical user interface section.

It is also convenient to set the size of the graphics area in XPP. The default plotting window size is from 0 to 20 for the x-axis and from −1 to 1 for the y-axis. This can be changed by using the Viewaxes menu or by adding the following line into the code:

```
@ xp=t, yp=x, xlo=0, xhi=15, ylo=0, yhi=20
```

This line sets the time variable to be assigned to the x-axis, with lower and upper axis limits defined by [0, 15] interval; x variable to be along the y-axis within [0, 20] interval.

The allocation of additional memory to store the simulation data can be done by using the @ maxstor command and equating it to the number that is bigger than the default value (5000 data points).

XPP Graphical User Interface Description

XPPAUT has many examples (.ode files) in the xppall/ode folder that will appear after XPPAUT installation. Follow the step-by-step installation instructions for your operating system which can be found in Ref. [1]. Example 1, described, above will be used to go over different options in the XPP graphical user interface. Create a text file named Example1 with the code shown above, and change its extension to .ode. When the .ode file is opened using the XPP program, the window should appear as shown in Fig. 1.

The top of the XPP window is called a title bar that displays the XPP program version and the .ode file name. Below the title bar, the XPP window has **ICs**, **BCs**, **Delay**, **Param**, **Eqns**, and **Data** buttons. A click on each of these buttons opens a new window that displays the following information: initial conditions (ICs), boundary conditions (BCs), delay (Delay), parameters (Param), equations (Eqns), and data (Data). Initial conditions and Parameters windows (see Fig. 2) are often used to change parameter values and initial conditions, so it is good to keep these windows open while working with XPP. Boundary conditions and Delay windows allow users to alter the boundary and delay initial data, respectively. The Equations window displays equations that cannot be edited in this window (File->Edit options from the left XPP menu can be used to edit equations; see description below). Data window allows users to save data points in a file, which then can be, for example, plotted in MS Excel or used in other data processing tools. The command line "Command:" is used to enter some information that is requested by XPP; for example, the total integration time or tolerance in nUmerics can be entered using the command line. In the middle, the XPP window has a graphics area where variables can be plotted as a function of time (see Fig. 3) or the phase plane with nullclines can be analyzed (see Fig. 4).

Along the left side, the XPP window contains the menu items among which Initialconds, Nullcline, Dir.field/flow, Graphic stuff, nUmerics, File, Erase, Sing pts, Viewaxes, Xi vs t, are the most often used for plotting variables as a function of time or for the phase plane analysis. Each of these options can open an additional menu with a list of other items (some of them will be described below). Note that the capital letter in the name of the menu item (or the letter enclosed in a pair of round brackets) indicates that it can be used as a "hot key". Thus, you could either click on that menu item or press the hot key. Also, the Esc key is used to quit or close a menu item.

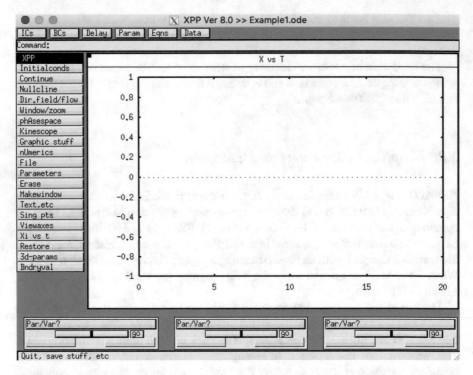

Fig. 1 Main XPP window that contains the title bar, the menu bar with buttons and the command line under the title bar, menu items along the left, a graphics area in the middle and an information area at the bottom

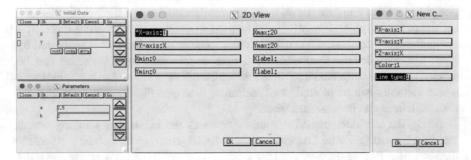

Fig. 2 XPP windows that are often used for plotting a solution in the graphic area. Initial data (top-left corner) window can be opened by clicking on ICs at the top of the XPP menu, Parameters window (bottom-left corner) is opened by clicking on the Param option at the top of the XPP menu, 2D View (in the middle) window can be opened by clicking Vewaxes->2D in the menu items along the left side of the XPP window, New C (right) window is opened by clicking Graphic stuff (in the left menu of XPP) and then (A)dd curve (inside the Graphic stuff menu)

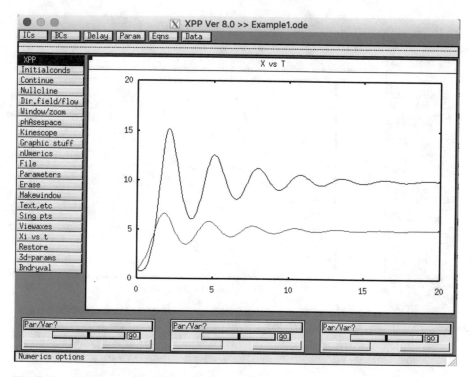

Fig. 3 The integration results of Example 1 equations shown in the XPP graphics area. The temporal evolution of both X (black curve) and Y (red curve) variables shows damped-oscillatory behavior

Initialconds is used to start integration of the model equations by clicking (G) o in the menu. The menu inside also has various other options to define initial conditions. (R) ange is used to create a range of initial conditions and generate several solution curves. (L) ast is used to apply the end values of the last integration as the starting point for a new integration. (M) ouse and m (I) se are used to select an initial condition by clicking in the graphics area using a computer mouse cursor, just once or multiple times, respectively. (N) ew prompts you for new initial data and then integrates the equations using that data. Initial data can be uploaded from a file using (F) ile, defined using a formula with form (U) la. There are also some other options in the Menu to specify initial values and initiate integration of model equations. Continue is used to continue integration for a longer time period. The longer integration period is specified in the command line, Xi vs t, and then Enter command is required to update numbers along the axis of the graphics area.

Nullcline is used to plot nullclines that are defined as curves where dx/dt = 0 and dy/dt = 0. It is useful for analysis of a system of two equations in the phase plane. Before plotting nullclines, we first have to set the phase plane in the graphics area by filling X in X-axis: and Y in Y-axis: spaces in the 2D View window

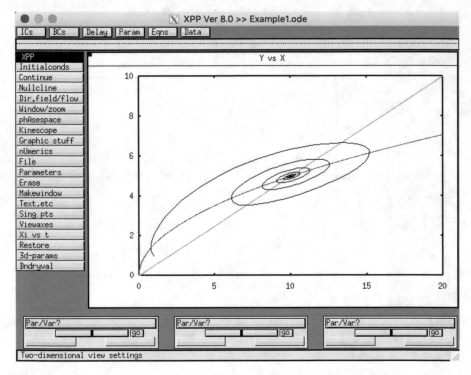

Fig. 4 Phase plane, nullclines and a solution. The red curve is X-nullcline, the green line is Y-nullcline, and the black curve is the solution corresponding to X = Y = 1 initial point

(see Fig. 2), which can be opened by clicking on Viewaxes and then 2D in the menu. Also set appropriate ranges for X and Y variables by defining Xmin, Ymin, Xmax and Ymax values in that 2D View window (for Example 1, set Xmin:0, Ymin:0, Xmax:20, Ymax:10). To plot nullclines, click Nullclines and (N)ew. In our Example 1, the red curve representing the x-nullcline where x variable does not change (dx/dt = 0), and the green line representing the y-nullcline where y variable does not change (dy/dt = 0), are plotted as shown in Fig. 4. The intersection of two nullclines is an equilibrium point where both x and y variables remain constant in time.

Dir.field/flow is used to draw the direction field (vectors) or flow (trajectories) started at different points of the phase-plane, defined by the grid. The grid size is defined by the user in the command line. If the grid size is set large, the plotting all trajectories can take a long time, densely populating the graphics area with trajectories. To abort any given trajectory, press Esc, and to stop the whole process of drawing trajectories, press / key.

Window/zoom is used to define the graphics area by setting X and Y ranges to zoom the region or to automatically fit the window so the entire curve is contained within it.

phAsespace is used to set a periodic domain and solve equations on a torus or cylinder.

Kinescope is used to capture the bitmap of the active window and then play it back.

Graphic stuff is often used to add another curve to the graphics area. The (A)dd curve option opens the New C window shown in Fig. 2, where the user can define the variable on each axis, the color of the curve (a number from 1 to 9 for the color (1 = red, 9 = violet)), and the line type (1—solid and 0—dotted). Other useful commands are (D)elete last, which removes the most recent curve, (R)emove all, which deletes all curves but the first, (E)dit curve, which is used to edit the first (0), the second (1), etc. curve, (P)ostscript is used to write a postscript representation of the current window, (F)reeze is used to create a permanent curve in the window, S(V)G to save, and exp(o)rt data to export data.

nUmerics is used to specify parameters of the numerical integration. For this book, the most used options are: Total, which is used to set the amount of time to integrate; Dt, which sets the integration step used by the fixed step integrator; Method, which is used to choose the integration method (e.g., (S)tiff integration method is required in some case studies in the book); and Bounds, which sets a global bound on the integrator. If any variable value exceeds the bound magnitude, the integration stops. The default Bounds are 100. Thus, the Bounds value must be changed for problems in which variables can take values bigger than 100, otherwise a message appears and the XPP program does not proceed with the integration.

File is used to exit XPP (Quit), to start the AUTO program (Auto), to view the code (Prt src), to create a file with numerics info and other parameters (Write set), to read the information from the file created by the Write set command (Read set), and to edit equations in the .ode file from XPP (Edit).

Parameters option is used to change the value of a parameter by typing a parameter name, and then adding its new value in the command line. However, the same can be conveniently done in the Parameters window that can be opened using the Param button from the top menu.

Erase command cleans the graphics area from curves.

Makewindow is used to create and destroy additional graphics windows. For example, (C)reate command creates a copy of the currently active graphics window. The graphs in this window can be modified without affecting the other windows. (D)estroy can be used to destroy the currently active window. The main window cannot be destroyed. Also, (K)ill all can be used to remove all graphics windows, keeping only the main graphics window.

Text,etc is used to place, delete, or edit text, arrows, or markers.

Sing pts is used to analyze stability of singular points (equilibrium points). The (M)ouse option is to use the mouse for an initial guess for a singular point location (click by the mouse cursor in the area near where two nullclines intersect). Answer (No) or (Yes) to the "Print Eigenvalues" question and (No) or (Yes) to the "Draw Invariant Sets" question. A small rectangle will appear in the phase-plane that shows the position of the equilibrium (in the phase plane, shown in

Fig. 4, the small rectangle will appear where x- and y-nullclines intersect). The stability window provides the following information: if the equilibrium point is stable or unstable, c+ and c− show the number of complex eigenvalues that have positive and negative real parts, respectively, r+ and r− indicate the number of positive and negative real eigenvalues, and im shows the number of imaginary eigenvalues.

In the Example 1 for $a = 2.5$, $b = 2$, the Equilibria window that appears after Sing pts displays c+ = 0, c− = 2, im = 0, r+ = 0, r− = 0, implying that there are only two complex eigenvalues having negative real parts. It also indicates that the equilibrium point is stable. For different values of a and b parameters the stability of the point may change. For the Example 1 with $a = 2$ and $b = 2.5$, the Sing pts will return c+ = 2 indicating two complex eigenvalues with positive real parts. In this case, the window will also show that the equilibrium point is unstable.

Viewaxes is used to select different types of graphs. For example, for 2D graphs, the user can specify two variables, limits for each variable, and labels (optional) for the axes in the 2D View window (see Fig. 2). The 2D View window can be opened by selecting the 2D option inside the Viewaxes menu.

Xi vs t works as a shortcut to the Viewaxes option and is used for faster choosing of the variable name to display it as a function of time in the graphics area.

Restore redraws the most recent data in the graphics area using the graphics parameters of the active window.

The **3d-params** option is used for 3D views to rotate the axes and perspective planes.

Bndryval is used to set parameters (e.g. variable starting and ending values) for solving boundary value problems.

Solving Equations and Displaying Solution

To start integration of Example 1 equations, open the Initial Data window (see Fig. 2) using the ICs button, set initial values X = 1 and Y = 1, then click OK and Go. The X curve will appear in the graphics area (the black curve in Fig. 3). Open Viewaxes->2D and set Ymin: 0, Ymax: 20 to have a better graphic view. To add Y solution to the graphics area, open Graphic stuff->(A) dd curve and type Y in Y-axis line and 1 in Color space, then click OK. The Y solution curve will appear in the graphics area (the red curve in Fig. 3).

To set the X–Y phase plane, open Viewaxes->2D, fill X in X-axis and Y in Y-axis, and set Xmin: 0, Ymin: 0, Xmax: 20, Ymax: 10. Nullclines can be computed by clicking Nullclines->(N) ew. However, to increase visibility of nullcline lines, go to Numerics->Ncline ctrl and enter ncline mesh 1000 in the command line, and replot nullclines using Nullclines->(N) ew again. Two nullclines should appear; the red curve representing the x-nullcline, where the x variable does not change ($dx/dt = 0$), and the green line representing the y-nullcline, where the y variable does not change ($dy/dt = 0$). See Fig. 4.

A solution can be added to the graphics area by clicking `Initialconds->`
`(G) o`, or using `m(I)ce` and then the mouse cursor by clicking on any starting point
in the graphics area (several times to have multiple solution curves). The solution
that corresponds to $X = 1$ and $Y = 1$ initial point is shown in the graphics area of
Fig. 4. The solution is converging to the stable steady state value $X = 10$, $Y = 5$ (see
also Fig. 3). Next, we will use the AUTO program to explore the behavior of the
system of differential equations in Example 1 by varying a and b parameter values.

AUTO Program

The AUTO program is integrated with XPP and is used to further investigate the
dynamic behavior of systems of differential equations and build parameter bifurcation diagrams. To start the AUTO program, open the `File menu in XPP` and
click `Auto`.

Make sure that the system is at a steady state before starting the AUTO program.
This is important because the last point of the simulation in XPP is passed to AUTO
as a starting point, from which AUTO continues analysis of the system. Thus, this
point should be a steady state point. For Example 1 that has an oscillating solution,
the integration should continue until the non-oscillating state is reached. Therefore,
the model parameters should also have values for which the stable steady state
solution can be reached. The `Initialconds->(L) ast` command in XPP can be
used to integrate the equations until a non-oscillating steady state is obtained.

The AUTO window has menu items along the left and a graphics area in the
middle, where bifurcation diagrams can be displayed (see Fig. 5). This menu
contains **Parameter**, **Axes**, and **Numerics** options that are used to open windows shown in Fig. 6, where the user has to provide necessary information for the
bifurcation analysis.

The **Parameter** menu is for listing 8 model parameters in order, see Fig. 6
(Parameters window). The default is the first 8 or fewer parameters in .ode file. `Par1`
must be the main parameter and `Par2` as the second parameter to be used in one- and
two- parameter bifurcation diagrams. Set `Par1:a` and `Par2:b` as the first and
second parameters for the bifurcation analysis of the system of differential equations
in Example 1.

Then, `hI-lo` in **Axes** menu (see Fig. 6 AutoPlot window) allows the user to set
axes for the variable (along Y-axis) and the main parameter (along X-axis) in the
one-parameter bifurcation diagram as shown in Fig. 5. To set axes as in Fig. 5, set
`Y-axis:X`, `Main Parm:a`, `Second Parm:b`. Use `Ymin`, `Ymax` and `Xmin`,
`Xmax`, to set axes ranges in the graphics area. For Example 1, we set `Ymin:0`,
`Ymax:20` and `Xmin:1`, `Xmax:3`.

The **Numerics** button opens the AutoNum window that is shown in Fig. 6. We
use **Numerics** to specify the step size and the range that are used by solver to
change the main parameter. The main parameter will be varied between `Par Min`
and `Par Max` limits by the solver, to obtain steady state solutions for different

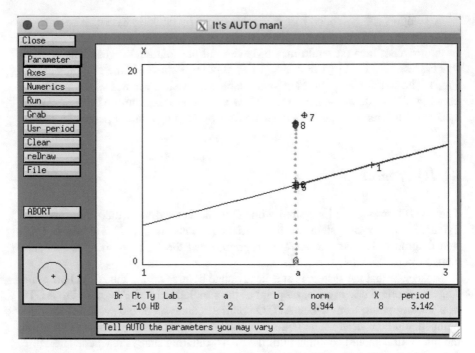

Fig. 5 AUTO window and one-parameter bifurcation diagram for Example 1. The point 1 ($a = 2.5$, $X = 10$) is where the construction of the diagram was started and then continued from $a = 2.5$ and $X = 10$ to $a = 3$, using positive $Ds = 0.02$ (setting the increasing direction for the parameter a value change), and from $a = 2.5$ and $X = 10$ to $a = 1$ and $X = 4$, using negative $Ds = -0.02$ (setting the decreasing direction for the parameter a value change). The stable steady state solution (red line) continues until the Hopf bifurcation point at $a = 2$ and $X = 8$, then the unstable steady state solution (black line) begins. The periodic solution is shown by the green dotted line

Fig. 6 Parameter, hI-lo (located in the Axes menu), and Numerics windows in the AUTO program. The first window titled Parameters appears when the Parameter option is clicked in the AUTO menu; the second window (AutoPlot) appears when either the hI-lo or Two par options are selected in the Axes menu; the AutoNum window appears when the Numerics option is selected

parameter values. The same range (Xmin: 1, Xmax: 3) as set in hI-lo (**Axes**) for the main parameter value also can be used to fill Par Min and Par Max spaces. Thus, we can set Par Min: 1 and Par Max: 3. Next, we suggest a value for Ds that is the initial step size for the bifurcation calculation. The AUTO program uses adaptive step size; thus, Ds is a proposed value to start with. Ds value can be positive or negative. The default Ds value is 0.02. The sign indicates the direction in which the parameter will be changed. Dsmin and Dsmax are the minimum step size (positive) and the maximum step size, respectively. If one-parameter bifurcation diagram lines appear to be broken, this may indicate that the Dsmax value is too big.

This information is sufficient to compute the bifurcation diagram shown in Fig. 5. The **Run** menu is used to start the integration. This menu contains Steady sate, Periodic, Bdry Value, Homoclinic, and hEteroclinic options. Also, when Two par is specified in the **Axes** menu, the Two param option will appear in the **Run** menu. This option will be used to compute the two-parameter bifurcation diagram.

To produce the one-parameter bifurcation diagram in Fig. 5, start with Ds = 0.02 (default value) and in the **Run** menu click Steady state, AUTO will compute the right side of the bifurcation diagram (the red line from initial point 1 that corresponds to $a = 2.5$ and $X = 10$ will continue to the point where $a = 3$ and $X = 12$, see Fig. 5). Click **Grab** to grab point 1 (use the mouse cursor to select point 1 and then click the Enter key). Then, in **Numerics**, set Ds to be negative 0.02 to perform the calculation in the opposite direction. Open the **Run** menu and click Steady state. The red line will continue towards $a = 2$, $X = 8$, and the black line will then continue to the point where $a = 1$. The red line indicates the stable steady state solution and the black line represents the unstable solution. Use **Grab** again to grab the Hopf bifurcation point (the point where a stable steady state solution becomes unstable, at $a = 2$ and $X = 8$ in this example). The information area at the bottom reports the Hopf bifurcation point as HB, and provides values of a and b parameters and the variable X. In the **Run** menu, click the Periodic option and this will print the periodic solution in the graphics area (the green dotted line in Fig. 5).

To compute the corresponding two-parameter bifurcation diagram, use **Grab** to grab the Hopf bifurcation point. Thus, the two-parameter bifurcation diagram will begin at this Hopf bifurcation point and the corresponding a and b parameter values. To compute the two-parameter bifurcation curve use the following two steps:

1. In the **Axes** menu choose the Two par option, and the same kind of window as the second window shown in Fig. 6 should appear. Set limits for a and b parameters using Xmin, Xmax, Ymin, Ymax spaces (for example, set Xmin: 1, Ymin: 1, Xmax: 2, Ymax: 2) in the window. Then, in the graphics area, the Y axis will represent parameter b and the X axis will represent parameter a (see Fig. 7).

2. In the **Run** menu, click the Two param option; the blue line that represents the two-parameter bifurcation diagram should appear in the graphics area (see Fig. 7). This blue line gives values of a and b parameters that correspond to the Hopf

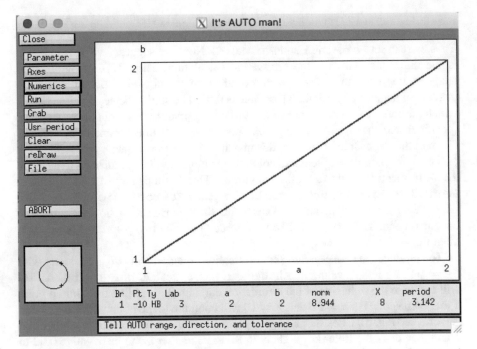

Fig. 7 Two-parameter bifurcation diagram for the Example 1 problem. The blue line shows a and b parameter values that correspond to the Hopf bifurcation. For all a and b parameter values below the blue line ($a > b$) the steady state solution is stable, and for all a and b parameter values above the blue line ($b > a$) the steady state solution is unstable

bifurcation. This line separates the two parameter regions that correspond to stable and unstable steady state solutions. In the region where $a > b$ (below the blue line in Fig. 7) the steady state solution is stable, and in the region where $a < b$ (above the blue line in Fig. 7) the steady state solution is unstable.

In this section, we followed just a basic set of steps that are used to build one- and two-parameter bifurcation diagrams. Setting other **Numerics** options (see the AutoNum window in Fig. 7) may be necessary for successful integration, so as to obtain the bifurcation diagrams. For example, if the integration is interrupted and does not continue, then reducing `Dsmin`, `EPSL` and `Dsmax` values by a factor of 10 could fix the problem.

Partial Differential Equations

A representative example of solving partial differential equations (PDEs) with XPPAUT is described by Tyson in Mitotic Cycle Regulation. II. Traveling Waves chapter of this textbook. To solve partial differential equations, the space is discretized and a set of N ordinary differential equations (ODEs) is generated using the method of lines. XPPAUT has a concise way to represent an array of ODEs in the ode file. For example, to define a set of 100 differential equations by the method of lines, we can write a line $v[1..100]'=...$, which describes the differential equations for $v1$, $v2,...,v100$ variables. A spatial partial derivative must be written in the numerical form. For example, a second derivative $D\frac{\partial^2 v}{\partial x^2}$ must be written as $(D/h^2)*(v[j+1]-2*v[j]+v[j-1])$. In XPPAUT, it is important that we use the letter "j" for the index. Also, if D/h^2 is used several times in the code it is often defined as a new parameter, in the Tyson's example (see Mitotic Cycle Regulation. II. Traveling Waves chapter) the name del was used. The parameter that depends on other parameters can be defined by a statement starting with the exclamation mark, !, for example, we can add the line $!del=D/h*h$ to the ode file that will define the del parameter.

The boundary conditions can be set by statements that specify the values of $v0$ and $v101$ variables. For example, the statements $v0 = v1$ and $v101 = v100$ will enforce 'no flux' boundary conditions at the two ends of the domain. We also don't need to type initial conditions for each variable vj. The initial conditions can be defined by a formula. To define initial conditions, click Initial Conds -> formUla and prompt for the 'variable', type $v[1..100]$ and hit Return key, at the 'formula' prompt, type the formula that specifies the initial values (for example heav(1-[j]) or $1/(1+exp(([j]-33)/10)))$, and hit Return key several times. XPP will then solve the set of ODEs within the defined domain. The result can be displayed by setting parameters in Viewaxes Array (see Table M in the Method section of Mitotic Cycle Regulation. II. Traveling Waves chapter).

MATLAB and Octave

MATLAB and **Octave** are numerical computing environments supporting the matrix-based programming language that allows users to express matrix and array mathematics directly. MATLAB has many advantages over other programing languages which allows users to store and represent data in the form of matrices, to test algorithms immediately without recompilation, and to use numerous tools that are built for many specific applications. The MATLAB syntax for writing programs is simple and easy to learn, implement and understand. The purpose of this section is to describe the basic MATLAB syntax and elements that are used in some codes written for case studies in this book.

MATLAB can take typed commands at MATLAB's >> command prompt and return the corresponding results. The sequence of commands can also be written in m-file, which is a file with file extension .m. Then, MATLAB executes each command in the m-file sequentially. A code in MATLAB can be written as a script (the simplest type of program with a sequence of commands) or using self-contained modules (functions) that perform a specific task. Functions provide more flexibility; each function can be used many times in the program, and can also be called and tested outside of the main program. Functions can take input values and return output values; in general, we can state:

```
function [y1, ..., yN] = fun1(x1, ..., xM)
```

where a function named fun1 takes M inputs x1, ..., xM and returns N outputs y1, ..., yN. Function names must begin with an alphabetic character, and can contain letters, numbers, or underscores. The declaration statement is the first line of the function that is executed. The function body ensues the declaration line and contains a collection of statements which defines what the function does with inputs to generate outputs. The function body ends with the "end" command. For example, the function that computes the squared difference between two input values can be written as:

```
function f = sq_diff(x1, x2)
f=(x1-x2)^2;
end
```

we can then call this function from the command line to get the squared difference between two numbers, for example as y=sq_ diff(7, 3) and get y = 16. To write this function in a function file, the name of the file must match the name of this function. For this example, the file containing this function must be named as sq_diff.m. A function file can contain several function definitions. Then, the name of the file must match the name of the first function in the file. By contrast, a script file which contains commands and function definitions cannot have the same name as a function in the file.

All ODE solvers in MATLAB use similar syntaxes and can solve systems of equations of the form y′=f(t,y). MATLAB recommends ode45 solver as the first choice for most problems. It is used to solve non-stiff differential equations. For stiff differential equations, MATLAB has a variable-step, variable-order solver called ode15s. To get a solution y(t) of the system of differential equations y′=f(t,y) we can use the following line:

```
[t, y]=ode45(rhs, t_range, y0);
```

where rhs defines right-hand sides of the ordinary differential equations, t_range defines the integration time interval, y0 represents initial conditions. For example, to integrate the system of two equations

$$\frac{dx}{dt} = y^2 - 2.5x$$

$$\frac{dy}{dt} = -x + 2y$$

in t_range = [0, 20] with initial values x(0) = y(0) = 1, the following code can be used

```
%Example 1

function integration

rhs= @(t,y) [y(2)^2-2.5*y(1); -y(1)+2*y(2)];

t_range = [0 20];
y0=[1;1];

[t, y]=ode45(rhs, t_range, y0);

plot(t, y(:,1), 'k','linewidth',2, t, y(:,2), 'r','linewidth',2)
```

This code must be saved as integration.m file. The first line that begins with symbol % is a comment line; all lines that begin with symbol % are ignored by MATLAB program. A semicolon indicates the end of the command or the end of the row (e.g. in a matrix). Importantly, MATLAB is case-sensitive, thus uppercase and lowercase letters are considered to be different.

In this code, y(1) and y(2) represent x and y variables in the system of two equations, correspondingly. Thus, in the MATLAB code, y is not just a single variable but an array or a vector generally consisting many variables. Then calling integration function in the MATLAB (or Octave) command prompt will produce Fig. 8 which is similar to Fig. 3 obtained by integrating the same system of equations with XXP solver (see Fig. 3 in the XPPAUT section). Figure 8a shows results obtained with ode45 solver and Fig. 8b shows results obtained by using the same code but changing ode45 to ode15s. We can notice that ode15s solver produces smoother curves. The plot is a function that creates a 2-D line plot. In our example, plot creates the 2-D plot of data for y(1) and y(2) variables versus time.

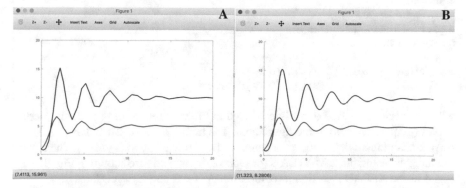

Fig. 8 Octave solutions for Example 1 problem generated by using ode45 (**a**) and ode15s (**b**) solvers with default settings. ode15s solver produces more accurate integration results

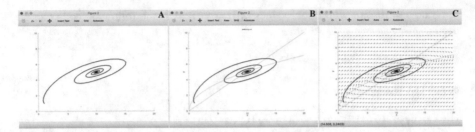

Fig. 9 Phase plane analysis using Octave. (**a**) A trajectory in the phase plane using the command plot. (**b**) x-nullcline (blue) and y-nullcline (orange) are added by using the command ezplot. (**c**) The direction field (red arrows) is added by using the command quiver

The colon notation in y(:, 1) means that it is the column of data points for the variable y(1) and y(:, 2) is the column of data points for the variable y(2). The plot function, in this example, also includes the color ('k' and 'r' correspond to black and red color codes) and line style (linewidth) information.

To produce trajectories in the phase plane, we can modify the plot function as

```
plot(y(:,1), y(:,2) 'k','linewidth',2)
```

This will produce a trajectory shown in Fig. 9a, which is similar to Fig. 4 produced using the XPPAUT software (see XPPAUT section).

Nullclines can be added to the phase plane by using ezplot command which is used to plot the solution to the implicit equations. In our example, nullclines are defined by the following implicit equations: $y^2 - 2.5x = 0$ and $-x + 2y = 0$. Thus, defining the right-hand side functions as

```
xnullc = @(x,y) y.^2-2.5*x;
ynullc = @(x,y) -x+2*y;
```

we can use ezplot command to plot solutions of xnullc=0 and ynullc=0 equations as

```
% plot x- and y-nullclines
figure(2); hold on

ezplot(@(x,y) xnullc(x,y), [0 20 0 10]);
ezplot(@(x,y) ynullc(x,y), [0 20 0 10]);
```

where figure(2); hold on line is used in order to retain all previously plotted curves (graphic objects) in the figure area. The numbers in [0 20 0 10] inside the ezplot command define the axes ranges as [xmin xmax ymin ymax]. The results produced by using these MATLAB (or Octave) commands are shown in Fig. 9b (see Fig. 4 in the XPPAUT section to compare MATLAB and XPPAUT results).

The direction filed can be produced by using the command quiver that displays velocity vectors as arrows with components (u,v) at the points (x,y). The length of the arrows (u,v) can be computed from the right-hand sides of the differential equations. To get the scaled direction field, each velocity vector should be normalized (divided) by its length. The (x,y) positions for the velocity arrows can be defined by the command `meshgrid` that creates a 2-D grid with uniformly spaced x-coordinates and y-coordinates in the given interval [a, b] with a step d. The following code can be used to add the direction field to the graphic area:

```
%produce the direction field
a = 0; % minimum axis limit
b = 20; % maximum axis limit
d = 0.5; % step for the mesh grid
[X, Y] = meshgrid(a:d:b, a:d:b/2); % define the mesh grid

u=Y.^2-2.5*X;
v=-X+2*Y;

len = sqrt(u.^2+v.^2); %length of the velocity vector

quiver(X, Y, u./len, v./len, 0.5, 'r'); % scaled by 0.5
```

The resulting direction filed is shown in Fig. 9c.

MATLAB does not provide build-in functions that could help us to perform a bifurcation analysis and obtain one- and two-parameter bifurcation diagrams as shown in Figs. 5 and 7 in the XPPAUT section. However, additional packages/ toolboxes for bifurcation analysis can be integrated into MATLAB. For example, Dynamical Systems Toolbox that is integration of AUTO bifurcation software into MATLAB (see Ref. [9] for all information about this toolbox) or Matcont package [10, 11].

MATLAB is also extensively used in this textbook for stochastic simulations. Gillespie's stochastic simulation algorithm is fully described by Chen in Stochastic Gene Expression chapter, by Miles & Mogilner in the Collective molecular motor transport chapter, and by Xing & Zhang in the Principle of Cooperativity in Olfactory Receptor Selection chapter. The corresponding MATLAB codes in these chapters represent a set of great examples of how Gillespie's stochastic simulation algorithm can be implemented in MATLAB to solve problems in Computational and Systems Biology fields.

Python

PyDSTool supports phase plane analysis, continuation and bifurcation analysis by providing access to the bifurcation analysis tool AUTO [7]. Therefore, PyDSTool can be used as an alternative to XPPAUT computational software. PyDSTool relies on numpy, scipy, MatPlotLib packages, which have to be installed over Python before using PyDSTool. The PyDSTool installation instructions, tutorial and tool-box documentations can be found in Ref. [12].

For example, we can use PyDSTool to analyze the system of two equations (1) from XPPAUT section. The following Python code describes the system of differential equations, then integrates the system and plots the solution:

```python
import PyDSTool as dst
import numpy as np
import matplotlib.pyplot as plt
import matplotlib


#create the object of the arg class and call it DSargs
DSargs = dst.args(name='Example1',
 varspecs={
 'x' : 'y^2-a*x',
 'y' : '-x+b*y',
 },
 pars={
 'a' : 2.5,
 'b' : 2.0,
 },
 ics={
 'x' : 1.0,
 'y' : 1.0, },
 tdata=[0,20])

DSargs.xdomain = {'x': [0, 20], 'y': [0, 10]}

#build a solver object
ode = dst.Vode_ODEsystem(DSargs)

#integrate and produce data for plotting
pts = ode.compute('Example1').sample()

plt.figure(figsize=(10,7)) # set figure size

plt.plot(pts['t'], pts['x'], 'k', pts['t'], pts['y'], 'r') #plot x vs
t, black color, and y vs t, red color

plt.xlabel('time')
plt.ylabel('x, y')

plt.savefig('./Fig1.tif')
```

This code will produce the plot shown in Fig. 10. The code can be executed in the Jupiter Notebook which is a web-based interactive computing notebook environment available in the ANACONDA navigator. Python code starts with import statement that is used to import modules and packages. import statement finds and initializes a module, and defines a name in the local namespace. If no matching

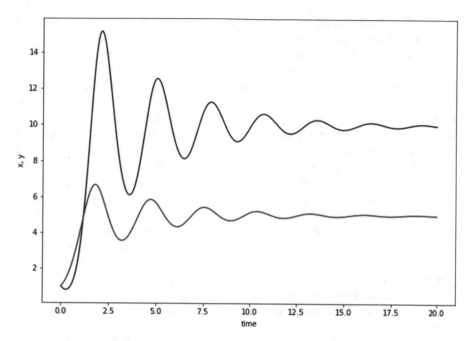

Fig. 10 The integration results of the system of two deferential equations using the Python code. Similar to XPP (see Fig. 3 in XPPAUT section) and MATLAB (see Fig. 8 in MATLAB section) results, the temporal evolution of both X (black curve) and Y (red curve) variables shows damped-oscillatory behavior

module is found, `ImportError` is returned. Thus, the installation of all listed packages has to be completed before running this code. "as `localname`" is used to create a shorter local name for an imported module. For example, in this code, `dst` will be used as a local name for `PyDSTool` and `plt` stays for `matplotlib.pyplot`.

Comments in Python begin with a hash mark # and continue to the end of the line. After importing all necessary packages, the `args` class is called in the code. The `args` is a class provided in PyDSTool to simplify the syntax for arguments, it includes structures similar to Python dictionaries. In `DSargs = dst.args()` line, we have created an object of the `args` class and named it `DSargs`. In our case, the `args` specifies all important information necessary for integrating the differential equations of the dynamical system. The `varspecs` defines the right-hand sides of ordinary differential equations for each variable of the system. The `pars` describes all parameters and assigns them initial values. The `ics` sets initial condition for each variable. The `tdata` specifies the time interval over which equations are integrated. The `xdomain` sets bounds on variables. This attribute is added separately to demonstrate another way of adding attributes to `DSargs`. The attributes of `DSargs` can be changed by calling `set` method of `DSargs`. For example, a parameter value can be changed by adding the following line `DSargs.set(pars={'a':`

1.3}). It is recommended to make all changes by calling set methods in order to set individual attributes without affecting the values of others.

Now we have all necessary information describing the system of differential equations with parameters, in order to build a solver object. In the ode = dst. Vode_ODEsystem(DSargs) line, we pass all information in DSargs to an initialization call of a PyDSTool class for an ODE solver known as Vode. A Vode generator object will be returned and given the name "ode" in the script, for subsequent use. You can give any Vode object identifier you like, but here it is chosen to be ode. To integrate the system ode.compute() method is called and .sample() is called afterwards in order to produce and store data for plotting. The resulting data is plotted and saved as an image file by using matplotlib. pyplot which includes a collection of functions that make matplotlib work like MATLAB. Using these functions, you can create a figure and a plotting area in the figure, plot some lines in the plotting area, decorate the plot with labels, etc. See the plot documentation in Ref. [13] for a complete list of line styles and format strings. In our code, only basic matplotlib.pyplot functions are used to set the figure size, plot variables vs. time, label axes and save the figure.

Phase plane analysis can be performed using methods in PyDSTool.Toolbox. The phase plane methods can be imported by adding the following line to the Python script:

```
from PyDSTool.Toolbox import phaseplane as pp
```

which imports phaseplane methods and calls it pp. For example, we can compute x- and y-nullclines by applying pp.find_nullclines() method to the ode object as

```
nulls_x, nulls_y = pp.find_nullclines(ode, 'x', 'y', n=3, eps=1e-8,
max_step=0.1)
```

where n value sets the number of starting points in the domain to find nullcline parts, eps value sets accuracy, and max_step defines a maximum step for the solver. The small number for n is used for a sufficiently simple phase plane geometry as in our system of two differential equations. To find the geometric structures, the n parameter specifies a sample grid size in the given domain; n=3 uses three starting points in the domain for numerical algorithms to find nullcline parts. Now, to plot x- and y-nullclines and also a trajectory we can use the plot method again as follows:

```
#plot x- and y-nullclines
plt.plot(nulls_x[:,0], nulls_x[:,1], 'r')
plt.plot(nulls_y[:,0], nulls_y[:,1], 'g')

#plot a trajectory
plt.plot(pts['x'], pts['y'], 'k', linewidth=2)
```

```
plt.xlabel('x')
plt.ylabel('y')
```

this part of the script produces Fig. 11a. To find the intersection of the nullclines (fixed point), pp_find_fixedpoints() method can be applied. A direction field can be computed by using vector field method as

```
pp.plot_PP_vf(ode, 'x', 'y')
```

which produces a vector filed as shown in Fig. 11b.

To compute the one-parameter bifurcation diagram, we can use PyCont that is a sub-package of PyDSTool which provides tools for numerical continuation and bifurcation analysis (see PyCont package description and tutorial in Ref. [14]). We can create PyCont instance by passing existing ode object to ContClass as PC = dst.ContClass(ode). Then, this name can be used to get access to PyCont methods.

Figure 11 shows that variables exhibit damped-oscillation dynamics for a = 2.5 and b = 2 parameter values. Therefore, the steady-state can be reached after sufficiently long time. By changing a model parameter, we can then search for a Hopf bifurcation point. The following code can be used to detect the Hopf bifurcation in our system of two differential equations:

```
#increase the integration time to reach a steady-state.
ode.set(tdata=[0,3000])

#Create an instance of PyCont
PC = dst.ContClass(ode)

# specify the bifurcation curve type and name it.
PCargs = dst.args(name='EQ1', type='EP-C')
```

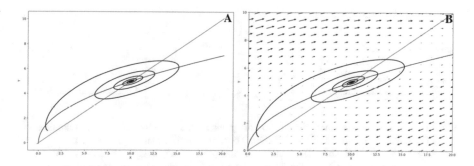

Fig. 11 Phase plane analysis results obtained using PyDSTool methods. (**a**) x-nullcline (red curve) and y-nullcline (green line) and a trajectory (black curve) shown in the (x, y) phase plane. (**b**) shows the vector field by the black arrows (whose size scales with the magnitude of the vector). These figures are similar to Fig. 4 obtained by using XPPAUT (see XPPAUT section) and Fig. 9b, c obtained by using MATLAB tools (see MATLAB section).

```
# indicate the bifurcation parameter
PCargs.freepars= ['a']


PCargs.StepSize = 1e-2
PCargs.MaxNumPoints = 50
PCargs.MaxStepSize = 1e-1


#set to search Hopf bifurcation point
PCargs.LocBifPoints = 'H'
PCargs.SaveEigen = True


# Declare a new curve based on the above criteria
PC.newCurve(PCargs)


PC['EQ1'].backward()


PC.display(['a','x'], stability=True, figure=3)
plt.savefig('./Fig3A.png', dpi=170)
```

this part of the script finds Hopf bifurcation point and produces the curve shown in
Fig. 12a. The curve type EP-C stands for Equilibrium Point Curve and the branch is
labeled EQ1.

PCargs.freepars=['a'] indicates the parameter that will be varied (our
bifurcation parameter). This parameter should be among parameters specified in
DSargs.pars. Then, PCargs.StepSize, PCargs.MaxStepSize,
PCargs.MaxNumPoints specify the step that will be used to change the parameter, the maximum step size and number of points. Usually, trial-and-error is used to
set values for the step sizes. Recall that similar information must be provided in the
AUTO program described in the XPPAUT section.

The PCargs.LocBifPoints = 'H' attribute tells PyCont to search for a
Hopf bifurcation point. It is also possible to compute all bifurcation points for the

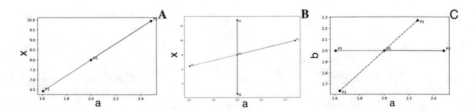

Fig. 12 One- and two-parameter bifurcation diagrams for the system of equations in the Example
1. (**a**) The stable steady state solution (solid line) continues until the Hopf bifurcation point at $a = 2$
and $X = 8$, then the unstable steady state solution (dotted line) begins. (**b**) shows the periodic
solution (vertical lavender line). The results shown in (**b**) agree with results obtained by using
XPPAUT computational software (see Fig. 5 in the XPPAUT section). (**c**) Two-parameter bifurcation diagram. The blue dashed line shows a and b parameter values corresponding to Hopf
bifurcation. This line separates the stable steady state solution ($a > b$) and the unstable steady
state solution ($b > a$) regions. This diagram agrees with the result obtained by using XPPAUT
software (see Fig. 7 in the XPPAUT section)

curve type by setting PCargs.LocBifPoints = 'all'. Then, by using the PCargs.info() method, we can get various information about the curve, such as special points that were detected. The PCargs.SaveEigen attribute is a Boolean variable that determines whether or not the eigenvalues of the equilibrium points should be saved along the curve. The True state of the variable allows us to obtain the stability information along the equilibrium curve.

To compute the bifurcation diagram, we have to first call the newCurve() method of the PyCont instance, passing the parameter dictionary as an argument. Then, to continue the curve in a specific direction, call forward() or backward() or both methods. The solution curves can be displayed by using the PC.display() method and the bifurcation diagram can be saved by calling plt.savefig().

When the Hopf bifurcation point is successfully found, the periodic solution that starts at the Hopf bifurcation point can be continued by using the following code:

```
PCargs = dst.args(name='LC1', type='LC-C')
PCargs.freepars = ['a']
PCargs.initpoint = 'EQ1:H1'

PCargs.MinStepSize = 0.000001
PCargs.MaxStepSize = .01
PCargs.StepSize = 0.00001
PCargs.MaxNumPoints = 420
PCargs.NumSPOut = 10000;
PCargs.LocBifPoints = []
PCargs.verbosity = 2
PCargs.SolutionMeasures = 'avg'
PCargs.SaveEigen = True

PC.newCurve(PCargs)

PC['LC1'].backward()
PC['LC1'].forward()

PC.display(['a','x'], stability=True, figure=3)
PC['LC1'].display(('a','x_min'), stability=True, figure=4)
```

which produces the periodic solution shown in Fig. 12b (vertical line). The curve type LC-C stands for Limit Cycle Curve and the curve is labeled LC1. To generate the branch, we set the same parameter as the bifurcation parameter PCargs.freepars = ['a']. PCargs.initpoint = 'EQ1:H1' indicates that the branch starts at the Hopf bifurcation point.

As in the previous code, PCargs.StepSize, PCargs.MaxStepSize, PCargs.MaxStepSize, PCargs.MaxNumPoints specify the integration step settings and usually are set after several trial simulations. NumSPout number allows us to get the complete information (output) every NumSPout steps.

The `PCargs.LocBifPoints = []` attribute tells PyCont to search for unspecified type of bifurcation points. `PCargs.verbosity` sets verbosity level of the solver. `PCargs.SolutionMeasures` specifies which solution measures to compute along the limit cycle curve. The solution measure `'avg'` sets it to the average of each cycle component. Other options are `'max'`, `'min'`, `'nm2'` that set the solution measures to maximum, minimum, and L2 norm of each cycle component, respectively. To continue the `'LC1'` curve both `PC['LC1'].forward()` or `PC['LC1'].backward()` methods are called.

To produce two-parameter bifurcation diagram we can use the following code:

```
PCargs.name = 'HO1'
PCargs.type = 'H-C2'
PCargs.initpoint = 'EQ1:H1'
PCargs.freepars = ['a','b']
PCargs.MaxNumPoints = 50
PCargs.MaxStepSize = 0.1
PCargs.LocBifPoints = []
PCargs.SaveEigen = True

PC.newCurve(PCargs)

PC['HO1'].forward()
PC['HO1'].backward()
plt.figure(figsize=(10,7))

# Plot two parameter bifurcation diagram
PC.display(('a','b'), stability=True, figure=3)
```

This code will produce the two-parameter bifurcation diagram shown in Fig. 12c. The curve type `H-C2` stands for Hopf point curve, method 2. The curve is labeled as HO1 here. `PCargs.initpoint = 'EQ1:H1'` indicates that the branch starts at the Hopf bifurcation point. `PCargs.freepars = ['a','b']` sets two bifurcation parameters to compute the two-parameter bifurcation diagram. All other PyDSTool methods and attributes in this code have been already described above.

In this section, we explored the computation of bifurcation diagrams that are typically computed for dynamical systems studied in Systems Biology classes. However, PyCont of PyDSTool package allows users to compute different curve classes (e.g. EP-C, LP-C, H-C1, H-C2, FP-C, LC-C) and also find various special points. More examples and documented demo scripts can be found in Ref. [14].

References

1. http://www.math.pitt.edu/~bard/xpp/xpp.html
2. Hoops S et al (2006) COPASI—a COmplex PAthway SImulator. Bioinformatics 22 (24):3067–3074
3. Moraru II et al (2008) Virtual Cell modelling and simulation software environment. IET Syst Biol 2(5):352–362

4. Faeder JR, Blinov ML, Hlavacek WS (2009) Rule-based modeling of biochemical systems with BioNetGen. Methods Mol Biol 500:113–167
5. Danos V, Feret J, Fontana W, Harmer R, Krivine J (2007) Rule-based modelling of cellular signalling. Lect Notes Comput Sci 4703:17–41
6. Lopez CF et al (2013) Programming biological models in Python using PySB. Mol Syst Biol 9:646
7. Clewley RH, Sherwood WE, LaMar MD, Guckenheimer JM (2007) PyDSTool, a software environment for dynamical systems modeling. http://pydstool.sourceforge.net
8. Ermentrout B (2002) Simulating, analyzing, and animating dynamical systems. SIAM
9. Etienne Coetzee (2020). Dynamical Systems Toolbox (https://www.mathworks.com/matlabcentral/fileexchange/32210-dynamical-systems-toolbox), MATLAB Central File Exchange.
10. MatCont is a Matlab software project for the numerical continuation and bifurcation study of continuous and discrete parameterized dynamical systems which is led by Willy Govaerts (Gent,B) and Yuri A. Kuznetsov (Utrecht,NL) and Hil G.E. Meijer (UT, Enschede, NL). MatCont can be downloaded at https://sourceforge.net/projects/matcont/
11. Dhooge A, Govaerts W, Kuznetsov YA, Meijer HGE, Sautois B (2008) New features of the software MatCont for bifurcation analysis of dynamical systems. MCMDS 14(2):147–175
12. https://pydstool.github.io/PyDSTool/FrontPage.html
13. https://matplotlib.org/3.3.0/api/_as_gen/matplotlib.pyplot.plot.html#matplotlib.pyplot.plot
14. https://pydstool.github.io/PyDSTool/PyCont.html#head-aba5d2b0f08788d3fdfe090a2241d990976e22d4

Index

© Springer Nature Switzerland AG 2021
P. Kraikivski (ed.), *Case Studies in Systems Biology*,
https://doi.org/10.1007/978-3-030-67742-8